AF575280

Springer Tracts in Mechanical Engineering

For further volumes:
http://www.springer.com/series/11693

Xin-Jun Liu • Jinsong Wang

Parallel Kinematics

Type, Kinematics, and Optimal Design

Xin-Jun Liu
Tsinghua University
Beijing
China, People's Republic

Jinsong Wang
Tsinghua University
Beijing
China, People's Republic

ISSN 2195-9862 ISSN 2195-9870 (electronic)
ISBN 978-3-642-36928-5 ISBN 978-3-642-36929-2 (eBook)
DOI 10.1007/978-3-642-36929-2
Springer Heidelberg New York Dordrecht London

Library of Congress Control Number: 2013941875

Printed on acid-free paper

Springer is part of Springer Science+Business Media (www.springer.com)

Preface

Parallel mechanisms (PMs) are the systems with closed-loop chains. Since they have advantages of compactness, high speed, high stiffness, high accuracy, high load-to-weight ratio, and low moving inertia, PMs have attracted an intensive attention from the academic and industrial communities. They have been and are being used in a wide variety of applications such as industrial robots, surgical robots, motion simulators, macro-, micro-, and nano-manipulators haptic devices, and even parallel kinematic machines (PKMs).

In recent years, the research and application have evolved from general six-DOF PMs to lower-DOF PMs. In particular, the kinematics of lower-DOF PMs has become a hot issue due to their inherit characteristics, which may lead to more successful applications than their 6-DOF counterparts. The relating topics include mobility analysis and type synthesis and kinematic analysis and optimal kinematic design. These fundamental issues, namely, parallel kinematics in short, are all the focus of this book.

This book is a summary and an extension of the work accomplished by the authors in the field of kinematic design of parallel mechanisms and parallel kinematics machines over the last 15 years.

The whole book includes three parts. The focused topic is type, kinematic analysis, and optimal design, respectively. Part I (Chaps. 1 and 2) presents a systematic classification and evolution-based-type synthesis of PMs available in practical applications. Part II (Chaps. 3, 4, 5, and 6) involves several fundamental issues on kinematic analysis of parallel mechanisms with 2–6 DOFs, including position, velocity Jacobian, singularity, and workspace. Part III (Chaps. 7, 8, and 9) presents kinematic synthesis and optimal design of parallel mechanisms in terms of different kinematic performance evaluation criteria, accompanied with a few typical design cases.

The main features of the book include:

- *This book not only includes the main aspects and important issues of conventional parallel kinematics but also presents many novel conceptions and approaches, i.e., type synthesis based on evolution, performance evaluation and optimization based on screw theory, and singularity model taking into account motion and force transmissibility.*

- *This book covers the systematic classification of PMs as well as providing a large number of mechanical architectures of PMs available to be used in practical applications.*
- *This book focuses on the kinematic design of parallel mechanisms. In particular, it selects parallel kinematic machines, one successful application of parallel mechanisms in the field of machine tools, as a design case.*
- *A large number of case studies and numerical analyses help the audience master the main ideas of the book at both theory and practice level.*

While this book is primarily intended for researchers and engineers working on parallel kinematic machines, parallel manipulators, parallel robots, and other parallel devices, we hope that it will also be of interest to a broader class of readers: (a) graduate students involved in the above areas since the methods proposed are mainly based on linear algebra and basic skills in kinematics, which they are familiar with, and (b) researchers in screw theory since the book acts as a successful application of screw theory in mechanism design. In brief, this book can be a textbook for graduate students as well as general scientific technique personnel.

This book would not have been possible without the help and involvement of many people. In particular, we would like to thank Prof. Jingjun Yu from Beihang University who contributed some very useful suggestions to this book, Dr. Chao Wu and Dr. Fugui Xie for their research on kinematic performance evaluation and optimal design during their doctoral period, and Prof. Feng Gao who is the supervisor of Prof. Xin-Jun Liu from 1994 to 1999 in Yanshan University. The authors also gratefully acknowledge the continuous financial support of the National Natural Science Foundation of China (NSFC), especially, under the grant 51135008 and the support of the National Basic Research Program (973 Program) of China under the grant 2013CB035400.

Tsinghua Yuan, Beijing
People's Republic of China
2 March 2013

Xin-Jun Liu
Jinsong Wang

Contents

Nomenclature

Kinematics: This refers to the study of motion without regard for forces.

Kinetics: This pertains to the study of forces on systems in motion.

Mechanism: A mechanism is a device that transforms motion into some desirable pattern and typically develops very low forces and transmits little power.

It is a system of elements arranged to transmit motion in a predetermined fashion. *A mechanism must have a positive DOF.*

Machine: A machine typically contains mechanisms that are designed to provide significant forces and transmit significant power.

It is a system of elements arranged to transmit motion and energy in a predetermined fashion.

Degree of freedom (DOF): DOF pertains to the number of independent parameters required to completely define configuration in space at any instant of time of a mechanism.

The DOF of an assembly of links completely predicts the character of the assembly. Only three possibilities exist: If the DOF is positive, it will be a *mechanism* and the links will exhibit relative motion. If the DOF is exactly zero, then it will be a *structure* and no motion is possible. If the DOF is negative, then it is a *preloaded structure*, which means that no motion is possible and some stresses may also be present at the time of assembly. (*Norton, R.L. 1999. Design machinery: an introduction to the synthesis and analysis of mechanisms and machines. McGraw-Hill: New York, pp.32*).

Mobility: Mobility is the number of independent input required to completely define the configuration of a mechanism.

The difference between DOF and mobility is that the DOF number cannot be more than six, but mobility can be any number. Therefore, the DOF of the mobile platform of a parallel mechanism is its mobility. No such relationship exists for a serial mechanism. For example, the mobility of a serial 6R mechanism is 6, but its DOF may be 3, 4, 5, or 6.

A **kinematic chain** is defined as an assemblage of links and joints, interconnected by means of providing controlled output motion in response to supplied input motion.

A **mechanism** is defined as a kinematic chain in which at least one link has been "grounded," or attached, to the frame of reference (which itself may be in motion). (*Norton, R.L. 1999. Design machinery: an introduction to the synthesis and analysis of mechanisms and machines. McGraw-Hill: New York, pp.27*).

A **machine** is defined as a combination of resistant bodies arranged to compel the mechanical forces of nature to do work accomplished by determinate motions.

A **link** (or **member**) is a rigid body that possesses at least two nodes, which are points for attachment to other links. (*Norton, R.L. 1999. Design machinery: an introduction to the synthesis and analysis of mechanisms and machines. McGraw-Hill: New York, pp.24*).

A **joint** (or **kinematic pair**) is a connection between two or more links (at their nodes), which allows for some motion, or potential motion, between the connected links. Therefore, a joint imposes some physical constraints on the relative motion between the two links.

The contact surface of a link is called a **pair element**. Two pair elements form a **kinematic pair**. A kinematic pair is called a **lower pair** if the two elements come into contact with each other with a substantial surface area. A kinematic pair is called a **higher pair** if the pair elements are in contact at a point or along a line.

A **revolute** (R) **joint** permits two paired elements to rotate with respect to each other about an axis that is defined by the geometry of the joint. Hence, the R joint imposes five constraints between the paired elements and is a 1-DOF joint.

A **prismatic** (P) **joint** enables two paired elements to slide with respect to each other along an axis that is defined by the geometry of the joint. Thus, the P joint imposes five constraints between the paired elements and is a 1-DOF joint.

A **cylindrical** (C) **joint** permits rotation about, and independent translation along, an axis that is defined by the geometry of the joint. Hence, the C joint imposes four constraints between the paired elements and is a 2-DOF joint.

A **spherical** (S) **joint** enables one element to rotate freely with respect to the other about the center of the sphere in all possible orientations. No translation between the paired elements is permitted. Thus, the S joint imposes three constraints on the paired elements and is a 3-DOF joint. Sometimes, an S joint is designed as the combination of three intersecting R joints.

A **universal** (U) **joint** is essentially a combination of two intersecting R joints. Hence, it is a 2-DOF joint.

Linkage: Linkages are made up of links and joints.

Pure rotation means that the body possesses one point (center of rotation), which exhibits no motion with respect to the "stationary" frame of reference. All other points on the body describe arcs about that center. A reference line drawn on the body through the center only changes the angular orientation of the body.

Pure translation indicates that all points on a body describe parallel paths. A reference line drawn on the body changes the linear position but not the angular orientation of the body.

Complex motion is defined as a simultaneous combination of rotation and translation. Any reference line drawn on the body changes both the linear position and angular orientation of the body. Points on the body travel nonparallel paths, and at every instant, a center of rotation that continuously changes location exists.

Number synthesis is the determination of the number and order of links and joints necessary to produce the motion of a particular DOF. Order in this context refers to the number of nodes per link.

Type synthesis refers to the definition of the proper type of mechanism best suited to a given problem.

The **dimensional synthesis** of a linkage is the determination of the proportions (lengths) of the links necessary to accomplish the desired motions or performance.

Pose is the position and orientation of a mobile platform.

Configuration refers to the combined positions and orientations of all links and a mobile platform.

Inverse kinematics pertains to the problem of identifying the input parameters for a given pose of a mobile platform. A similar term is inverse kinematic problem.

Direct kinematics pertains to the problem of identifying the pose of a mobile platform for specified input. Similar terms are direct kinematic problem and forward kinematics.

Working mode refers to one of several solutions to the inverse kinematic problem.

Assembly mode pertains to one of several solutions to the direct kinematic problem.

Part I
Mechanism Type

Chapter 1
Classification of Parallel Mechanisms

Abstract This chapter provides a systematic classification of parallel mechanisms based upon the latest research on the architectures of parallel mechanism. The list will be classified by increasing numbers of degrees of freedom, from 2 to 6, companied with the motion pattern of the mobile platform. Various mechanical architectures of parallel mechanism will be exposed, as well as typical examples of applications.

Keywords Architectures • Classification • DOF • Parallel mechanism • Complex legs

1.1 Definition and Characteristics

Mechanical systems that allow a rigid body to move with respect to a fixed base play a highly important role in numerous applications. A rigid body in space can move in various ways, in translation or rotation. The motions are called *degrees of freedom* (DOF). The total number of DOFs of a rigid body in space cannot exceed six (e.g., three translational motions along the Cartesian x-, y-, and z-axes and three rotary motions about these axes, as shown in Fig. 1.1). As soon as it is possible to control several DOFs of the end-effector via a mechanical system, this system can be called a *robot*.

The last few decades have witnessed an important development in the use of robots in the industrial world; this increased dependence on robots is primarily due to their flexibility. However, the mechanical architecture of the most common robots is nonadaptive to certain tasks. Other types of architectures have therefore recently been studied and are being increasingly used regularly within the industrial realm. This holds true for the parallel manipulator, in which the system that converts the motions of several bodies into constrained motions of other bodies is referred to as the parallel mechanism.

X.-J. Liu and J. Wang, *Parallel Kinematics: Type, Kinematics, and Optimal Design*, Springer Tracts in Mechanical Engineering, DOI 10.1007/978-3-642-36929-2_1,

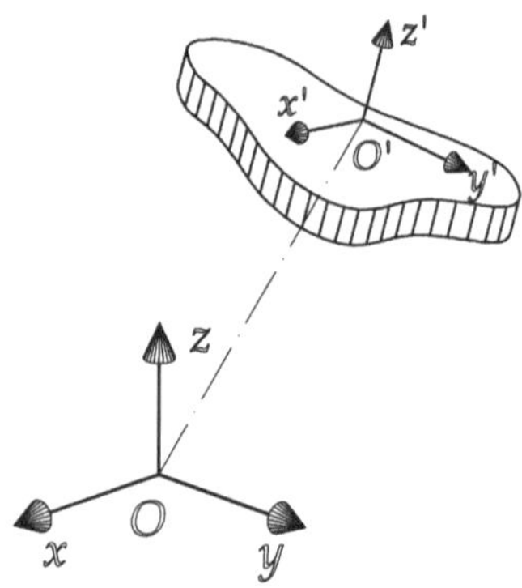

Fig. 1.1 Rigid body in the Cartesian space

A parallel mechanism is a closed-loop system that is made up of an end-effector (or mobile platform, moving platform) with n ($n > 1$) DOFs and a fixed base, linked together by m ($m > 1$ and may be greater or less than n) independent kinematic chains (or legs). These kinematic chains are assemblages of links and joints. Actuation takes place through k ($k \geq n$) simple actuators in chosen joints. According to the definition, several cases apply:

(1) $k = n$, non-redundantly driven parallel mechanism, which is a general case of parallel mechanisms
(2) $k > n$, redundantly driven parallel mechanism
(3) $m > n$, a parallel mechanism with redundant legs
(4) $m < n$, a parallel mechanism with one or more actuated joints in each leg; referred to as a hybrid parallel mechanism
(5) $n = m = k$, fully parallel mechanism

For case (3), there is usually only one redundant leg, in which no actuated joint exists. For such a mechanism, the DOF of the mobile platform is dependent on the mobility of the passive leg.

A parallel mechanism may be symmetrical or asymmetrical. It is said to be symmetrical if it satisfies the following conditions:

(a) It is a fully parallel mechanism.
(b) The arrangement of the joints attached to the mobile platform and the fixed base complies with a specified rule.
(c) The corresponding fixed-length links in all the legs have the same length.
(d) The type and number of joints in all the legs are arranged in an identical pattern.
(e) The number and location of the actuated joints in all the legs are the same.

When the conditions above are not satisfied, the mechanism is asymmetrical.

The parallel mechanism is interesting for the following reasons:

- A minimum of two legs enables the distribution of the load on the legs.
- When the actuated joints are locked, the mechanism remains in its position, an important safety aspect for certain applications.

Because the external load can be shared by actuators, parallel mechanisms tend to have a large load-carrying capacity. Parallel mechanisms are always presented as exhibiting excellent performance in terms of accuracy, rigidity, and ability to manipulate large loads. They have been used in a substantial number of applications ranging from astronomy to flight simulation and are becoming increasingly popular in the machine tool industry.

1.2 Joints and Legs

A parallel mechanism is a closed-loop system made up of joints and links with a specified pattern. The joints can be generally classified as simple joints and combined joints. Figure 1.2 shows examples of simple joints, such as the revolute (R), prismatic (P), cylinder (C), and spherical (S) joints. Figure 1.3a shows a typical combined joint, i.e., the universal (U) joint. Note that if an S joint is designed as the combination of three intersecting R joints, it becomes a combined joint (Fig. 1.3b).

The kinematic chains (legs) used mostly in parallel mechanisms are shown in Table 1.1. These are called simple legs. In this context, a 6-DOF leg means its end-effector has three translations and three rotations that are independent in the Cartesian space. To improve the performance or constrain the specified DOF of

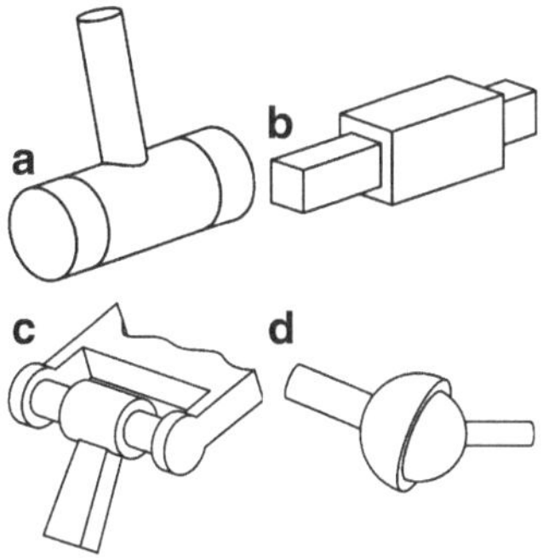

Fig. 1.2 Some simple joints: (**a**) revolute (R) joint, (**b**) prismatic (P) joint, (**c**) cylinder (C) joint, and (**d**) spherical (S) joint

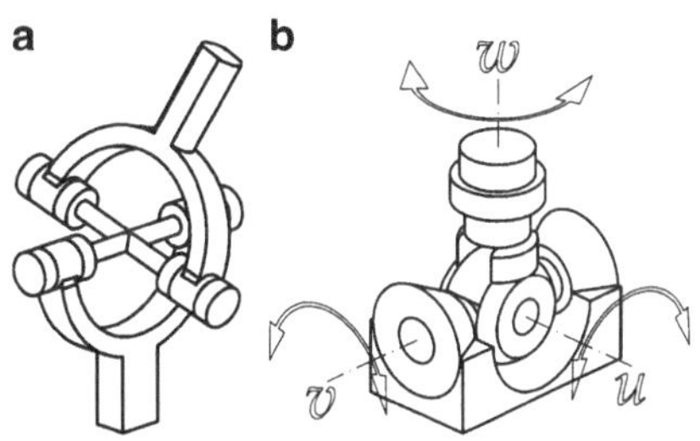

Fig. 1.3 Two combined joints: (**a**) universal (U) joint and (**b**) spherical joint

Table 1.1 Some simple legs typically used in parallel mechanisms

Leg DOF	Joints	Leg example	Figure illustration
2	R, R	RR	Fig. 1.4a
	R, P	RP, PR	Fig. 1.4b, c
	P, P	PP	
3	R, R, R	RRR	Fig. 1.5a, b
	R, P, R	RPR, PRR	Fig. 1.5c
	P, C	PC	
	R, C	RC, CR	
4	P, R, U	PUR, PRU, UPR, RPU	Fig. 1.6a
	P, R, C	PRC, RPC, CPR	Fig. 1.6b
	P, S	PS	Fig. 1.6c
	P, R	PRRR	
	R, S	RS	
	R, C	CRR, RRC	Fig. 1.6d
5	R, R, S	RRS, RSR	Fig. 1.7a
	R, P, S	RPS, PRS, SPR, PSR	Fig. 1.7b, c
	P, S	PPS	
	P, C, U	PCU	
	R, U	RUU, URU, RRRU	Fig. 1.7d
	R, C	RRCR	
	P, U	PUU, UPU	
	R, P, U	RPUR (specific condition needed)	Fig. 1.31
6	P, S	PSS, SPS	Fig. 1.8a
	P, U, S	PUS, UPS, SPU	Fig. 1.8b
	R, S	RSS, SRS	
	U, R, S	RUS, URS, SRU	
	P, R, S	PPRS, PPSR	Fig. 1.8c

parallel mechanisms, some simple mechanisms, particularly parallelograms, are used as one part of a kinematic chain. Such a kinematic chain is referred to as a complex leg. Table 1.2 shows some complex legs. Simple mechanisms are typically not used in a leg with six DOFs.

1.3 Architectures

Given that the number of DOFs of a rigid body in space cannot exceed six, the number of DOFs of a parallel mechanism can be any number between two and six. From the birth of the first design of a parallel mechanism, a large number of mechanical designs for parallel mechanisms with two to six DOFs have been proposed. Some typical parallel mechanisms are introduced in the succeeding sections.

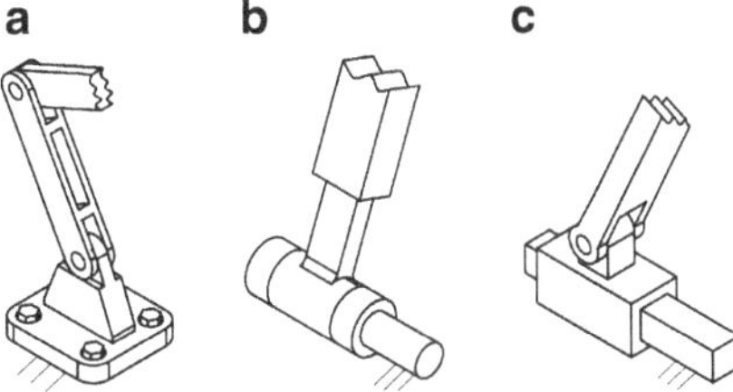

Fig. 1.4 Three kinds of simple legs with two DOFs: (**a**) RR leg, (**b**) RP leg, and (**c**) PR leg

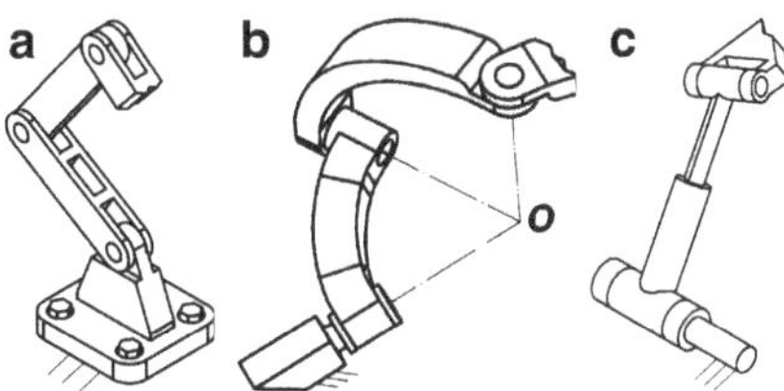

Fig. 1.5 Three kinds of simple legs with three DOFs: (**a**) planar RRR leg, (**b**) spherical RRR leg, and (**c**) RPR leg

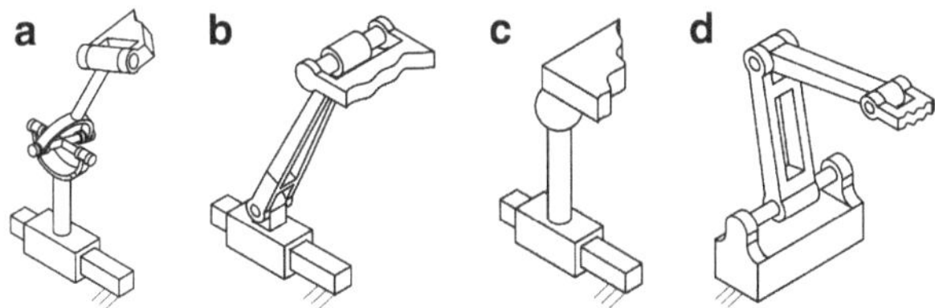

Fig. 1.6 Four kinds of simple legs with four DOFs: (**a**) PUR leg, (**b**) PRC leg, (**c**) PS leg, and (**d**) CRR leg

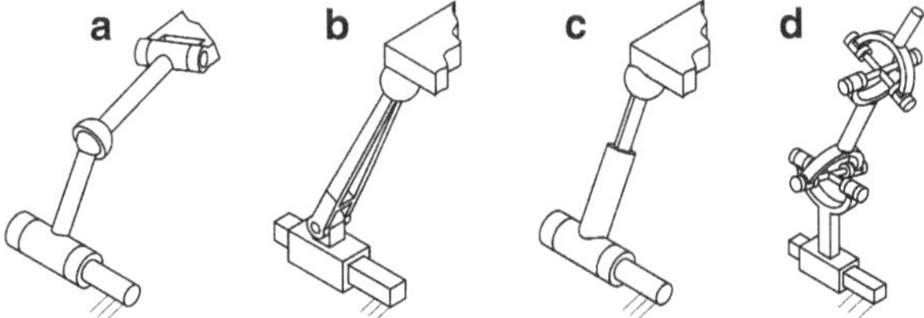

Fig. 1.7 Four kinds of simple legs with five DOFs: (**a**) RSR leg, (**b**) PRS leg, (**c**) RPS leg, and (**d**) PUU leg

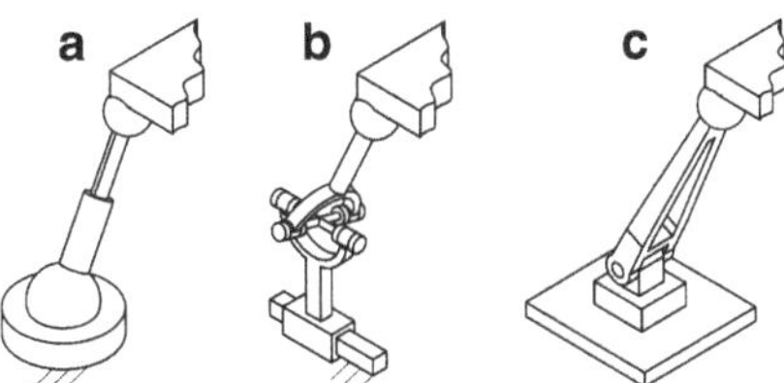

Fig. 1.8 Three kinds of simple legs with six DOFs: (**a**) SPS leg, (**b**) PUS leg, and (**c**) PPRS leg

Table 1.2 Some complex legs used in parallel mechanisms

Leg DOF	Joints and simple mechanisms	Leg example	Figure illustration
2	R, Pa	R(Pa), (Pa)R	
	P, Pa	P(Pa)	Fig. 2.24
	Pa	(Pa)(Pa)	Fig. 2.23
3	R, Pa	R(Pa)R	
	P, Pa, R	P(Pa)R	
	Pa, U	(Pa)U	
	P, P^P	$P(P^P)$	Fig. 1.9
4	R, Pa	RR(Pa)R	Fig. 1.10a
	P, Pa, R	PR(Pa)R	Figs. 1.10b and 2.12
	C, Pa, R	C(Pa)R	
	S, Pa	(Pa)S	
	P, P^P, R	$P(P^P)R$	Fig. 1.10c
	P^P, U	$(P^P)U$	
5	P, Pa, S	P(Pa)S, (Pa)PS	Fig. 1.11a, b
	R, R, Pa, U	RR(Pa)U	Fig. 1.11c
	P, R, Pa, U	PR(Pa)U	Fig. 1.11d
	P, P^P, U	$P(P^P)U$	Fig. 1.11e
	P^P, S	$(P^P)S$	
6	P^{5R}, R, S	$(P^{5R})SR$, $(P^{5R})RS$	Fig. 1.38d

Note: (Pa), (Ps), (P^P), and (P^{5R}) denote planar parallelogram, spatial parallelogram with S joint or U joint, a mechanism in which two platforms are connected by three UU chains, and planar 5R parallel mechanism, respectively

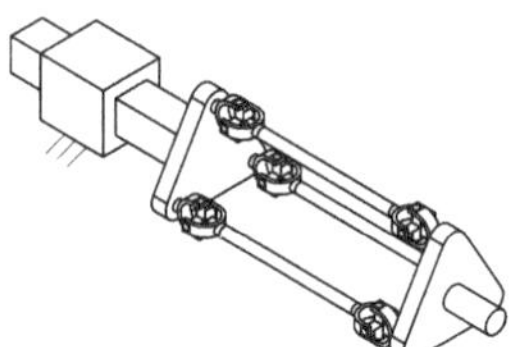

Fig. 1.9 Complex leg with three DOFs: the $P(P^P)$ leg

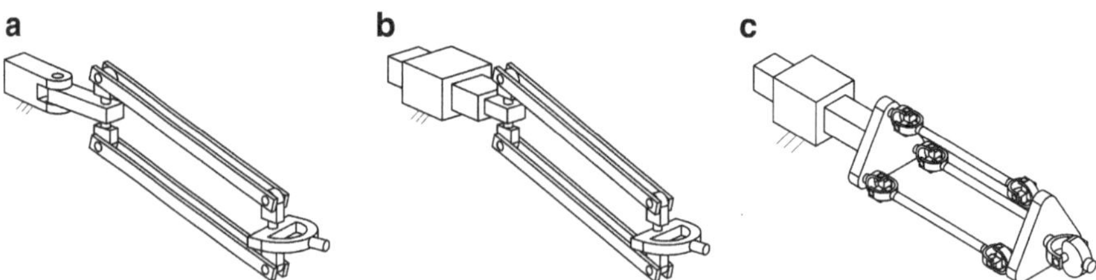

Fig. 1.10 Three kinds of complex legs with four DOFs: (**a**) RR(Pa)R leg, (**b**) PR(Pa)R leg, and (**c**) $P(P^P)R$ leg

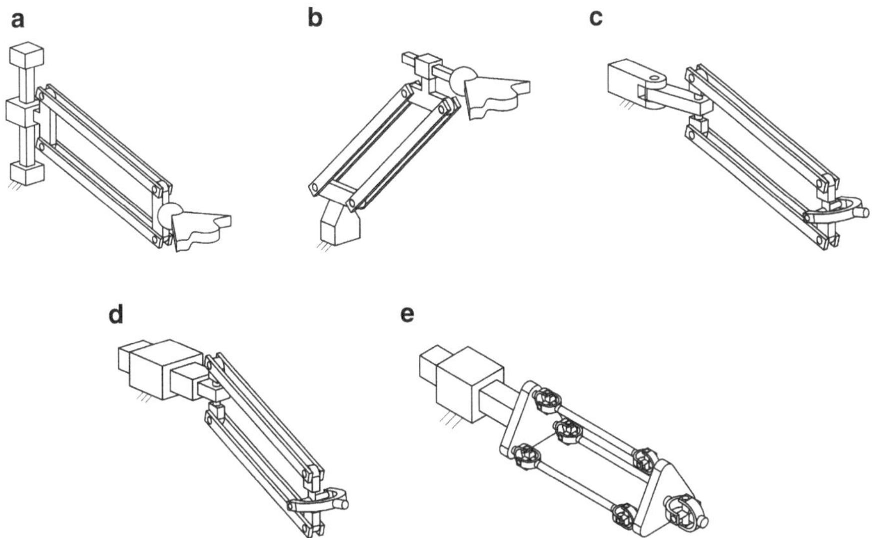

Fig. 1.11 Five kinds of complex legs with five DOFs: (**a**) P(Pa)S leg, (**b**) (Pa)PS leg, (**c**) RR(Pa)U leg, (**d**) PR(Pa)U leg, and (**e**) P(P^P)U leg

1.3.1 Two-DOF Parallel Mechanisms

Most existing 2-DOF parallel mechanisms are planar manipulators with two translational DOFs. In those designs, only prismatic and revolute joints are used. McCloy (1990) showed that 20 different combinations exist if there are only five bars. This number is reduced to six (Fig. 1.12) if the actuators are assumed attached to the ground, that there is no passive prismatic joint, and that no actuator supports the weight of another actuator. Among these mechanisms, the 5R symmetrical parallel mechanism is the most extensively studied (Gao et al. 1998; Cervantes-Sánchez et al. 2000, 2001; Liu et al. 2006b, c; Macho et al. 2008); the PRRRP mechanism usually has the advantage of iso-stiffness along the actuation direction. It is typically used in machine tools (Stengele 2002).

Aside from the mechanisms shown in Fig. 1.12, some other 2-DOF parallel mechanisms are shown in Fig. 1.13. This figure illustrates the mechanism with three RRR kinematic chains. Each leg features an active R joint; thus, this is a redundant mechanism (Kock and Schumacher 1998). Figure 1.13b shows that the mechanism has two RRR legs and one RR leg, in which only the former are active while the latter is passive. This mechanism has the mobility of the RR leg, i.e., two positional DOFs in a plane. Figure 1.13c illustrates an interesting mechanism, whose mobile platform has two translational motions along the x- and y-axes (Chen et al. 2007). Furthermore, the leg with the PRRR chain has no stiffness along the z-axis. A motion-decoupled mechanism, which consists of two PC or PP kinematic chains, is illustrated in Fig. 1.13d. Given that the two actuated prismatic joints are arranged

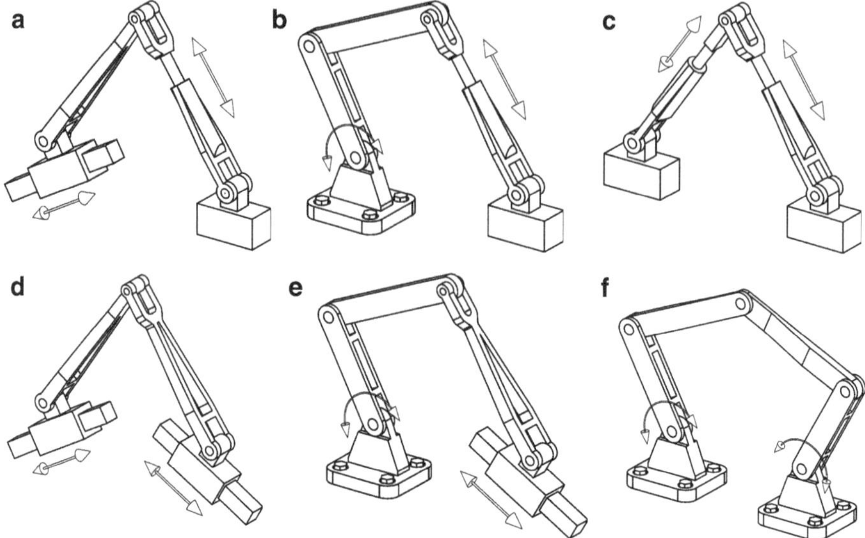

Fig. 1.12 Two-DOF parallel mechanisms: (**a**) PRRPR mechanism, (**b**) RRRPR mechanism, (**c**) RPRPR mechanism, (**d**) PRRRP mechanism, (**e**) RRRRP mechanism, and (**f**) 5R mechanism

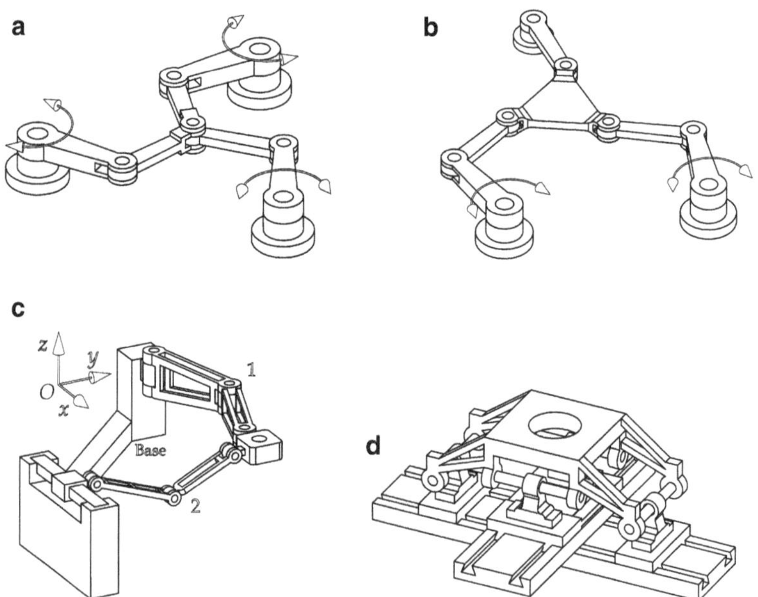

Fig. 1.13 Some 2-DOF parallel mechanisms: (**a**) 3-RRR mechanism with a redundant active leg, (**b**) 2-RRR&1-RR mechanism with a redundant leg, (**c**) decoupled RRR&PRRR mechanism, and (**d**) decoupled 2-PC(or 2-PP) mechanism

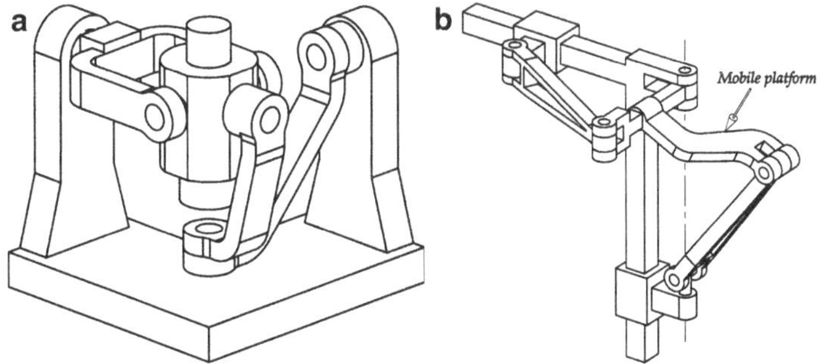

Fig. 1.14 Two orientational 2-DOF parallel mechanisms: (**a**) 5R parallel mechanism and (**b**) PRRURUP mechanism

in a mode wherein they are perpendicular to each other and the two passive P joints are also in this mode, the two translations of the mobile platform are decoupled.

Some other interesting 2-DOF parallel mechanisms are introduced in Sect. 2.3.1.

The mobile platforms of the mechanisms shown in Fig. 1.13c, d have two translational DOFs. Figure 1.14 shows two parallel mechanisms with two orientational DOFs. The mechanism shown in Fig. 1.14a is a five-bar spherical mechanism (Gosselin and Caron 1999). In Fig. 1.14b, three joints, i.e., two prismatic joints and one revolute joint of the mechanism, are fixed to the base. The mobile platform is linked to the output link of a slider-rocker mechanism by a revolute joint and is connected to the base by a PUR chain. The two revolute joints attached to the mobile platform are parallel to each other and are also parallel to one revolute joint in the universal joint. The rotational axis of the revolute joint attached to the base is collinear to that of another revolute joint in the U joint. At any moment, the parallel mechanism can be considered the combination of two slider-rocker mechanisms. The two orientations of the mechanism illustrated in Fig. 1.14b are decoupled when the two prismatic joints are active (Carricato and Parenti-Castelli 2004).

1.3.2 Three-DOF Parallel Mechanisms

Many 3-DOF parallel mechanisms are available, with some popular ones introduced in this chapter. An example is the planar 3-RRR (R stands for revolute joint) parallel mechanism (Gosselin and Angeles 1988) shown in Fig. 1.15. The mobile platform has three planar DOFs, which are two translations along the x- and y-axes and a rotation around the axis perpendicular to the O-XY plane. Figure 1.16 shows some other parallel mechanisms with three planar DOFs. Among these mechanisms, that shown in Fig. 1.16d is an actuation-redundant version of the mechanism in Fig. 1.16c. This way, singularity is avoided and the orientational workspace of the mobile platform is improved (Wu et al. 2007). Figure 1.17 shows a planar 3-DOF parallel mechanism with decoupled motions (Yu et al. 2008). The mobile platform

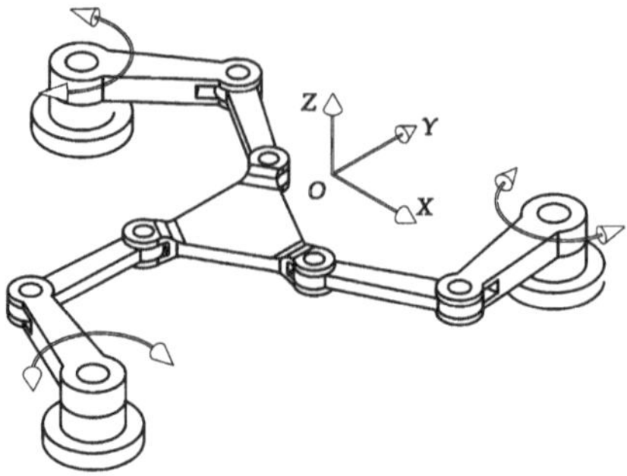

Fig. 1.15 Planar 3-RRR parallel mechanism

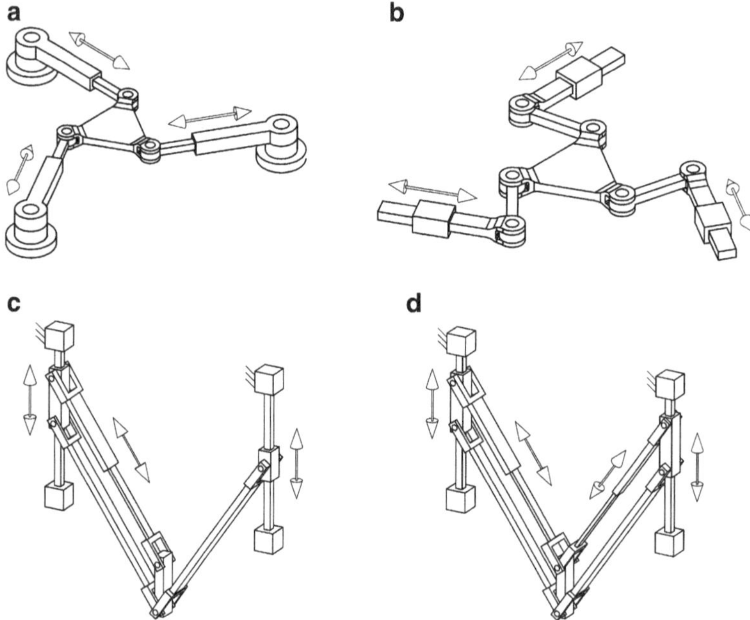

Fig. 1.16 Four planar 3-DOF parallel mechanisms: (**a**) 3-RPR mechanism, (**b**) 3-PRR mechanism, (**c**) 2-PRR&1-RPR mechanism, and (**d**) 2-PRR&2-RPR mechanism

is connected to the base by PPRP, PR, and PRP chains, where the axes of two C joints are collinear and three P joints are actuated.

Another example with 3-RRR chains is the spherical parallel mechanism (Gosselin and Angeles 1989; Liu et al. 2000) shown in Fig. 1.18a. In this design, all the joint axes intersect at a common point. The motion of any point in the mechanism is the rotation about the point. Moreover, the mobile platform has only orientational DOFs with respect to the base. In each RRR chain, the relative angle between adjacent R joints can be varied. For example, when the angle between the

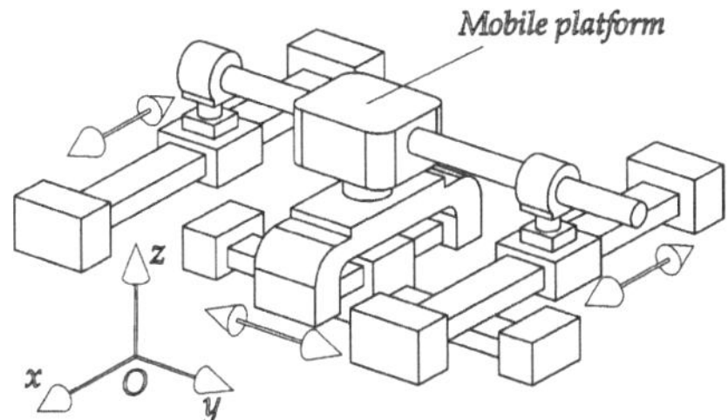

Fig. 1.17 Planar 3-DOF parallel mechanism with decoupled motions

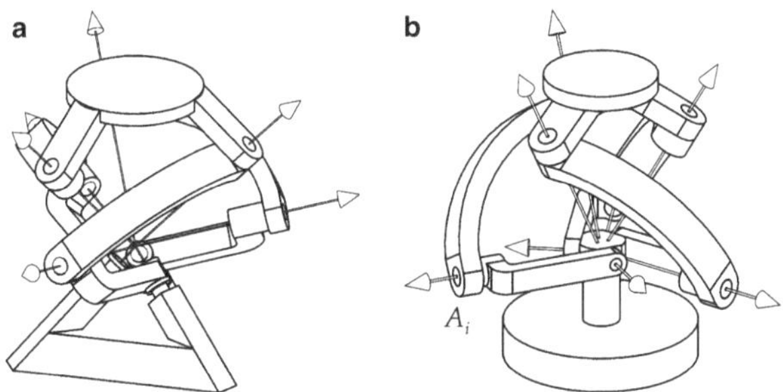

Fig. 1.18 Spherical 3-RRR parallel mechanisms: (**a**) a general case and (**b**) a special case

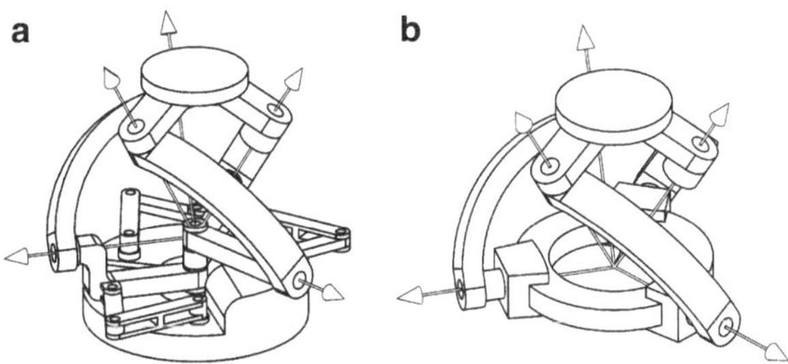

Fig. 1.19 Revised version of the spherical 3-RRR parallel mechanism (**a**) with a planar four-bar mechanism as the input and (**b**) with 3-PRR chains

R joint attached to the base and its adjacent R joint is 90°, the mechanism will be that shown in Fig. 1.18b. Furthermore, its input can be given by a planar four-bar mechanism, shown in Fig. 1.19a. The mechanism in Fig. 1.18b shows that the locus of the end point A_i of the input link is a circle. Then, the input link can be mounted on a circular guide, and the spherical 3-RRR parallel mechanism can be varied to a spherical 3-PRR mechanism (see Fig. 1.19b).

In the family of 3-DOF parallel mechanisms, a highly important group has drawn considerable attention. They have the same characteristic, which is that the mobile platform has complex DOFs, three independent DOFs (one translation and two

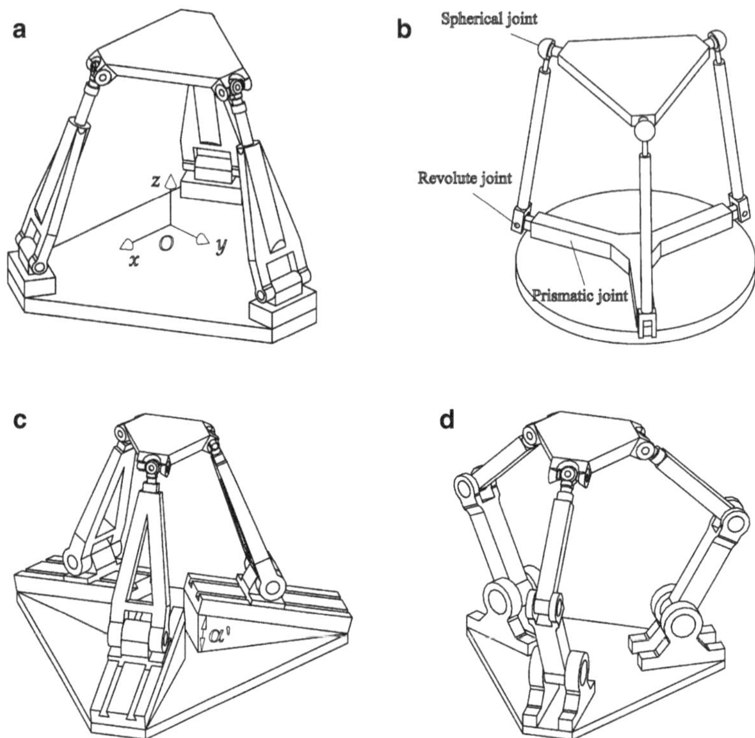

Fig. 1.20 Some 3-[PP]S parallel mechanisms: (**a**) 3-RPS mechanism, (**b**) 3-PRS mechanism, (**c**) inclined 3-PRS mechanism, and (**d**) 3-RRS mechanism

rotations), and parasitic motions. When different orientation description methods are used, the parasitic motions also vary. They are one rotation and two translations when three Euler angles are used (Carretero et al. 2000) and two translations when *azimuth* and *tilt* angles are employed (Liu and Bonev 2008). These mechanisms are called 3-[PP]S parallel mechanisms (Bonev 2002), whose three spherical joints move in vertical planes intersecting at a common line. Such mechanisms are referred to as zero-torsion mechanisms (Bonev 2002). Abundant literature on 3-[PP]S parallel mechanisms has been published. The 3-RPS parallel mechanism (see Fig. 1.20a) presented by Hunt (1983) is such a mechanism and has been studied by many other researchers (Lee and Arjunan 1991; Fang and Huang 1997). Another [PP]S parallel mechanism with three PRS kinematic chains (Fig. 1.20b) was analyzed by Carretero et al. (1998); an inclined 3-PRS parallel mechanism (Fig. 1.20c) was examined by Pond and Carretero (2004); a 3-RRS mechanism (Fig. 1.20d) was investigated by Li et al. (2002). Finally, a 3-PCU (see Fig. 2.50) was proposed by Liu et al. (2004). Among these mechanisms, the 3-PRS mechanism shown in Fig. 1.21 is worth mentioning. DS Technologie in Germany developed a new machining tool head (Wahl 2000), the Sprint Z3, which is based on the 3-PRS

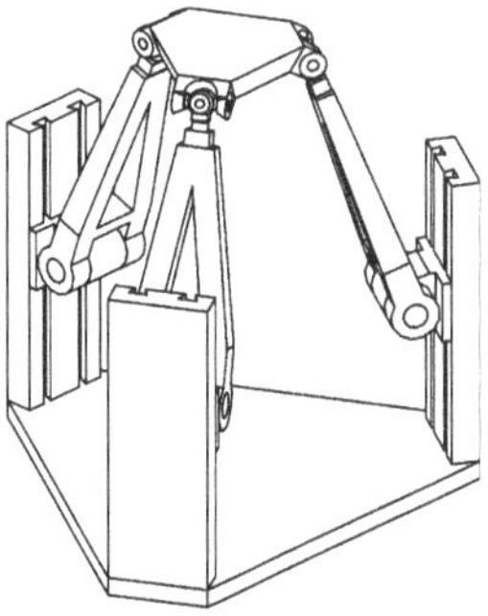

Fig. 1.21 3-PRS mechanism applied to Sprint Z3 by DS Technologie of Germany (Wahl 2000)

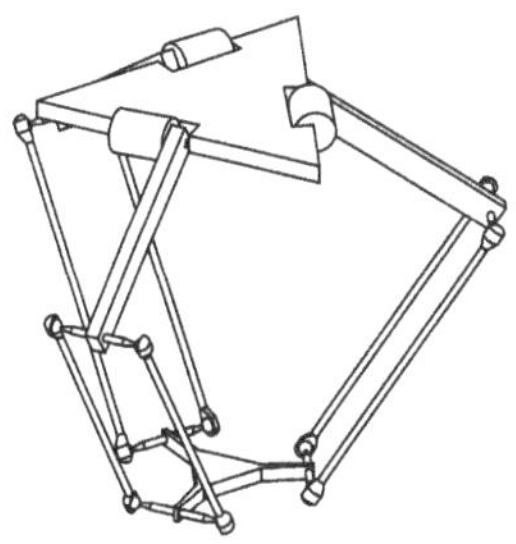

Fig. 1.22 DELTA mechanism

mechanism. The Z3 tool head was designed to endow performance improvements to thin wall machining applications for aluminum aerospace structural components. In May 2000, Cincinnati Machine and DS Technologie announced a strategic alliance through which machine tools, services, and support from both companies are offered to the aerospace industry in North and South America through Cincinnati Machine. Five-axis machine tools equipped with Z3 have achieved huge success in industry. The experience with the Z3 head shows that a machine tool with a zero-torsion tool head has advantages in terms of efficiency and accuracy. Furthermore, a zero-torsion mechanism has a simpler kinematic model (Liu and Bonev 2008). However, its kinematics, calibration, and control are relatively difficult because parasitic motions exist in a 3-[PP]S parallel mechanism.

The most famous mechanism with three translations is DELTA (Fig. 1.22), proposed by Clavel (1986) and marketed by the Demaurex Company and ABB under the name IRB 340 FlexPicker. In DELTA, the mobile platform is connected to the base by three legs with PR(Ps) chains, where (Ps) stands for the spatial four-bar parallelogram with four spherical joints, P prismatic joint, and R revolute joint, where the R joints are actuated. DELTA has been regarded by the industry as a highly attractive innovation and is covered by a family of 36 patents (Bonev 2001). When considering three translational parallel mechanisms, the mechanism proposed by Tsai (Tsai and Stamper 1996; Fig. 1.23) is worth mentioning. In this design, the three legs are equipped with PR(Pa)R chains, where (Pa) denotes the planar four-bar parallelogram with four revolute joints. Although the manipulator proposed by Tsai has translations identical to those of DELTA, this manipulator is

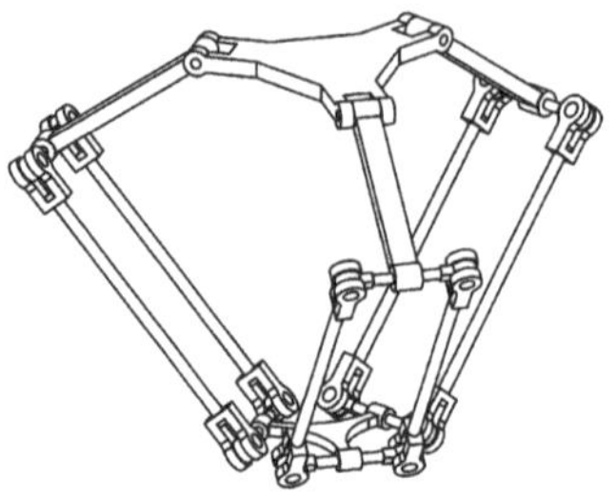

Fig. 1.23 Tsai's mechanism

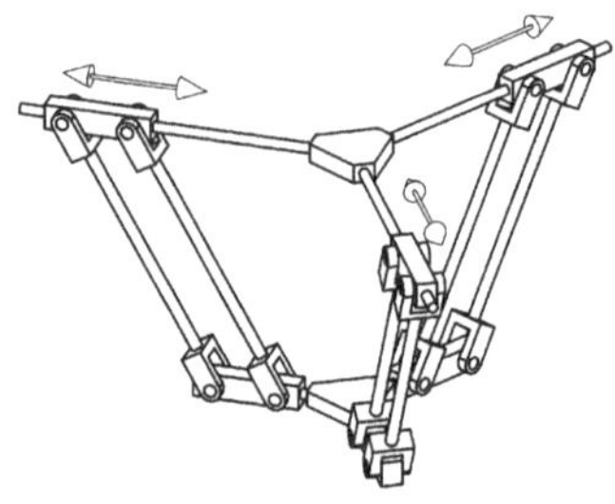

Fig. 1.24 StarLike mechanism

not a version of DELTA. It is probably the first design to solve the problem of the UU chain (see Sect. 2.3.1.1). Another parallel mechanism with three translational DOFs is StarLike (Fig. 1.24), designed by Hervé (1992) on the basis of group theory. This type of parallel mechanism has wide applications in the industrial world; examples include pick-and-place applications, parallel kinematic machines, and medical devices (Bonev 2001). Although these three parallel mechanisms have the same capability, DELTA was more successful in the light industry than the mechanism proposed by Tsai and StarLike because of its easy assembly.

DELTA and Tsai's mechanism have the advantages of high velocity and high acceleration. When the input mode is varied, the performance levels of the two mechanisms are correspondingly different. The most commonly used method is replacing the revolute joint and input link with a prismatic joint and slider, respectively. The linear DELTA and Tsai's mechanism are shown in Fig. 1.25. These mechanisms may earn an identical performance on workspace sections along the actuation direction and a relatively simple kinematic design (Liu 2006). Furthermore, if the directions of the three linear actuations are collinear with three Cartesian axes (see Fig. 1.26), some performance levels will improve. For example, the mechanisms are isotropic at their original points, and their kinematics will be very simple (Liu et al. 2003).

Some 3-DOF translational parallel mechanisms have recently been presented. Figure 1.27 illustrates typical mechanisms; that shown in Fig. 1.27a consists of three RRC chains (Zhao 2000), that shown in Fig. 1.27b is made up of three CRR (or PRRR) chains (Kim and Tsai 2004), and the mechanism in Fig. 1.27b is composed of three PRRR chains with the axes of all R joints intersecting at one common point (Kong and Gosselin 2004a). In the 3-RRC and 3-CRR mechanisms, the axes of

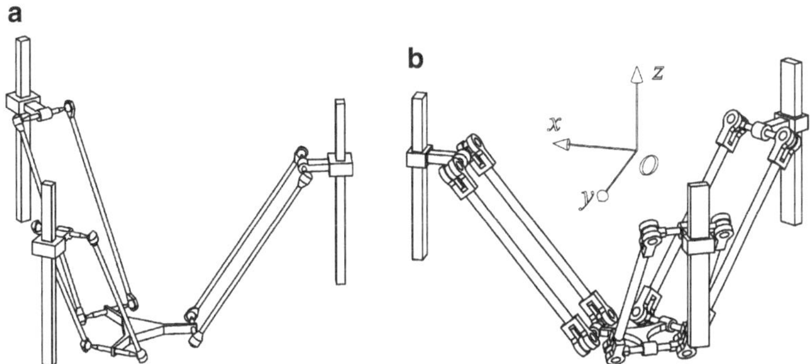

Fig. 1.25 Two 3-DOF translational parallel mechanisms: (**a**) linear DELTA and (**b**) linear Tsai's mechanism

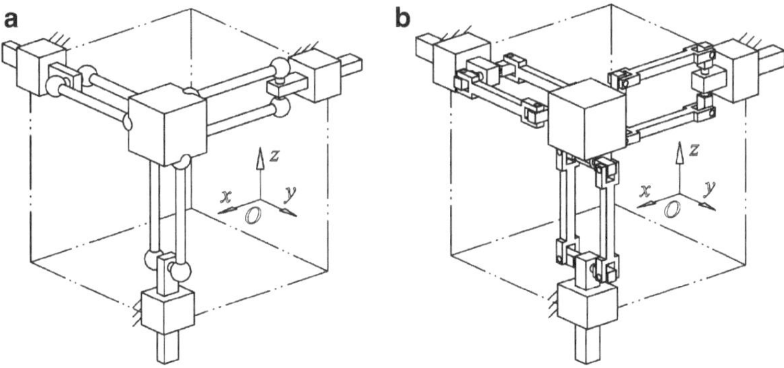

Fig. 1.26 Two 3-DOF parallel cube mechanisms: (**a**) cube-DELTA and (**b**) cube-Tsai's mechanism

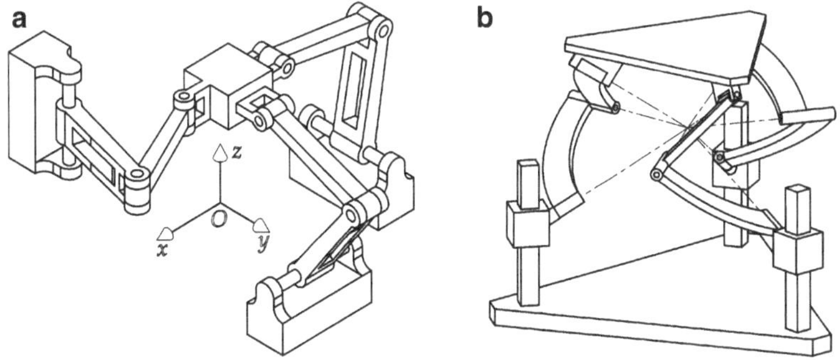

Fig. 1.27 Three kinds of translational parallel mechanisms: (**a**) with 3-CRR (or 3-PRRR) chains (in each leg, the axes of all the R joints are parallel to one another and are orthogonal to those of the other two legs) and (**b**) with 3-PRRR chains (the axes of all the R joints intersect at one point)

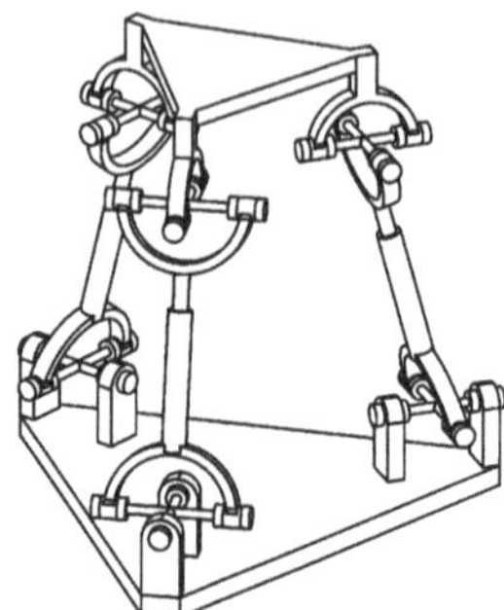

Fig. 1.28 3-UPU parallel mechanism

all the R joints in each leg are parallel to one another and are orthogonal to those of the other two legs. Generally, the 3-CRR mechanism is better than the 3-RRC mechanism in the workspace because passive prismatic motion exists in the latter. Some other 3-DOF translational parallel mechanism can be found in the reference (Carricato and Parenti-Castelli 2003).

Some 3-DOF translational parallel mechanisms have two universal joints in their legs. For example, a different architecture of a 3-DOF 3-UPU translational parallel mechanism (see Fig. 1.28) was proposed in 1996 by Tsai (1996). In this design, the mobile platform is connected to the base through three identical legs. Each leg comprises a prismatic actuator with two U joints at the ends. In the mechanism, the two inner revolute joints are parallel to each other; the same is true for the two outer revolute joints. The special arrangement of the axes of the universal joints (Fig. 1.28) provides the necessary constraint that keeps the orientation of a mobile platform constant. In practical applications, however, guaranteeing these conditions is difficult, if not impossible; usually, constraint singularity exists in a kinematic chain with two universal joints (Zlatanov et al. 2002). The 3-UPU parallel mechanism developed in Seoul National University loses control when the mobile platform plane moves away from its initial configuration (Zlatanov et al. 2002). For this case, therefore, a symmetrical 3-DOF translational parallel mechanism with two U joints in each of its legs is not recommended. To solve this problem, a parallelogram can be used in the legs, an idea introduced in Sect. 2.3.1.1.

In another type of 3-DOF parallel mechanism, the mobile platform is connected to the base through four legs, in which the fourth leg is a passive one and is also the leading leg, indicating that the leg determines the motion of the mobile platform. A spherical coordinate parallel mechanism with 3-PUS&1-UP chains, a similar parallel mechanism with 3-UPS&1-UP chains (Siciliano 1999), a pure rotational parallel mechanism with 3-UPS&1-S chains (Wang and Gosselin 1999), and a pure translational parallel mechanism with 3-UPS&1-R(Pa)(Pa) chains (Schoppe et al. 2002) are shown in Fig. 1.29a–d, respectively. They have common characteristics, i.e., each of the actuated legs has six DOFs and the mobile platform has the DOFs of the passive leg. The mechanism in Fig. 1.29a was applied to the machine tool design

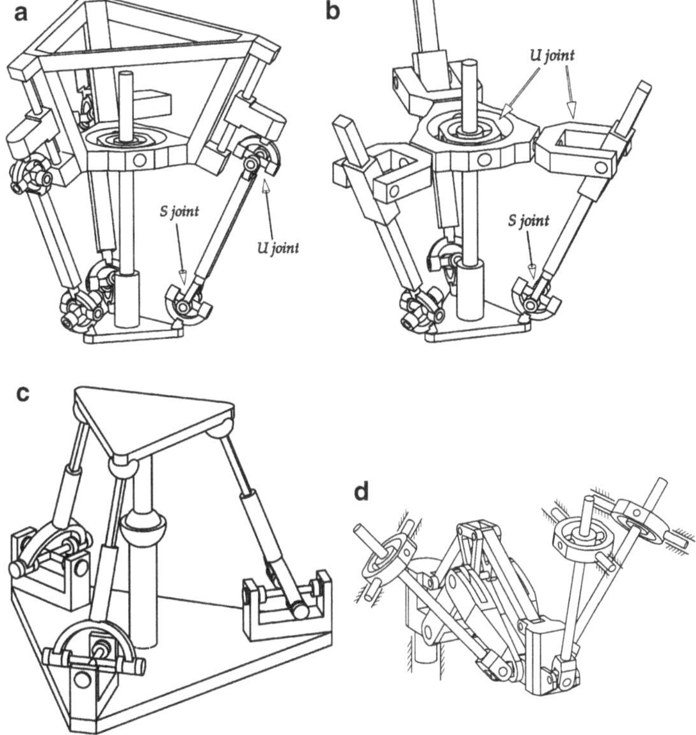

Fig. 1.29 Four 3-DOF parallel mechanisms with four legs: (**a**) 3-PUS&1-UP mechanism, (**b**) 3-UPS&1-UP mechanism (the parallel mechanism of the Tricept series machines (Neumann 2006)), (**c**) 3-UPS&1-S mechanism, and (**d**) 3-UPS&1-R(Pa) (Pa) mechanism (the parallel mechanism of SKM 400 (Schoppe et al. 2002))

of the IFW of the University of Hannover. The machine designed with the second mechanism and the tool head attached to its mobile platform is the Tricept (see http://www.neosrobotics.com/), which has also achieved great success in the fields of machining and industrial robotics. The mechanism in Fig. 1.29d is a parallel device in which the spindle is fixed by the SKM 400 machine tool (Schoppe et al. 2002) developed by Heckert Company in Germany.

No spatial fully parallel mechanism with two translational DOFs and one rotational DOF has existed until 2001 (Fig. 2.15; Liu et al. 2001a). After this, researchers proposed similar mechanisms that are based on the units of single opened chains (Yang 2004) and screw theory (Kong and Gosselin 2005). Some other mechanisms are introduced in Sect. 2.3.

Figure 1.30 shows another concept of a spatial 3-DOF parallel mechanism, that of Exechon (Neumann 2006). The mobile platform is connected to the base by three legs, two of which have UPR chains with the third having an SPR chain. The mechanism is said to have a very large workspace. With a two-axis tool head

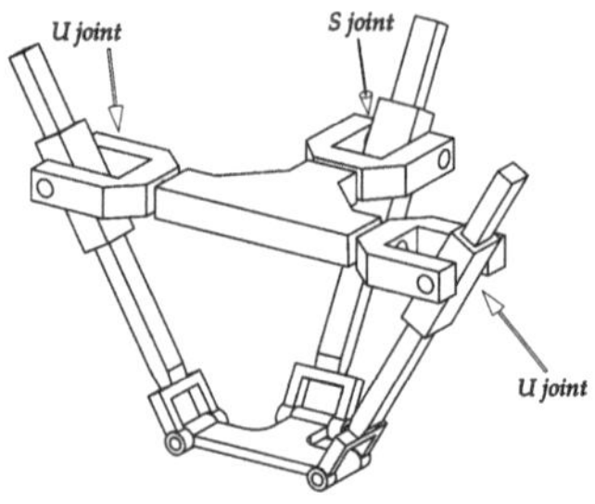

Fig. 1.30 Spatial 3-DOF parallel mechanism with 2-UPR&1-SPR chains

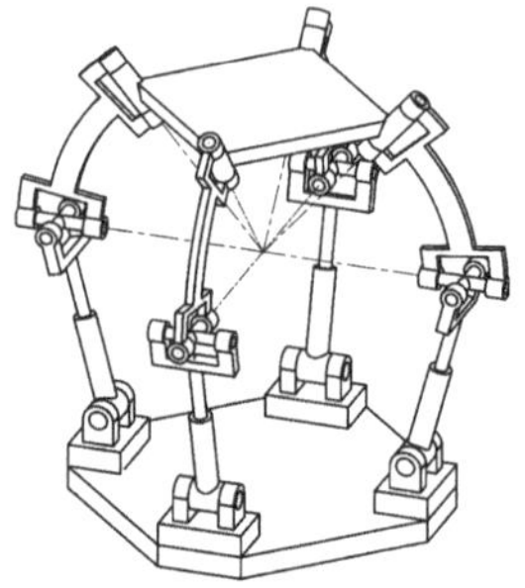

Fig. 1.31 Four-DOF parallel mechanism with four RPUR chains

attached to the mobile platform, Exechon has the capability of agile machining, single setup machining, compound angle machining, and multipath blending and tool-drag elimination. It is also capable of five-face machining with one setup.

1.3.3 Four-DOF Parallel Mechanisms

Few 4-DOF mechanisms exist in the family of parallel mechanisms. In particular, designing a 4-DOF symmetrical parallel mechanism with two links and three joints in each of its legs is very difficult to accomplish. On the basis of screw theory, some symmetrical parallel mechanisms have recently been proposed (Fang and Tsai 2002; Huang and Li 2003; Kong and Gosselin 2004b). Each leg of most of these mechanisms is usually equipped with three links and more than three joints. Figure 1.31 shows a 4-DOF parallel mechanism presented in Huang and Li (2003). Each of its legs is equipped with three links and four joints (one R joint, one P joint, one U joint, and one R joint in sequence from the base platform to the mobile one). The mechanism has three orientational DOFs and one translational DOF if every R joint attached to the mobile platform and one R joint of every U joint intersect at one common point. More links and joints are placed in each leg; thus, such a mechanism is relatively disadvantaged in terms of accuracy.

H4 (Pierrot et al. 2001) is the most popular among the 4-DOF parallel mechanisms. Figure 1.32 presents the architecture of H4, in which the mobile platform is connected to two movable links by revolute joints, and each of the said

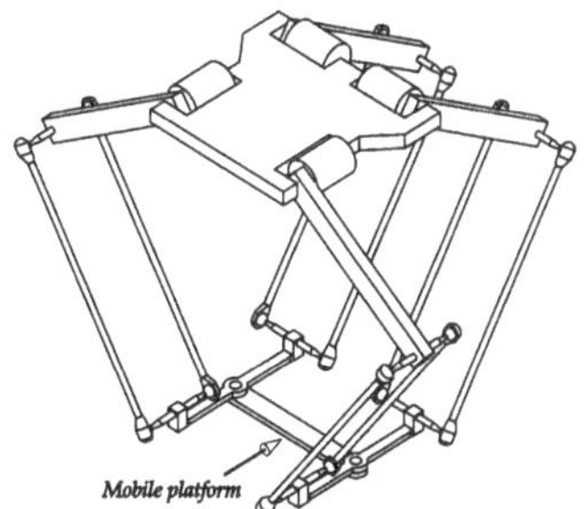

Fig. 1.32 Four-DOF parallel mechanism, H4

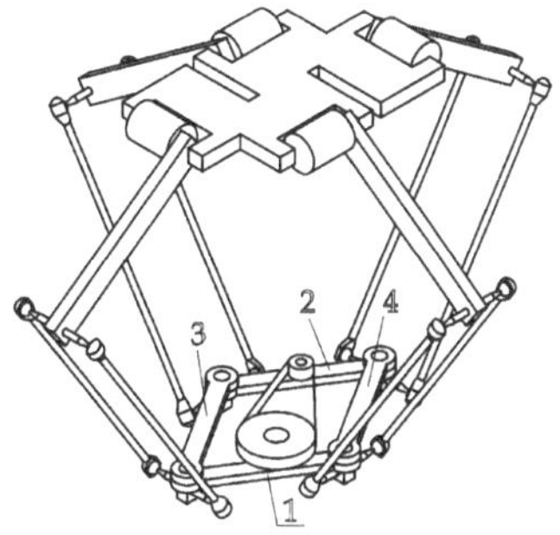

Fig. 1.33 Improved version (Par4) of H4

links is connected to the base by two identical R(Ps) chains. In Company et al. (2006), the arrangement of four chains was discussed. On the basis of this concept, several versions of the mechanism, such as I4 (Krut et al. 2003a), Eureka (Krut et al. 2003b), Par4 (Nabat et al. 2005), Heli 4 (Krut et al. 2006), and Dual 4 (Pierrot et al. 2006), have been proposed. Par4 (Fig. 1.33) is the improved version of I4 and H4. The articulated traveling plate of Par4 is composed of four parts: two main parts (1 and 2) linked by two rods (3 and 4) with revolute joints (Fig. 1.33). An amplification system, which can comprise a gear or belt between two main parts, is added to obtain a complete turn: $\pm\pi$. Using this mechanism as basis, Adept Technology released the Quattro (see http://www.adept.com/products/robots/parallel/quattro-s650/general), which, at 240 cycles per minute, is the industry's fastest pick-and-place robot.

1.3.4 Five-DOF Parallel Mechanisms

The proposal of a 5-DOF fully parallel mechanism with symmetrical architecture is also challenging. Recently, some 5-DOF parallel mechanisms have been synthesized on the basis of screw theory (Fang and Tsai 2002; Huang and Li 2003). One of the 5-DOF parallel mechanisms presented by Fang and Tsai (2002) is illustrated in Fig. 1.34. The mobile platform is connected to the base by five RPUR chains, and the P joints can be actuated. In this mechanism, the following conditions should be satisfied: the R joints connected to the base should be parallel to the adjacent R joints

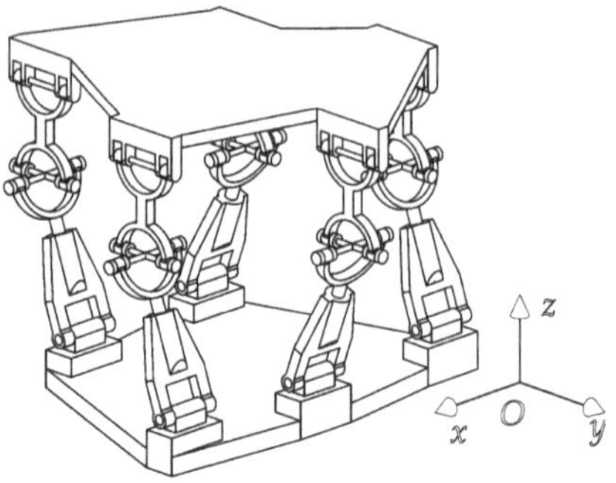

Fig. 1.34 Five-DOF parallel mechanism with five RPUR chains (Fang and Tsai 2002)

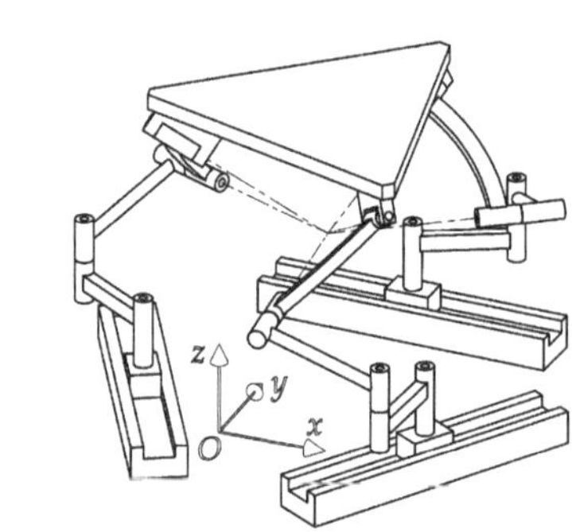

Fig. 1.35 Five-DOF parallel mechanism with three PRRRR chains (Huang and Li 2003)

of the U joints, and the R joints attached to the mobile platform should be parallel to the adjacent R joints of the U joints. The mechanism has three translational and two rotational DOFs, and the rotation about the *z*-axis is constrained (Fig. 1.34). Figure 1.35 shows such a mechanism with three PRRRR chains (Huang and Li 2003). In this mechanism, the six upper R joints should intersect at one common point, and the six lower R joints should be parallel to the *z*-axis. The mobile platform of the mechanism has three rotations and two translations with respect to the base, and the translation along the *z*-axis is constrained (see Fig. 1.35). However, this mechanism is asymmetrical because two of the three legs have two actuators. Moreover, each leg of these mechanisms is equipped with three links and more than three joints. This type of mechanism is more complex in terms of kinematic and dynamic analyses. They require considerable improvement before they are applied in industry.

1.3.5 Six-DOF Parallel Mechanisms

Six-DOF parallel mechanisms are the most popular manipulators and have been studied by more researchers. The architecture shown in Fig. 1.36 is a classical 6-DOF parallel mechanism with six UPS kinematic chains. Most of the 6-DOF parallel mechanisms consist of six legs, which can theoretically be arranged

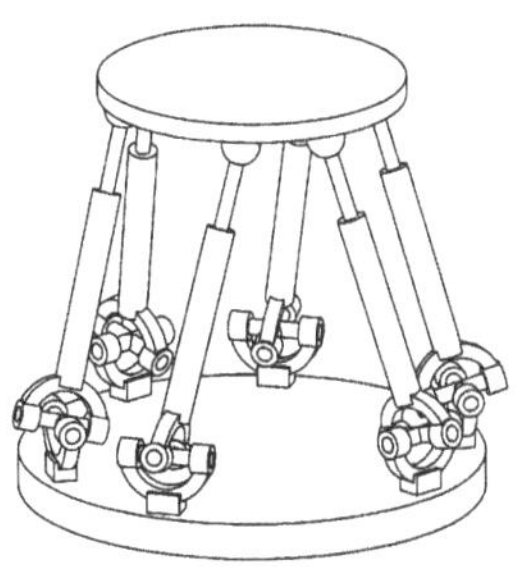

Fig. 1.36 General Stewart platform with six DOFs

arbitrarily. Examples are 6-6 (six joints on the base and six on the mobile platform) (Sreenivasan et al. 1994), 6-3 (Hunt 1983), 5-5 (Hunt and Primrose 1993), 5-4 (Innocent and Parenti-Castelli 1993), 4-4 (Lin et al. 1992), 3-2-1 (Bruyninckx 1997), or cubic-type mechanisms (each two of the six legs are settled at the side of a cube) (Dafaoui et al. 1998). Such a parallel mechanism possesses the advantages of high stiffness, low inertia, and large payload capacity. However, they suffer from relatively small useful workspaces and design difficulties. Furthermore, their direct kinematics is a highly difficult problem.

A rigid body in space has, at most, six DOFs. A 6-DOF parallel mechanism indicates that its mobile platform is fully free in the Cartesian space. From the perspective of type synthesis, therefore, parallel mechanisms with six DOFs are much easier to design than mechanisms with less than six DOFs. Any kinematic chain (such as UPS, PUS, RUS, SPS, RSS, PSS, PPRS, PRPS, and PPSR chains), whose end-effector has six DOFs (NOT mobility), can be the leg of a 6-DOF parallel mechanism because this type of kinematic chain imposes no constraint on the mobile platform. Thus, the 6-DOF parallel mechanisms have the largest number of DOFs in the family of parallel mechanisms. Figure 1.37 illustrates some typical 6-DOF parallel mechanisms.

Legs cannot be arranged arbitrarily. A rule that must be complied with is that the hexagon formed by the joints fixed to the base should not be similar to that formed by the joints attached to the mobile platform. Otherwise, the parallel mechanism belongs to architectural singularity (Ma and Angeles 1991) and will lose its control in the workspace.

Some exotic chain mechanisms are also available. The mechanism is actuated by a planar mechanism (e.g., a five-bar mechanism), or two actuators are in each leg; in this case, the mechanism usually consists of three legs. Figure 1.38a shows a mechanism with three PRPS mechanisms, in which the two P joints are active. The mechanisms in Fig. 1.38b, c have PPRS and PPSR chains, respectively, and both are actuated by planar motors. Figure 1.38d illustrates a mechanism with 3-(P^{5R})SR chains, in which the position of each S joint is determined by a five-bar parallel mechanism.

Parallel mechanisms with six legs usually have a limited tilting angle of the mobile platform. Actuation redundancy is one solution to improving the angle. Eclipse series mechanisms (Kim et al. 2001; Kim et al. 2002) have redundant

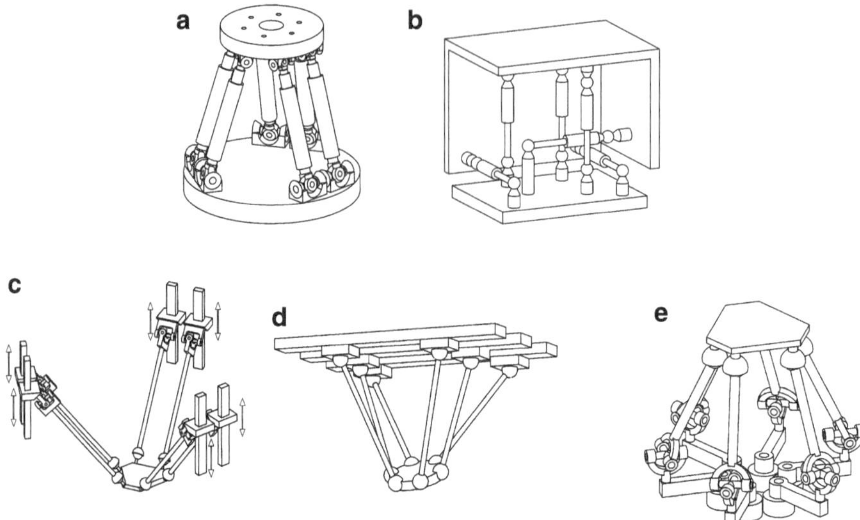

Fig. 1.37 Some typical 6-DOF parallel mechanisms: (**a**) a 6-UPS mechanism with two adjacent legs parallel to each other at its initial configuration, (**b**) a 3-2-1 type of mechanism with six UPS chains, (**c**) a 6-PUS parallel mechanism, (**d**) the Hexaglide mechanism (Honegger et al. 2000), and (**e**) a 6-RUS mechanism

actuations. In particular, the mobile platform of Eclipse II can tilt to 360° about three orthogonal axes. Figure 1.39 shows the mechanism of Eclipse I, in which the mobile platform is connected to the base by three PPRS chains. In this mechanism, the six P joints are all actuated. To obtain a high tilting angle of the mobile platform, two additional actuators $A1$ and $A2$ (Fig. 1.39) are attached to two of the three revolute joints. This way, the developed machine that uses the Eclipse I mechanism can perform five-face machining (Kim et al. 2001).

Given that parallel mechanisms with six DOFs can simulate any composite motion of a rigid body in space, they have been extensively applied in motion simulators, damping devices, positioning machines, and any other device that requires six DOFs. However, many experiments indicate that such a mechanism is disadvantageous in the application of machine tools because of potential calibration difficulty and lower tilting angles. Since the end of the twentieth century, serial-parallel machine tools have drawn considerable interest because these machines offer the advantages of both serial and parallel mechanisms. As a result, 2- and 3-DOF parallel mechanisms have been increasingly proposed and studied. Some have been used in industry, and the exploration of such mechanisms is foreseen to continue into the future. So we will discuss an issue on type synthesis of parallel mechanisms with lower DOF (two to five) in detail.

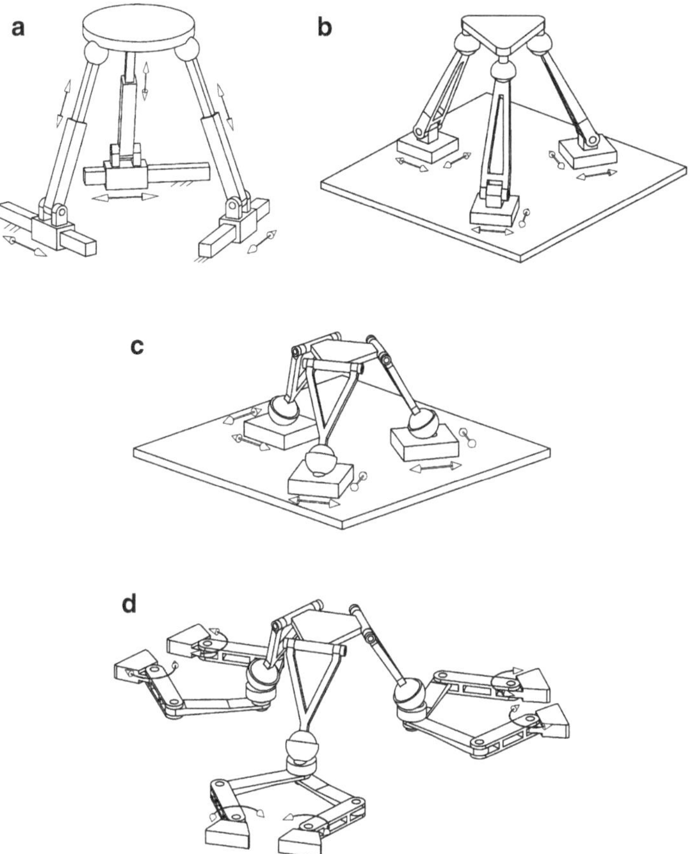

Fig. 1.38 Some 6-DOF parallel mechanisms with three legs: (**a**) 3-PRPS mechanism, (**b**) 3-PPRS mechanism, (**c**) 3-PPSR mechanism, and (**d**) 3-(P^{5R})SR mechanism

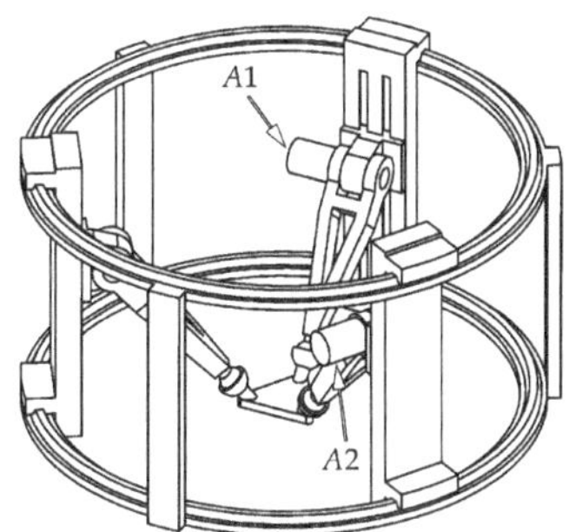

Fig. 1.39 Six-DOF parallel mechanism with eight actuators, Eclipse I

References

Bonev I (2001) The Delta parallel robot-the story of success. http://www.parallelmic.org/Reviews/Review002p.html

Bonev IA (2002) Geometric analysis of parallel mechanisms. Ph.D. thesis, Laval University, Quebec

Bruyninckx H (1997) The 321-HEXA: a fully-parallel manipulator with closed-form position and velocity kinematics. In: Proceedings of IEEE international conference on robotics and automation. IEEE Computer Society Press, Washington, DC, pp 2657–2662

Carretero JA, Nahon M, Podhorodeski RP (1998) Workspace analysis of a three DOF parallel mechanism. In: Proceedings of the IEEE/RSJ international conference on intelligent robots and systems, IEEE press, Piscataway, N.J., Victoria, pp 1021–1026

Carretero JA, Podhorodeski RP, Nahon MA, Gosselin CM (2000) Kinematic analysis and optimization of a new three degree of freedom parallel manipulator. J Mech Des 122(1): 17–24

Carricato M, Parenti-Castelli V (2003) A family of 3-DOF translational parallel manipulators. ASME J Mech Des 125(2):302–307

Carricato M, Parenti-Castelli V (2004) A novel fully decoupled two-degrees-of-freedom parallel wrist. Int J Robot Res 23(6):661–667

Cervantes-Sánchez JJ, Hernández-Rodríguez JC, Angeles J (2001) On the kinematic design of the 5R planar, symmetric manipulator. Mech Mach Theory 36:1301–1313

Cervantes-Sánchez JJ, Hernández-Rodríguez JC, Rendón-Sánchez JG (2000) On the workspace, assembly configurations and singularity curves of the RRRRR-type planar manipulator. Mech Mach Theory 35:1117–1139

Chen C, Angeles J (2007) Generalized transmission index and transmission quality for spatial linkages. Mech Mach Theory 42:1225–1237

Clavel R (1986) Device for displacing and positioning an element in space. WIPO Patent, WO87/03528

Company O, Krut S, Pierrot F (2006) Internal singularity analysis of a class of lower mobility parallel manipulators with articulated traveling plate. IEEE Trans Robot 22(1):1–11

Dafaoui E-M, Amirat Y, Pontnau J, Francois C (1998) Analysis and design of a six-DOF parallel mechanism, modeling, singular configurations, and workspace. IEEE Trans Robot Autom 14(1):78–92

Fang Y, Huang Z (1997) Kinematics of a three-degree-of-freedom in-parallel actuated manipulator mechanism. Mech Mach Theory 32(7):789–796

Fang Y, Tsai L-W (2002) Structure synthesis of a class of 4-DOF and 5-DOF parallel manipulators with identical limb structures. Int J Robot Res 21(9):799–810

Gao F, Liu X-J, Gruver WA (1998) Performance evaluation of two-degree-of- freedom planar parallel robots. Mech Mach Theory 33(6):661–668

Gosselin CM, Angeles J (1988) The optimum kinematic design of a planar three-degree- of-freedom parallel manipulator. J Mech Transm Autom Des 110(1):35–41

Gosselin CM, Angeles J (1989) The optimum kinematic design of a spherical three-degree- of-freedom parallel manipulator. J Mech Transm Autom Des 111(2):202–207

Gosselin CM, Caron F (1999) Two degree-of-freedom spherical orienting device. US Patent, No. 5 966 991

Hervé JM (1992) Group mathematics and parallel link mechanisms. In: Proceedings of IMACS/SICE international symposium on robotics, mechatronics, and manufacturing systems, International Association for Mathematics and Computers in Simulation (IMACS), Kobe, pp 459–464

Huang Z, Li QC (2003) Type synthesis of symmetrical lower mobility parallel mechanisms using the constraint-synthesis method. Int J Robot Res 22(1):59–79

Hunt KH (1983) Structure kinematics of in parallel actuated robot arms. J Mech Transm Autom Des 105:705–712

Hunt KH, Primrose EJF (1993) Assembly configurations of some in-parallel actuated manipulators. Mech Mach Theory 28(1):31–42

Innocent C, Parenti-Castelli V (1993) Direct kinematics in analytical form of a general 5–4 fully-parallel manipulators. In: Angeles J, Kovacs P, Hommel G (eds) Computational kinematics, Schloss Dagstuhl, Germany, Kluwer Academic Publishers, pp 141–152

Kim HS, Tsai L-W (2004) Design optimization of a Cartesian parallel manipulator. J Mech Des 125(1):43–51

Kim J, Hwang JC, Kim JS et al (2002) Eclipse II: a new parallel mechanism enabling continuous 360-degree spinning plus three-axis translational motions. IEEE Trans Robot Autom 18(3):367–373

Kim J, Park FC, Ryu SJ et al (2001) Design and analysis of a redundantly actuated parallel mechanism for rapid machining. IEEE Trans Robot Autom 17(4):423–434

Kock S, Schumacher W (1998) A parallel x-y manipulator with actuation redundancy for high-speed and active-stiffness applications. In: Proceedings of the IEEE international conference on robotics and automation, IEEE press, Piscataway, N.J., Leuven, pp 2295–2300

Kong X, Gosselin CM (2004a) Type synthesis of 3-DOF translational parallel manipulators based on screw theory. ASME J Mech Des 126(1):83–92

Kong X, Gosselin CM (2004b) Type synthesis of 3T1R 4-DOF parallel manipulators based on screw theory. IEEE Trans Robot Autom 20(2):181–190

Kong X, Gosselin CM (2005) Type synthesis of 3-DOF PPR-equivalent parallel manipulators based on screw theory and the concept of virtual chain. ASME J Mech Des 127:1113–1121

Krut S, Company O, Benoit M, Ota H, Pierrot F (2003a) I4: a new parallel mechanism for SCARA motions. In: Proceedings of IEEE international conference on robotics automation, IEEE press, Piscataway, N.J., Taipei, pp 1875–1880

Krut S, Company O, Rangsri S, Pierrot F (2003b) Eureka: a new 5-degree-of-freedom redundant parallel mechanism with high tilting capabilities. In: Proceedings of IEEE/RSJ international conference on intelligent robots systems, IEEE press, Piscataway, N.J., Las Vegas, pp 3575–3580

Krut S, Company O, Nabat V, Pierrot F (2006) Heli4: a parallel robot for scara motions with a very compact traveling plate and a symmetrical design. In: Proceedings of the IEEE/RSJ international conference on intelligent robots and systems, IEEE press, Piscataway, N.J., Beijing, pp 1656–1661

Lee K-M, Arjunan S (1991) A three-degrees-of freedom micromotion in-parallel actuated manipulator. IEEE Trans Robot Autom 7(5):634–641

Li J, Wang J, Liu X-J (2002) An efficient method for inverse dynamics of the kinematic defective parallel platforms. J Robot Syst 19(2):45–61

Lin W, Duffy J, Griffis M (1992) Forward displacement analysis of the 4–4 Stewart platform. ASME J Mech Des 114:444–450

Liu X-J, Bonev IA (2008) Orientation capability, error analysis, and dimensional optimization of two articulated tool heads with parallel kinematics. ASME J Manuf Sci Eng 130(1), Article Number: 011015

Liu X-J (2006) Optimal kinematic design of a three translational DOFs parallel manipulator. Robotica 24(2):239–250

Liu X-J, Jeong J, Kim J (2003) A three translational DOFs parallel cube-manipulator. Robotica 21(6):645–653

Liu X-J, Jin Z-L, Gao F (2000) Optimum design of 3-DOF spherical parallel Manipulators with respect to the conditioning and stiffness indices. Mech Mach Theory 35:1257–1267

Liu X-J, Pruschek P, Pritschow G (2004) A new 3-DOF parallel mechanism with full symmetrical structure and parasitic motions. In: Proceedings of the international conference on intelligent manipulation and grasping, IEEE press, Piscataway, N.J., Genoa, pp 389–394

Liu X-J, Wang J, Gao F, Wang L-P (2001) On the analysis of a new spatial three degrees of freedom parallel manipulator. IEEE Trans Robot Autom 17(6):959–968

Liu X-J, Wang J, Pritschow G (2006b) Kinematics, singularity and workspace of planar 5R symmetrical parallel mechanisms. Mech Mach Theory 41(2):145–169

Liu X-J, Wang J, Pritschow G (2006c) Performance atlases and optimum design of planar 5R symmetrical parallel mechanisms. Mech Mach Theory 41(2):119–144

Ma O, Angeles J (1991) Optimum architecture design of platform manipulators. In: Proceedings of the fifth international conference on advanced robotics, IEEE press, Piscataway, N.J., Pisa, pp 1130–1135

Macho E, Altuzarra O, Pinto C et al (2008) Workspaces associated to assembly modes of the 5R planar parallel manipulator. Robotica 26(3):395–403

McCloy D (1990) Some comparisons of serial-driven and parallel driven manipulators. Robotica 8:355–362

Nabat V, Company O, Krut S, Rodriguez M, Pierrot F (2005) Par4: very high speed parallel robot for pick-and-place. In: Proceedings of IEEE international conference on intelligent robots and systems, IEEE press, Piscataway, N.J., Edmonton, pp 553–558

Neumann K-E (2006) Exechon concept. In: Proceedings of the 5th Chemnitz parallel kinematics seminar. Verlag Wissenschaftliche Scripten, Zwickau, pp 787–802

Pierrot F, Company O, Krut S, Nabat V (2006) Four-dof PKM with articulated travelling-plate. In: Proceedings of the 5th Chemnitz parallel kinematics seminar. Verlag Wissenschaftliche Scripten, Zwickau, pp 677–693

Pierrot F, Dauchez P, Fournier A (1991) Towards a fully-parallel 6 d.o.f. robot for high speed applications. In: Proceedings of IEEE international conference on robotics & automation, Sacramento, IEEE press, Piscataway, N.J., pp 1288–1293

Pierrot F, Marquet F, Company O, Gil T (2001) H4 parallel robot: modeling, design and preliminary experiments. In: Proceedings of the 2001 IEEE international conference on robotics and automation, Seoul, IEEE press, Piscataway, N.J., pp 3256–3261

Pond GT, Carretero JA (2004) Kinematic analysis and workspace determination of the inclined PRS parallel manipulator. In: Proceedings of 15th CISM-IFToMM symposium on robot design, dynamics, and control, Montreal, Paper Rom04-18

Schoppe E, Pönisch A, Maier V et al (2002) Tripod machine SKM 400 design, calibration, and practical application. In: Proceedings of the 3rd Chemnitz parallel kinematics seminar. Verlag Wissenschaftliche Scripten, Zwickau, pp 579–594

Siciliano B (1999) The Tricept robot: inverse kinematics, manipulability analysis and closed-loop direct kinematics algorithm. Robotica 17:437–445

Sreenivasan SV, Waldron KJ, Nanua P (1994) Closed-form direct displacement analysis of a 6–6 Stewart platform. Mech Mach Theory 29(6):855–864

Stengele G (2002) CROSS HÜLLER SPECHT Xperimental, a processing center with new hybrid kinematics. In: Proceedings of the 3rd Chemnitz parallel kinematics seminar. Verlag Wissenschaftliche Scripten, Zwickau, pp 609–627

Tsai L-W (1996) Kinematics of a three-DOF platform with extensible limbs. In: Lenarcic J, Parenti-Castelli V (eds) Recent advances in robot kinematics. Kluwer Academic Publishers, Dordrecht/Boston, pp 401–410

Tsai LW, Stamper R (1996) A parallel manipulator with only translational degrees of freedom. In: Proceedings of the ASME design engineering technical conference, Irvine, 96-DETC-MECH-1152, ASME press, New York

Wahl J (2000) Articulated tool head. WIPO Patent No. WO 00/25976

Wang J, Gosselin CM (1999) Static balancing of spatial three-degree-of-freedom parallel mechanisms. Mech Mach Theory 34:437–452

Wu J, Wang J, Li T, Wang L (2007) Performance analysis and application of a redundantly actuated parallel manipulator for milling. J Intell Robot Syst 50(2):163–180

Yang T-L (2004) Topology structure design of robot mechanisms (in Chinese). China Machine Press, Beijing

Yu A, Bonev IA, Zsombor-Murray P (2008) Geometric approach to the accuracy analysis of a class of 3-DOF planar parallel robots. Mech Mach Theory 43(3):364–375

Zhao TS (2000) Some theoretical issues on analysis and synthesis for spatial imperfect-DOF parallel robots (in Chinese). Ph.D thesis, Yanshan University, Qinhuangdao

Zlatanov D, Bonev IA, Gosselin CM (2002) Constraint singularities of parallel mechanisms. In: Proceedings of the 2002 IEEE international conference on robotics and automation, IEEE press, Piscataway, N.J., Washington, DC, pp 496–502

Chapter 2
Type Synthesis of Parallel Mechanisms

Abstract This chapter presents several general methods to achieve type synthesis of parallel mechanisms. In particular, an evolution-based approach is used for type synthesis and comprehensive enumeration of parallel mechanisms with parallelogram, as given the number of DOF ranging from 1 to 6. A number of novel mechanical architectures can be obtained correspondingly to improve the kinematic performances of traditional parallel mechanisms. Representative types of parallel mechanisms will then be chosen to be studied more specifically in the remainder of this book.

Keywords Architectures • Type synthesis • DOF analysis • Parallel mechanism • Evolution • Parallelogram

2.1 DOF Analysis

In this chapter, we consider a parallel mechanism to have n DOFs. That is, the mechanism has n DOFs in its normal configuration, excluding any singularity. The mechanism may lose or gain one or more DOFs in its singularity. This will be discussed in detail in Chap. 5.

Perhaps the foremost concern in studying the kinematics of mechanisms is the number of DOFs. The DOF of a mechanism is the number of independent parameters or input needed to completely specify the configuration of the mechanism. However, defining a general mobility criterion for closed-loop kinematic chains is difficult, as Hunt (1978) and Lerbet (1987) noted. Classical mobility formulas can cause the disregard of some DOFs. Grübler-Kutzbach's formula (Kutzbach 1933) is nevertheless generally used, and it may be written as

$$M = d\,(n - g - 1) + \sum_{i=1}^{g} f_i \tag{2.1}$$

X.-J. Liu and J. Wang, *Parallel Kinematics: Type, Kinematics, and Optimal Design*, Springer Tracts in Mechanical Engineering, DOI 10.1007/978-3-642-36929-2_2,

where

M: the mobility (DOFs of a system)
d: the order of the screw system that applies ($d = 3$ for planar and spherical motion, $d = 6$ for spatial motion)
n: the number of links including the frame
g: the number of joints
f_i: the DOFs associated with the ith joint

We can calculate the DOF number of a mechanism using Eq. (2.1). Nevertheless, this equation is not the universal equation for all parallel mechanisms. Sometimes, it fails when it is applied to a parallel mechanism with excessive constraints. For example, when it is applied to Tsai's mechanism (Fig. 1.23), it yields $M = -3$.

To precisely calculate the number of DOFs of a parallel mechanism, many scholars elaborated on Eq. (2.1) or proposed a new criterion. In 2003, Huang and Li (2003) provided a DOF criterion for some lower mobility parallel mechanisms with closed loops in each of their legs using constraint analysis to aid DOF analysis by the Grübler-Kutzbach's formula. In 2005, Gogu (2005) provided a brief presentation and conducted critical analysis of 35 approaches/formulas presented in literature during the last 150 years for DOF calculation and their origins, similarities, and limitations. In his article, he also explained why these formulas do not work for certain mechanisms and proposed a new formula for the rapid DOF calculation of parallel mechanisms with elementary legs. Dai et al. (2006) have recently presented a DOF criterion for overconstrained parallel platforms that employ both screws and reciprocal screws. Rico et al. (2006) proposed a criterion based on an analysis of the subalgebras of Lie algebra, se(3), also known as screw algebra, of the Euclidean group, SE(3). The criterion is said to provide the correct number of DOFs for a wider class of parallel mechanisms.

However, the objective of DOF analysis is never the DOF number itself, but both the DOF number and type, i.e., how many DOF numbers are there and what kind are they classified under. These targets indicate that the DOF formula is far from perfect. DOF analysis based on screw theory and Lie algebra may provide both the DOF number and type. Here, we introduce some methods of DOF analysis without the use of any mathematics.

2.1.1 Observation Method

This method can be applied to the DOF analysis of simple parallel mechanisms, some parallel mechanisms with a passive leg, and parallel mechanisms with six DOFs.

In the DOF analysis of a mechanism, the joint is the most important component. The observation method is based on the comprehensive understanding of the free motion of a joint. For example, with an R joint, the link may only rotate about the

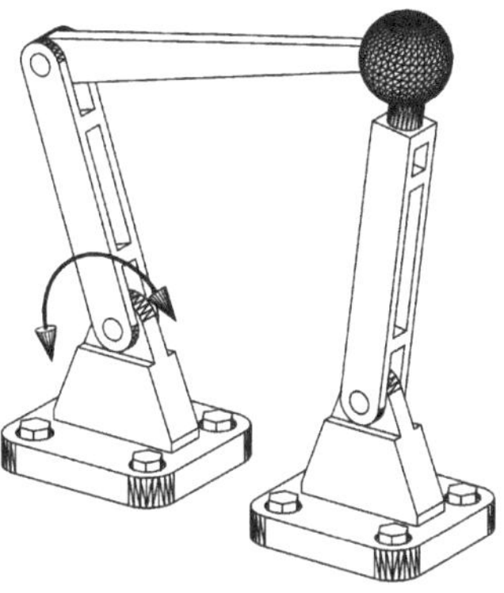

Fig. 2.1 RRSR four-bar mechanism

joint axis and remain in a specified plane; any point at the link describes a circle path centered at the point on the joint axis. If a link is linked to another by a prismatic joint, the loci of all the points at the link are straight lines that are parallel to one another. With a spherical joint, the link can rotate with respect to any line passing through the center of the joint.

Let us first consider planar parallel mechanisms. Only a 1-DOF joint normally exists in a planar parallel mechanism. For a planar mechanism with multi-DOF joints, one or more DOFs of the joint will lose its motion function but will play a mechanical role (e.g., guaranteeing the free motion of the mechanism if an error or deformation occurs). In this case, we say that there is an idle (or passive) DOF in the mechanism. A four-bar mechanism with an RRSR chain is shown in Fig. 2.1. Kinematically, only one DOF of the spherical joint is effective; the remaining two DOFs are idle. With the S joint, the planar four-bar mechanism still has one DOF. However, idle DOFs are necessary in some machines, especially in heavy machines (e.g., the heavy forming machine that works under a very large payload). In this chapter, we consider only the parallel mechanism without idle DOFs.

Revolute and prismatic joints are typically used in a planar parallel mechanism (see Figs. 1.12, 1.13, 1.15, and 1.16). The observation of the RRRPR parallel mechanism in Fig. 1.12b shows that it consists of two legs, i.e., the RR and RP chains, which are connected together by a common revolute joint, referred to as the end-effector of the mechanism. Our concern is the DOF of this end-effector. Because reference points P_1 and P_2 (Fig. 2.2) of the RR and RP chains have the same DOFs, i.e., two translational DOFs in the O-xy plane, the end-effector of the RRRPR parallel mechanism has the two DOFs. Similarly, we may infer that the end-effectors of the other parallel mechanisms shown in Fig. 1.12 have two translational DOFs. Given that the end-effector (the mobile platform) of each parallel mechanism shown in Figs. 1.15 and 1.16 is connected to the base by three planar kinematic chains, each mechanism consists of three single-DOF joints. The end-effector of such a chain has three DOFs in a plane. Any leg of the parallel mechanisms imposes no kinematic constraint on two others; thus, their mobile platforms have three DOFs, i.e., two translations and one rotation in a plane.

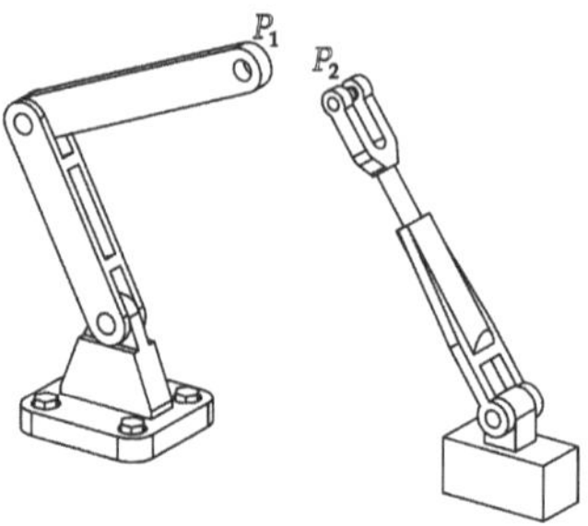

Fig. 2.2 RR and RP chains

Now, we consider the mechanism shown in Fig. 3.2c. The mobile platform of the mechanism is connected to the base by the RRR and PRRR chains. The axes of the RRR chains are all parallel to the others and are orthogonal to those in the PRRR chains. This arrangement shows that with the RRR chain, the mobile platform has three planar DOFs, i.e., two translations in the *O-xy* plane and one rotation about the *z*-axis. However, the rotation is constrained by the PRRR chain. Therefore, the mobile platform has only two translational DOFs.

In the family of parallel mechanisms, some mechanisms with n DOFs usually consist of n identical actuated legs with six DOFs and one passive leg with n DOFs connecting the mobile platform and base. This means that the DOF of the mechanism is dependent on the DOF of the passive leg. Such a mechanism has an advantage: its rigidity can be improved through optimization of the link rigidities to reach maximal global stiffness and precision (Zhang and Gosselin 2002). The Tricept parallel mechanism (Fig. 1.29b) is such a mechanism. In the Tricept, the fourth leg is a UP chain, which has two rotational DOFs and one translational DOF. Therefore, the parallel mechanism has the said DOFs, and identifying the DOFs of this group of parallel mechanisms is very easy.

The end-effector of a 6-DOF chain has six DOFs, i.e., three translations and three rotations. If the mobile platform (not an end point) is connected to the base through several 6-DOF chains, the platform is guaranteed to have six DOFs. If the chain number equals six, one DOF in each leg is actuated; if the number is less than six, at least one leg will have more than one actuator. Thus, analyzing the DOF of a parallel mechanism that has all 6-DOF chains may be the easiest approach.

2.1.2 Evolution Method

DELTA has achieved significant success in industry because of its rapid performance and easy setup. The mechanism itself, however, seems complex. An issue is how many DOFs DELTA has, or why it has three translational DOFs when it is first encountered. Perhaps the design conceived by Clavel for DELTA did not originate from the 6-DOF parallel mechanism. If we assume such an origin, the DOF analysis of DELTA will be easy. Figure 2.3 shows Pierrot's 6-DOF parallel mechanism with

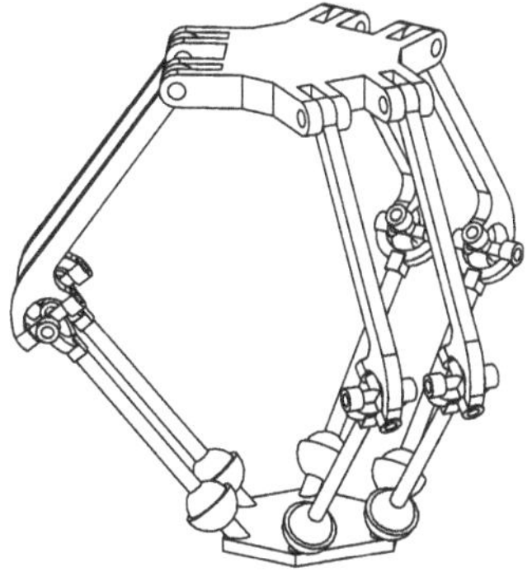

Fig. 2.3 Pierrot's 6-DOF parallel mechanism (Pierrot et al. 1991)

six RUS chains (Pierrot et al. 1991), in which every two legs are parallel to each other when the mechanism is in its initial configuration. If the input of two parallel legs is the same at every moment, the DOFs of the mobile platform are therefore three translations. In this case, the two parallel input links can be replaced with a single link and six actuators become three. The modified mechanism is DELTA, indicating that DELTA has three pure translational DOFs in space.

This DOF analysis method is useful only if the original mechanism of a parallel mechanism can be found. Sometimes, this is not an easy task.

2.1.3 Kinematic Analysis Method

Some mechanisms such as the 3-[PP]S parallel mechanisms shown in Fig. 1.20 have complex kinematic chains. Completely identifying the DOFs of these mechanisms is impossible using the methods mentioned above. For example, the mechanism in Fig. 1.20a has three RPS chains, each with five DOFs, i.e., two translations and three rotations. This kind of kinematic chain has one less DOF than the PRPS chain of the mechanism in Fig. 1.38a, which has six actuators in three legs and six DOFs. On this basis, we can conclude that the 3-PRS mechanism has three DOFs. However, we cannot identify exactly what the DOFs are. To this end, we can first analyze its kinematics by assuming that the mobile platform has six DOFs.

Figure 2.4 illustrates the kinematic model of a 3-[PP]S mechanism, in which points P_1, P_2, and P_3 of the mobile platform remain at planes Π_1, Π_2, and Π_3, respectively. The three points can be connected to the base by any one of the RR, RP, PR, and PP chains or other chains similar to these.

A kinematic model of the mechanism is developed, as shown in Fig. 2.4. The vertices of the output platform are denoted as platform joints P_i $(i = 1, 2, 3)$, and the vertices of the base joints are denoted as B_i $(i = 1, 2, 3)$. A fixed global reference system $\Re$: O-XYZ is located at the intersecting point of three planes Π_1, Π_2, and Π_3 with the z-axis coincident with the intersecting line and x-axis in plane Π_3. Another reference frame, called moving frame $\Re'$: o-xyz, is located at the center

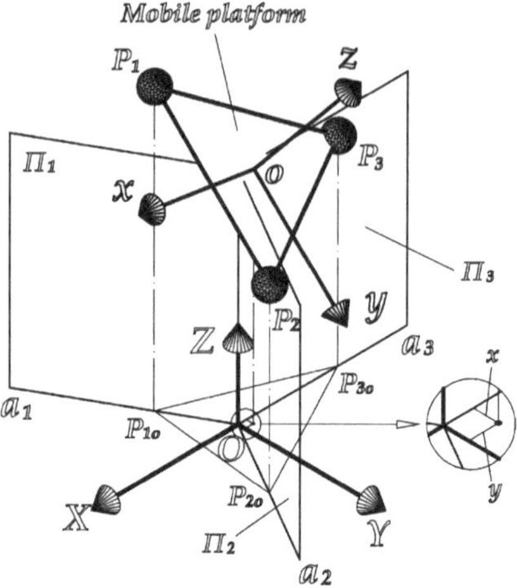

Fig. 2.4 Kinematic model of a 3-[PP]S mechanism

of regular triangle $P_1P_2P_3$. The z-axis is perpendicular to mobile platform $P_1P_2P_3$, x-axis directed along the P_3o direction, and y-axis in the mobile platform.

We assume that the mobile platform has six DOFs. Its pose is represented by a vector $\boldsymbol{X} = \begin{bmatrix} x & y & z & \psi & \theta & \phi \end{bmatrix}^{\mathrm{T}}$, where ψ, θ, and ϕ are three Euler angles. The position of the mobile platform in frame $\Re$ is given by the position vector $(\boldsymbol{o})_{\Re}$ of point o and the orientation given by matrix $\boldsymbol{T}$. Moreover, there are

$$(\boldsymbol{o})_{\Re} = \begin{bmatrix} x & y & z \end{bmatrix}^{\mathrm{T}} \tag{2.2}$$

$$\begin{aligned} \boldsymbol{T} = \boldsymbol{T}_{ZXY} &= \boldsymbol{R}_Y(\theta)\,\boldsymbol{R}_X(\psi)\,\boldsymbol{R}_Z(\phi) \\ &= \begin{bmatrix} c_\theta c_\phi + s_\psi s_\theta s_\phi & -c_\theta s_\phi + s_\psi s_\theta c_\phi & c_\psi s_\theta \\ c_\psi s_\phi & c_\psi c_\phi & -s_\phi \\ -s_\theta c_\phi + s_\psi c_\theta s_\phi & s_\theta s_\phi + s_\psi c_\theta c_\phi & c_\psi c_\theta \end{bmatrix} \end{aligned} \tag{2.3}$$

where c stands for cosine and s for sine. Vectors $(\boldsymbol{P}_i)_{\Re'}$ are defined as the position vectors of points P_i in frame $\Re'$, and

$$(\boldsymbol{P}_i)_{\Re'} = \begin{bmatrix} P'_{ix} & P'_{iy} & P'_{iz} \end{bmatrix}^{\mathrm{T}} = \begin{bmatrix} r\cos\varphi_i & r\sin\varphi_i & 0 \end{bmatrix}^{\mathrm{T}} \tag{2.4}$$

where $\varphi_i = (i-1)\,2\pi/3$ and r is the radius of the mobile platform. Then, the position vectors of points P_i in frame $\Re$ can be written as

$$(\boldsymbol{P}_i)_{\Re} = \begin{bmatrix} P_{ix} & P_{iy} & P_{iz} \end{bmatrix}^{\mathrm{T}} = \boldsymbol{T}\,(\boldsymbol{P}_i)_{\Re'} + \boldsymbol{O}' \tag{2.5}$$

That is,

$$P_{ix} = x + T_{11}P'_{ix} + T_{12}P'_{iy} + T_{13}P'_{iz} \tag{2.6}$$

$$P_{iy} = y + T_{21}P'_{ix} + T_{22}P'_{iy} + T_{23}P'_{iz} \tag{2.7}$$

$$P_{iz} = z + T_{31}P'_{ix} + T_{32}P'_{iy} + T_{33}P'_{iz} \tag{2.8}$$

where $T_{jk}(j, k = 1, 2, 3)$ is the jth row and kth column element in matrix $\boldsymbol{T}$.

Figure 2.4 shows that each point of P_i $(i = 1, 2, 3)$, which are vertices of the mobile platform, is always in a fixed plane. Therefore, we can write three constraint equations as follows:

$$P_{1y} = 0 \tag{2.9}$$

$$P_{2y} = -\sqrt{3}\,P_{2x} \tag{2.10}$$

$$P_{3y} = \sqrt{3}\,P_{3x}. \tag{2.11}$$

That is,

$$y + T_{21}r = 0 \tag{2.12}$$

$$y - \frac{r}{2}T_{21} + \frac{\sqrt{3}\,r}{2}T_{22} = -\sqrt{3}x + \frac{\sqrt{3}\,r}{2}T_{11} - \frac{3r}{2}T_{12} \tag{2.13}$$

$$y - \frac{r}{2}T_{21} - \frac{\sqrt{3}\,r}{2}T_{22} = \sqrt{3}x - \frac{\sqrt{3}\,r}{2}T_{11} - \frac{3r}{2}T_{12}. \tag{2.14}$$

From Eqs. (2.12), (2.13), and (2.14), we can obtain the following equations:

$$T_{21} = T_{12} \tag{2.15}$$

$$x = \frac{r}{2}\left(T_{11} - T_{22}\right). \tag{2.16}$$

Equations (2.12), (2.15), and (2.16) can be rewritten as

$$y = -rc_{\psi}s_{\phi} \tag{2.17}$$

$$-c_{\theta}s_{\phi} + s_{\psi}s_{\theta}c_{\phi} - c_{\psi}s_{\phi} = 0 \tag{2.18}$$

Then,

$$\phi = \tan^{-1}\left[\frac{s_\psi s_\theta}{(c_\psi + c_\theta)}\right] \tag{2.19}$$

$$x = \frac{r}{2}\left(c_\theta c_\phi + s_\psi s_\theta s_\phi - c_\psi c_\phi\right). \tag{2.20}$$

We can conclude that in vector $\boldsymbol{X} = \left[x\ y\ z\ \psi\ \theta\ \phi\right]^{\mathrm{T}}$, only z, ψ, and θ are independent, and three others, x, y, and ϕ, are dependent on ψ and θ. That is, a 3-[PP]S parallel mechanism has three independent DOFs, which are the translation along the Z-axis and two rotations about the X- and Y-axes.

However, these three methods cannot be applied to the DOF analysis of any parallel mechanism. The successful identification of DOF number and type can be implemented using screw theory or Lie algebra.

2.1.4 Method Based on Screw Theory

In the screw theory (Ball 1900), a unit screw $\boldsymbol{\$}$ is defined by a straight line with an associated pitch, as illustrated in Fig. 2.5, and is expressed as in a Plücker coordinate form

$$\boldsymbol{\$} = (\boldsymbol{s}\,;\boldsymbol{s}^0) = (\boldsymbol{s}\,;\boldsymbol{s}_0 + p\boldsymbol{s}) = (\boldsymbol{s}\,;\boldsymbol{r}\times\boldsymbol{s} + p\boldsymbol{s}) \tag{2.21}$$

where $\boldsymbol{s}$ is a unit vector pointing in the direction of the screw axis, $\boldsymbol{s}_0 = \boldsymbol{r}\times\boldsymbol{s}$ defines the moment of the screw axis about the origin of a reference frame, $\boldsymbol{r}$ is the position vector of any point on the screw axis with respect to the reference frame, and p is the pitch of the screw. If p is equal to zero, the screw reduces to

$$\boldsymbol{\$} = (\boldsymbol{s}\,;\boldsymbol{s}_0) = (\boldsymbol{s}\,;\boldsymbol{r}\times\boldsymbol{s}) \tag{2.22}$$

On the other hand, if p is infinite, the screw can be defined as

$$\boldsymbol{\$} = (\boldsymbol{0}\,;\boldsymbol{s}) \tag{2.23}$$

The screw expressed by Eq. (2.22) can represent a rotation in kinematics or a force in statics, and the screw expressed by Eq. (2.23) can represent a translation in kinematics and a couple in statics. We call the screw a twist if it represents the instantaneous rotation or translation of a rigid body and a wrench if it denotes a force or couple acting on a rigid body. In this regard, the first three components of a twist represent the unit angular velocity, and the last three components represent the unit linear velocity of a point in the rigid body that is instantaneously coincident with

Fig. 2.5 A general screw

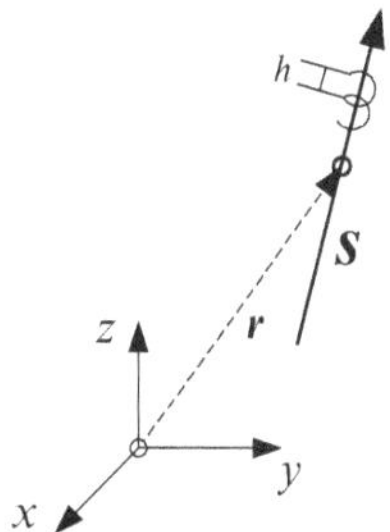

the origin of a reference frame. In similar, the first three components of a wrench represent the unit resultant force, and the last three components represent the unit resultant moment about the origin of a reference frame. A wrench of zero pitch represents a pure force, whereas a wrench of infinite pitch denotes a pure couple. On the other hand, a revolute joint can be represented by the screw expressed by Eq. (2.22), and a prismatic joint can be represented by the screw expressed by Eq. (2.23).

A unit screw, $\$^r = (\boldsymbol{s}^r;\ \boldsymbol{s}^{0r})$, and a set of unit screws, $\$_1, \$_2, \ldots, \$_n$, are said to be reciprocal if they satisfy the condition

$$\$^r \circ \$_i = \boldsymbol{s}_i \cdot \boldsymbol{s}^{0r} + \boldsymbol{s}_i^0 \cdot \boldsymbol{s}^r = 0 \quad (i = 1, 2, \cdots n) \tag{2.24}$$

where "∘" represents the reciprocal product and $\$_i$ is the ith screw of the screw set.

In an n-DOF serial manipulator, the twists associated with all the joints form a screw system of order n, called an n-system, provided that all the joint screws are linearly independent. For a spatial manipulator, if $n = 6$, there exists no screw reciprocal to the twist screw system. If $n < 6$, there exists 6-n linearly independent wrenches that form a (6-n)-system. Each wrench in the (6-n)-system is reciprocal to the n-system of twists, namely,

$$\$_j^r \circ \$_i = 0 \quad (i = 1, 2, \cdots n \textit{ and } j = 1, 2, \cdots 6 - n) \tag{2.25}$$

Equation (2.25) can be used to solve for a (6-n)-system of reciprocal wrenches. On the other hand, given 6-n linearly independent wrenches, Eq. (2.25) can be used to find an n-system of twists.

For a parallel manipulator illustrated in Fig. 2.6, when each kinematic pair in a limb is expressed as a unit screw, all these unit screws form a *limb twist system* (LTS). The linearly independent joint screws of a limb form an n-system. The reciprocal screws form a (6-n)-system, called a *limb wrench system* or a *limb constraint system* (LCS). In a low-DOF parallel manipulator, we note that a single limb exerts some restrictions on the moving platform. The overall constraint imposed on the moving platform, which is called a *platform wrench system* or a *platform constraint system* (PCS), is obtained by a combination of all LCSs. The PCS represents the restricted DOFs of the moving platform. Thus, the DOF

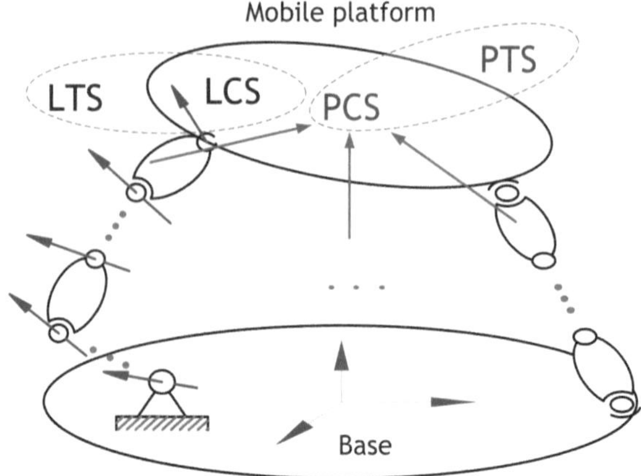

Fig. 2.6 Several pairs of screw systems in parallel mechanisms

of the moving platform can be determined by the *platform twist system* (PTS), which is reciprocal to the manipulator constraint system. In other words, the motion characteristics of the moving platform are completely determined by the nature of the PCS. For instance, for a general 3-DOF translational parallel manipulator (TPM), the PCS consists of three linearly independent constraint couples:

$$\begin{aligned} \boldsymbol{S}_1^r &= (0\,0\,0;1\,0\,0) \\ \boldsymbol{S}_2^r &= (0\,0\,0;0\,1\,0) \\ \boldsymbol{S}_3^r &= (0\,0\,0;0\,0\,1) \end{aligned} \tag{2.26}$$

which restricts three rotations of the moving platform (Fig. 2.6).

In general, when using the method based on the screw theory to determine the degrees of freedom of a low-DOF parallel manipulator, a whole analysis procedure is described as follows. First, select a single limb from the parallel manipulator. For the purpose of a convenient analysis, all multiple-DOF pairs in this limb are deposed into the equivalent basic joints. Thus, these 1-DOF kinematic pairs in the limb form a LTS. Next, an LCS reciprocal to the LTS, representing the constraint forces or couples acting on the moving platform, is derived by means of Eq. (2.25). In similar, the constraint forces or constraint couples provided by the other limbs are also obtained. All these derived screws compose a PCS, which represents the overall constraints for the parallel manipulator. Finally, by calculating the reciprocal product of the PCS, a PTS reflecting all instantaneous motions of this manipulator is obtained, and the DOF of the manipulator is determined accordingly.

Therefore, in order to achieve the type synthesis of a class of parallel manipulators, the process described above can provide us with a good inspiration. In other words, the screw theory is still effective during the synthesis process. The concrete synthesis process is realized through the following steps:

Step 1: *Describe all constraint screws acting on the movable platform.*

For any TPM, its moving platform has only three-dimensional translations and loses three rotational DOFs; thus, there exist three linearly independent CCs acting on the moving platform. These three constraint couples form a PCS of the TPMs, as described in Eq. (2.26).

Step 2: *Analyze the geometric condition of the LCS corresponding to a specific PCS.*

Different PCSs in a parallel manipulator result from the combination of all the LCS under different geometric conditions. In this sense, it is very necessary to analyze the geometric condition of the LCS corresponding to a specific PCS in order to obtain an available limb structure.

The family of the TPMs can be grouped into two categories: manipulators with independent constraints and manipulators with repeated constraints (also called overconstrained manipulators). In an overconstrained manipulator, its limb has less than five basic joints, which means fewer joints and simpler architecture than an independent-constraint manipulator with the same kinematic characteristics. In this regard, by analyzing the geometric condition of the LCS in terms of TPMs with independent constraints and overconstrained TPMs, respectively, we can find the corresponding PCS.

Step 3: *Make a solution to the LTS reciprocal to the LCS.*

According to the screw theory, each of screws in a LTS is reciprocal to the LCS. Therefore, once the screws in the LCS given, the screws in the LTS can be calculated by means of Eq. (2.25). A simple approach of determining a LTS is to find a base of this LTS. Then by a linear combination of these basic twists, other types of the LTS reflecting the limb structure can be obtained.

Step 4: *Analyze the geometric condition of the LTS corresponding to different LCS.*

In a TPM, each limb may provide couples with different numbers varied from zero to three to the moving platform; therefore, we should analyze the geometric condition met by the LTS corresponding to any kind of LCS such that we can find the geometric relationship among all kinematic pairs in the LTS and further construct this limb. Furthermore, to make sure that the moving platform could achieve a finite motion instead of an instantaneous motion, the arrangement of the kinematic pairs in each limb has to meet some additional geometric conditions, which is an important issue to be discussed.

Step 5: *Construct and allocate available the connecting chains in series.*

Once we have obtained the LTS satisfying some geometric conditions, we can construct a limb according to the screw expressions reflecting this LTS. On the other hand, note that the sequence of each screw in the LTS is changeable, which means the arrangement of the kinematic pairs may also be alterable.

Step 6: *Construct a desired parallel mechanism according to Steps 2–5.*

2.2 Evolution of Parallel Mechanisms

In the field of parallel mechanisms, an interesting problem is the identification of a method for designing a mechanical architecture for a parallel mechanism being given its number and type of DOF. After Gough established the basic principles of a mechanism with a closed-loop kinematic structure in 1947, many other parallel mechanisms with a specified DOF number and type have also been proposed. However, some parallel mechanisms have inherent relationships with others. A parallel mechanism may have evolved from another. We introduce the typical evolution of parallel mechanisms. As mentioned in Sect. 3.2.2, the evolution of parallel mechanisms may help us identify their DOFs.

Figure 1.36 shows the general architecture of 6-DOF parallel mechanisms. Theoretically, the arrangement of the six legs of the mechanism can be done at will, which leads to some potential 6-DOF parallel mechanisms, such as the mechanism shown in Fig. 1.37b, in which the legs are arranged in 3.2.1 style. This architecture presents application advantages in a micromotion system (Pernette and Clavel 1996) and shake table. The disposal of six legs (as shown in Fig. 1.37c, d) causes the mechanisms to move freely along a specified direction; this feature is very useful in industrial applications (Honegger et al. 2000).

Applying parallelity between every two legs of the six-leg mechanism in Fig. 1.36 generates the 6-DOF parallel mechanism presented in Fig. 1.37a. The number of DOFs of the mechanism differs if the two most adjacent legs have identical input. This is a topological mechanism of DELTA with linear actuators. Replacing the active P joints of the mechanism in Fig. 1.37a with R joints and fixing them to the base result in the modified mechanism with revolute actuators presented by Pierrot (1991; Fig. 2.3). Similarly, the mechanism can evolve to a DELTA robot. The fact that the two nearby revolute actuators in Fig. 2.3 always have the same output generates a different mobile platform output. This feature characterizes the design of the well-known DELTA robot (Fig. 1.22).

The supposed design conceived for the above-mentioned mechanisms not only aids the understanding of the mechanisms but also provides us new concepts for the design of a mechanism. For example, the 6-DOF parallel mechanism (Fig. 2.7) described in Dafaoui et al. (1998) can be taken as the topological architecture of the mechanism shown in Fig. 1.37a by rearranging the six legs. This inspires the design of new parallel mechanisms (see Fig. 1.26a, b) through the rearrangement of the three legs of DELTA and Tsai's parallel mechanism. In the two new designs, the three actuators are arranged according to the Cartesian axes, indicating that the actuation directions are normal to one another. In addition, the joints connected to the mobile platform are located on three sides of a cube; for this reason, we call this type of mechanism the parallel cube-mechanism. The mechanisms also have three translational DOFs, as in the cases of DELTA and Tsai's mechanism. However, some of their important characteristics differ from those of DELTA and Tsai's mechanism, making the designs novel. One of the most important study results of Liu et al. (2003) shows that a compliance center exists in the workspace,

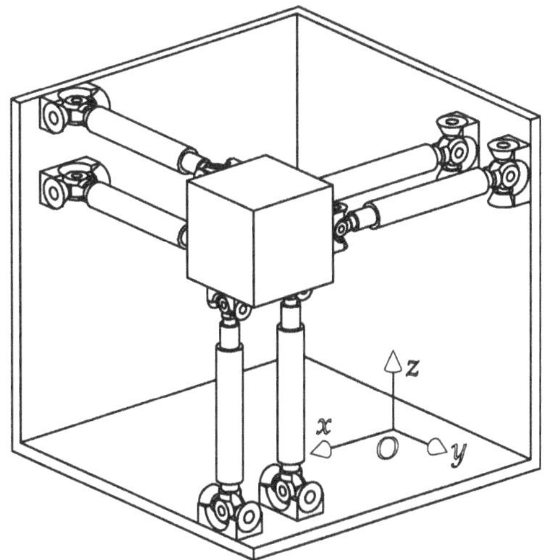

Fig. 2.7 Six-DOF parallel cube-mechanism with six UPS chains

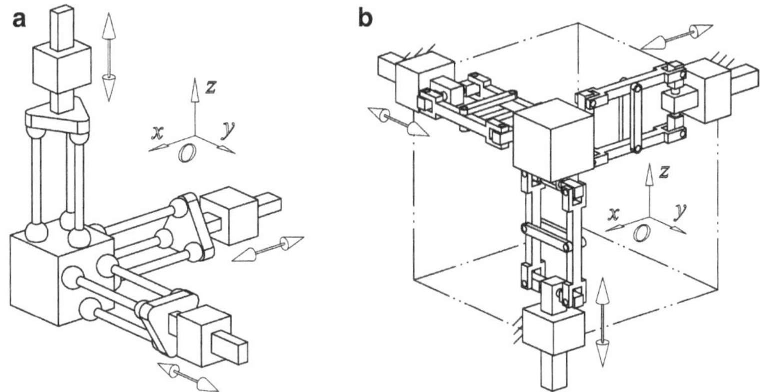

Fig. 2.8 Two pure translational parallel cube-mechanisms (Liu et al. 2003) with redundant constraint: (**a**) modification of DELTA and (**b**) modification of Tsai's mechanism

which is the initial position. At this point, the parallel cube-mechanisms behave in a manner similar to both a velocity and stiffness isotropic setup. These characteristics make the parallel cube-mechanism adaptive to such applications as micromotion mechanisms (Liu et al. 2001b; Ohya et al. 1999), remote center compliance devices (Kim et al. 1997), precision assembly machines (Dafaoui et al. 1998), and parallel kinematic machines (Huang et al. 2002). For application in a parallel kinematic machine, the stiffness of the system is one of the most important problems. The advantage of the presented parallel mechanisms is that the stiffness can be improved by increasing the redundant constraints (Liu et al. 2003), as in the architectures shown in Fig. 2.8. Figure 2.8a shows the modified version of the design in Fig. 3.14a,

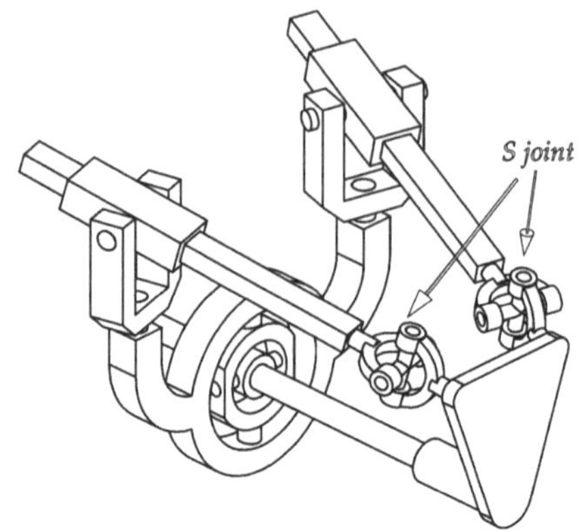

Fig. 2.9 Three-DOF parallel mechanism with 2-UPS&1-UP chains (the parallel mechanism of TriVariant (Huang et al. 2005a))

in which there are three spherical-bar-spherical or universal-bar-universal chains instead of two such chains in each leg. As seen in Fig. 2.8b, each parallelogram in Fig. 1.26b is divided into two or more parallelograms. The kinematic model of each of the mechanisms is identical to that of each of the mechanisms shown in Fig. 1.26. Furthermore, the revision has no negative influence on the kinematics, workspace, and other performance-related functions of the mechanism. The more redundant the constraint of the leg, the higher the stiffness of the mechanism and, in relative terms, the higher the fabrication accuracy it requires.

As shown in Fig. 1.29b, the parallel mechanism of the Tricept is equipped with the 3-UPS&1-UP chains, in which the P joints of the three UPS chains are active and the P joint of the UP chain is passive. On the basis of this mechanism, Huang and colleagues (2005a) designed a modified version with 2-UPS&1-UP chains (see Fig. 2.9), in which the three P joints are all active. Using this mechanism, Tianjin University (China) developed a 5-DOF hybrid machine called TriVariant. The modification not only inherits most of the advantages of the Tricept but also allows for cost-effectiveness compared with the Tricept because of the savings gained from forgoing one active limb. As analyzed in Huang et al. (2005a), the kinematic performance of TriVariant is comparable with that of the Tricept for the same task workspace and a set of similar geometrical parameters.

Again, considering the SKM 400 parallel mechanism shown in Fig. 1.29d, three P joints in the UPS chains are active, but the R joint in the R(Pa)(Pa) leg is passive. A 3-DOF pure translational parallel mechanism was proposed (Huang et al. 2005b), in which one UPS leg from the SKM 400 mechanism is excluded and the R joint is actuated (Fig. 2.10).

Although the design concepts from so many parallel mechanisms provide us more concepts for the design of a new mechanism, additional work remains to be done especially for designing lower-DOF parallel mechanisms that combine translational and rotational DOFs. Few spatial three-DOF parallel mechanisms combine two spatial translations and one rotation; these are presented in the following chapter.

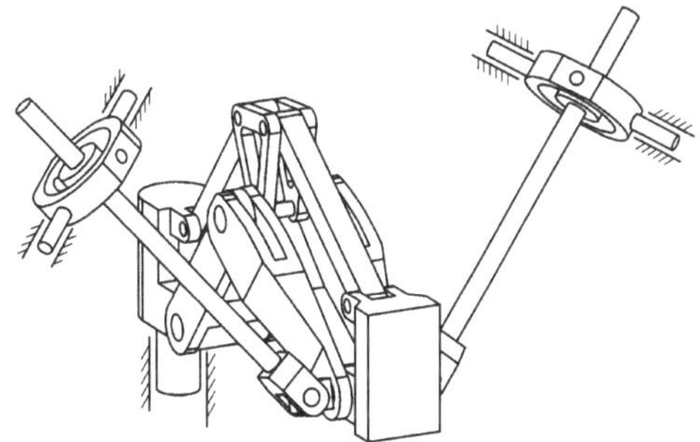

Fig. 2.10 Three-DOF pure translational parallel mechanism with 2-UPS&1-R(Pa)(Pa) chains (Huang et al. 2005b)

2.3 Type Synthesis

In the field of parallel mechanisms, one of the most important and interesting problems is architecture design, i.e., type synthesis. Numerous parallel mechanisms with a specified number and type of DOF have been proposed (see Merlet (2000) for examples). Most recently, several systematic approaches have been proposed for the type synthesis of parallel mechanisms, such as the methods based on displacement group theory (Hervé 1999) and screw theory (Zhao 2000; Huang and Li 2002; Fang and Tsai 2002; Kong and Gosselin 2004a, b), as well as the units of single opened chains (Yang 2004), vector approach (Carricato and Parenti-Castelli 2003), and Lie subgroups and submanifolds of the special Euclidean group, SE(3) (Meng et al. 2007). Using these methods as bases, researchers proposed many new parallel mechanisms that have not been previously conceived of. These substantially enrich the theory of parallel mechanisms. In this paper, we introduce the type synthesis of some parallel mechanisms.

2.3.1 Type Synthesis of Parallel Mechanisms with Parallelogram

2.3.1.1 Motivation

Few spatial parallel mechanisms with two translations and one rotation exist in the family of parallel mechanisms, and few fully parallel mechanisms have high rotational capability. To these ends, we design a mechanism that satisfies these requirements.

Equation (2.1) indicates that in a symmetrical spatial parallel mechanism, each of the three legs should have five DOFs. The 3-PRS parallel mechanism is a typical example. Then, we consider the mechanism shown in Fig. 2.11, in which the mobile platform is connected to the base through two PRS chains and one PUU chain.

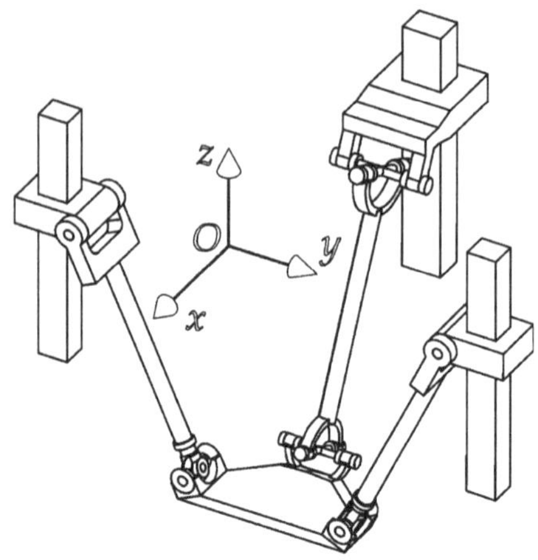

Fig. 2.11 Parallel mechanism with a 2-PRS&1-PUU chain

In this design, the R joints of two PRS legs are parallel to each other, and the legs are in a common plane. With this architecture, the mechanism should have three DOFs. With the two PRS legs, the mobile platform cannot move along the x-axis and cannot rotate about the z-axis, leaving four DOFs, two translations, and two rotations on the mobile platform. For example, endowing the mobile platform with three spatial DOFs necessitates two translations and one rotation; the third leg with the PUU chain must constrain the rotational DOF of the mobile platform about the x-axis. Whether this can be accomplished depends on the UU chain. A question arises as to the ability of the UU chain to constrain the rotational DOF. To resolve this issue, we first investigate the kinematic characteristics of the UU chain.

Figure 2.12a shows a UU chain, in which the axes of four revolute joints are denoted as y_1, z_1, y_2, and z_2. At any moment, the y_1-axis is always parallel to the y_2-axis. The configuration when links 1, 2, and 3 are collinear is referred to as the initial configuration. In this chain, there are four DOFs that are four rotations about the four axes. In movement, if the z_1-axis is always parallel to the z_2-axis, we call this translational motion; otherwise, it is referred to as spatial motion. In translational motion, axes y_1 and y_2 and axes z_1 and z_2 are always parallel to each other (see Fig. 2.12b). The transformation matrix of link 3 with respect to link 1 can be written as

$$\boldsymbol{Q}_1 = \boldsymbol{R}_{(z_1,\,\theta_1)}\boldsymbol{R}_{(y_1,\,\theta_2)}\boldsymbol{R}_{(y_2,-\theta_2)}\boldsymbol{R}_{(z_2,\,\theta_3)} = \boldsymbol{R}_{(z_1,\,\theta_1)}\boldsymbol{R}_{(z_2,\,\theta_3)} \tag{2.27}$$

where $\boldsymbol{R}_{(z_1,\,\theta_1)}$ stands for the transformation matrix when it rotates at angle θ_1 about the z_1-axis. The other matrices follow this rule. Equation (2.27) indicates that link 3 applies no self-rotation during translational motion.

When the movement of link 3 is of spatial type with respect to link 1 (Fig. 2.12c), the transformation matrix of link 3 with respect to link 1 is

$$\boldsymbol{Q}_2 = \boldsymbol{R}_{(z_1,\,\theta_1)}\boldsymbol{R}_{(y_1,\,\theta_2)}\boldsymbol{R}_{(y_2,\,\theta_3)}\boldsymbol{R}_{(z_2,\,\theta_4)} \tag{2.28}$$

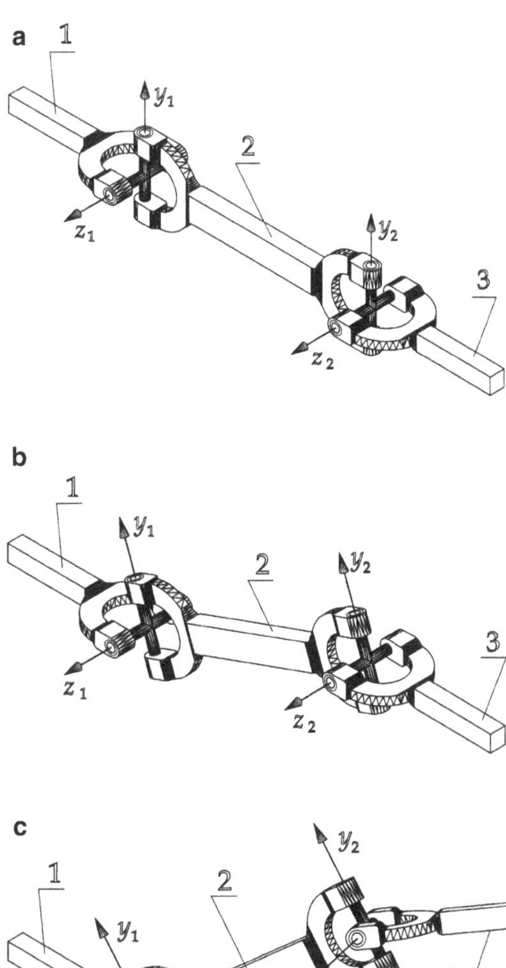

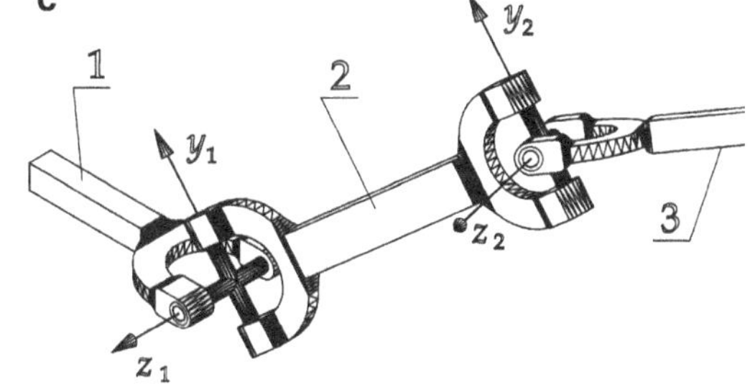

Fig. 2.12 Kinematic characteristic analysis of a UU chain: (**a**) initial configuration, (**b**) configuration when links 1 and 3 are parallel to each other, and (**c**) configuration when links 1 and 3 are not parallel to each other

Given that the y_1-axis is always parallel to the y_2-axis, Eq. (2.28) can be rewritten as

$$Q_2' = \boldsymbol{R}_{(z_1,\,\theta_1)}\boldsymbol{R}_{(y',\,\theta_2\,+\,\theta_3)}\boldsymbol{R}_{(z_2,\,\theta_4)} \tag{2.29}$$

which is actually the transformation matrix with Euler-angle z-y-z convention. Therefore, when the UU chain changes its initial configuration to the configuration that links 1 and 3, and these links are not parallel to each other, link 3 has three rotations with respect to link 1, including a self-rotation.

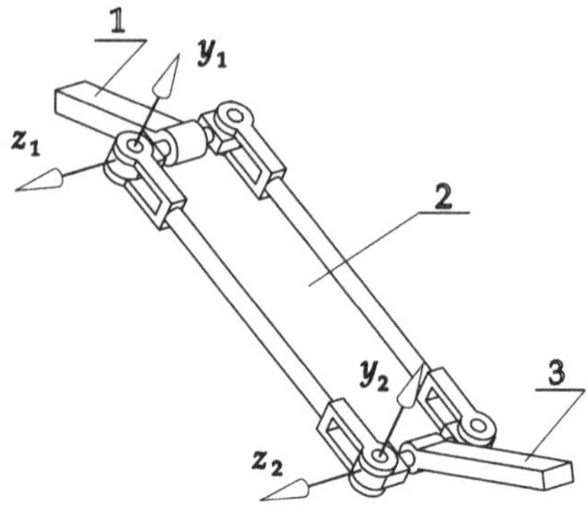

Fig. 2.13 Substitute of the UU chain

Thus, we draw the following conclusions: (a) When the UU chain is at its initial configuration, it can constrain the self-rotation of link 3. (b) When the movement of link 3 is translational with respect to link 1, link 3 does not apply self-rotation. (c) If the movement of link 3 is spatial with respect to link 1, the self-motion of link 3 occurs.

The analysis above shows that the PUU chain in the mechanism of Fig. 2.11 cannot completely constrain the rotation of the mobile platform about the x-axis, indicating that the parallel mechanism has four DOFs at certain configurations. Figure 2.12 and Eq. (2.28) show that to constrain the rotation, the two inner revolute joints of the UU chain should be parallel to each other; the same applies to the two outer revolute joints. Only in this geometric condition can the mechanism shown in Fig. 2.11 have three spatial DOFs, i.e., two translations and one rotation.

In practice, however, the geometric condition cannot be satisfied at any moment. *Is there any substitute mechanism for the UU chain?*

To solve the problem of the UU chain, we provide a solution (Fig. 2.13), in which the UU combination is replaced by a revolute joint, planar parallelogram, and another revolute joint. The axes of the two revolute joints are parallel to each other and are orthogonal to those of the four revolute joints in the parallelogram. This arrangement guarantees constant parallelism between axes y_1 and y_2 and axes z_1 and z_2. The solution is called the variation of the UU chain, referred to as the R(Pa)R chain. With this variation, when link 3 moves with respect to link 1, self-rotation no longer occurs.

The PUU chain in the mechanism in Fig. 2.11 can be replaced by the PR(Pa)R chain shown in Fig. 2.14. The new parallel mechanism (Liu et al. 2001a) is illustrated in Fig. 2.15. It has three DOFs, which are the translations in the O-yz plane and the rotation about the line passing through the two S joints attached to the mobile platform.

2.3.1.2 Concept of a Parallelogram

Why can a parallelogram solve the problem of the UU chain? The concept of the parallelogram was used in some other parallel mechanisms. In the DELTA

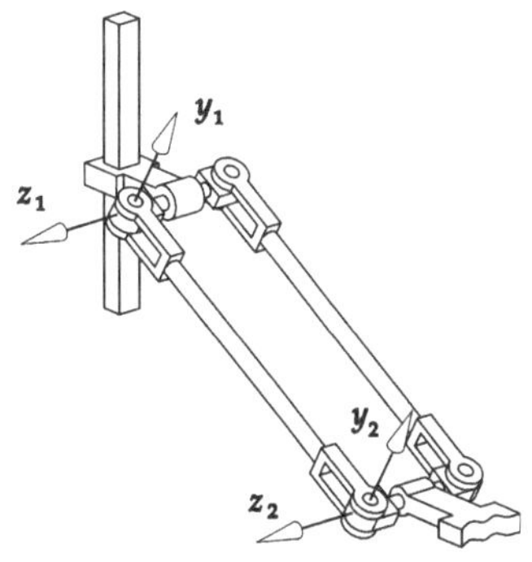

Fig. 2.14 Variation of the PUU chain, a PR(Pa)R chain

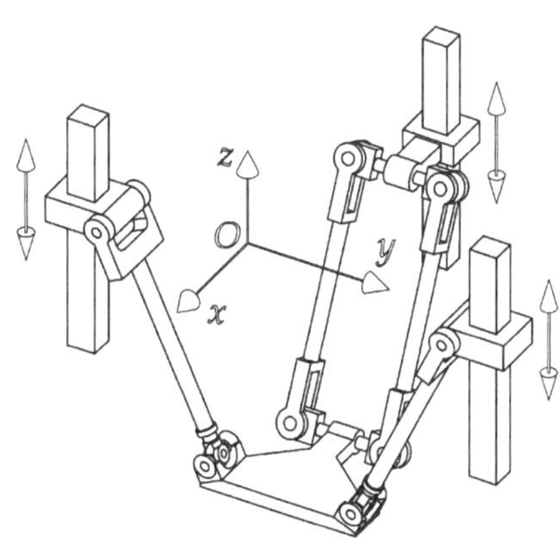

Fig. 2.15 Spatial 3-DOF parallel mechanism with a 2-PRS&1-PR(Pa)R chain (Liu et al. 2001a)

mechanism (Fig. 1.22), each of the three legs consists of a spatial four-bar mechanism. The four bars are connected end to end by spherical joints. Although it is a spatial mechanism, each two of the four bars should be parallel to each other at every instant, for which DELTA has three translational DOFs. The design concept was extended to a new family of 4-DOF parallel mechanisms, H4 (Fig. 1.32), in 1999. Such a concept is highly important to parallel mechanism design in which a planar four-bar parallelogram is used.

The mechanism of the planar four-bar parallelogram lies in four bars connected end to end by revolute joints. Figure 2.16 shows that links 1 and 3, as well as links 2 and 4, are identical in link length. Links 1 and 2 can always have identical orientations with links 3 and 4, respectively. The parallelogram was first applied to the design of the StarLike mechanism (Fig. 1.24) in 1992 by Hervé; this mechanism also has three translational DOFs, used later in the design of another parallel mechanism (Fig. 1.23) with three translational DOFs by Tsai in 1996. From then on, the concept of the parallelogram has attracted the attention of many researchers. Application examples include TURIN (Sorli et al. 1997) with six DOFs and CaPaMan (Ceccarelli 1997) with three DOFs.

Although the locus of any point on the output link is a circle, a planar parallelogram enables an output link to maintain a fixed orientation with respect to

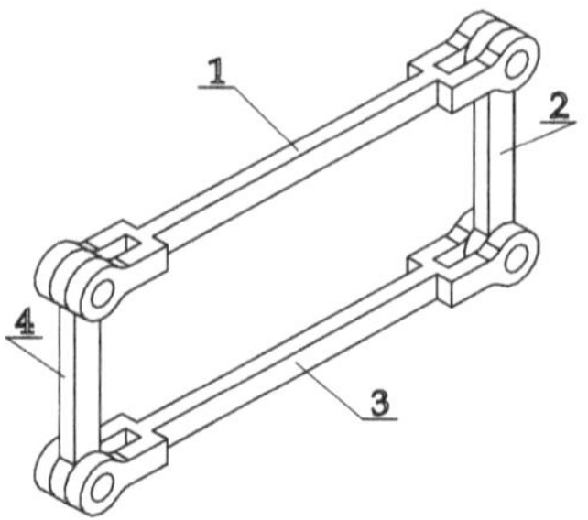

Fig. 2.16 Planar parallelogram

an input link, which is why the concept of the parallelogram is used in these parallel manipulators. In particular, addressing the UU chain (as shown in Sect. 4.1.1) is highly useful. In the above-mentioned designs, the parallelogram guarantees desired output, such as no rotational DOFs for StarLike and Tsai's mechanisms.

Conversely, the planar parallelogram, being a leg, can improve the kinematic performance of a system. Typically, the constant link in a leg for most parallel mechanisms is only a single bar. If the leg is constructed by a simple mechanism, a parallel mechanism will have better performance in terms of kinematics and dynamics. The first design that employed a simple mechanism in leg kinematics is a micromotion mechanism proposed by Hudgens and Tesar (1988). In this mechanism, each leg is driven by a four-bar mechanism mounted on the base, an approach that improved the positional resolution of the mechanism. Tahmasebi (1992) presented a 6-DOF parallel mechanism by means of the planar 2-DOF five-bar mechanism; the positional resolution, stiffness, and force control of the mechanism were improved. Frisoli et al. (1999) developed a novel tendon-driven five-bar linkage with a large isotropic workspace and applied it to the legs of a 6-DOF haptic device, thereby generating good kinematic isotropy and acceleration. Moreover, Chung and Lee (2001) proposed a new 2-DOF parallel mechanism, in which each leg includes a four-bar mechanism, showing advantages in terms of good kinematic performance and static balance (Wang and Gosselin 1999).

Another advantage to designing a mechanism with a parallelogram acting as the constant link in each leg is that leg stiffness can be increased to a large extent. A parallelogram has higher stiffness with respect to a single bar. Moreover, leg stiffness can be improved further by increasing redundant constraints. For instance, the parallelogram can be designed as in Fig. 2.17, in which the number of parallelograms is increased. The more parallelograms the leg has, the higher the stiffness of the mechanism and, relatively, the higher the fabrication accuracy required because of the redundant constraints.

For most parallel mechanisms, rotational capability reaches its maximum at the original point of the workspace. The maximum value is usually limited for some parallel mechanisms with spherical joints because of the limited swing angle of such joints. Moreover, the titling angle of the mobile platform becomes increasingly smaller from the point to the boundary. A parallelogram enables the output link to remain at a fixed orientation with respect to an input link in such a way that attaching

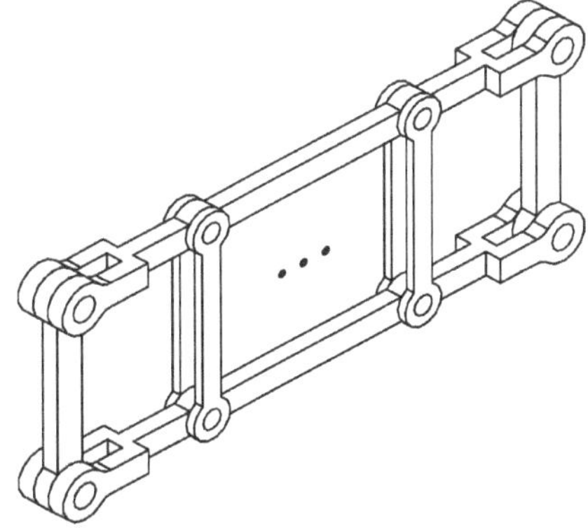

Fig. 2.17 Parallelogram with redundant constraints

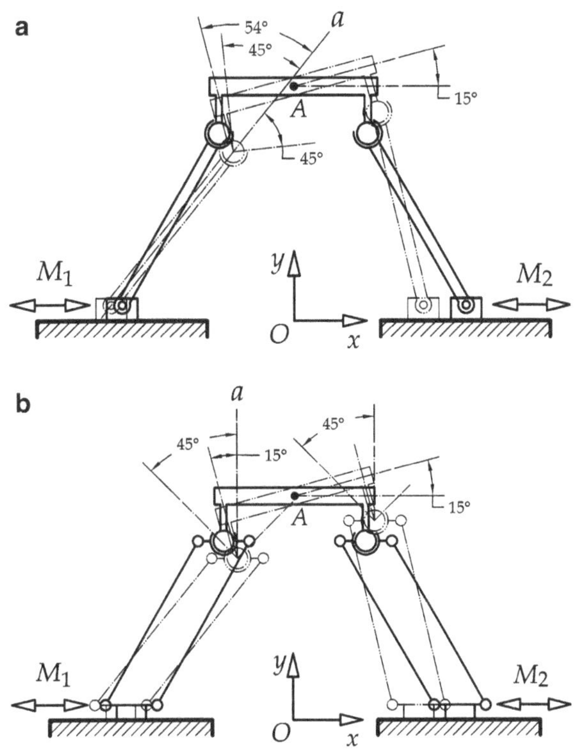

Fig. 2.18 Rotational capability comparison of two kinds of leg mechanisms: (**a**) with a single bar and (**b**) with a planar four-bar parallelogram

one end of a ball-and-socket joint to the output link can relatively increase the swing range of the joint. The method eventually improves the rotational capability of the mobile platform. For example (refer to Fig. 2.18), suppose that in some kind of parallel mechanism, the combination of actuations M_1 and M_2 enables the mobile platform to rotate about point A. In Fig. 2.18a, each of the two legs is a traditional one, which consists of a length-fixed link. The link is connected to the mobile platform by a ball-and-socket joint. In Fig. 2.18b, each leg consists of a planar four-bar parallelogram, which is also connected to the mobile platform by a ball-and-socket joint. The socket is fixed to the output link of each parallelogram. In Fig. 2.18, line a is the reference line, from which the tilting angle for the joints is limited to $\pm 45°$. Let each mobile platform rotate 15° about original point A from zero orientation. Figure 2.16 shows that the orientation of line a in Fig. 2.18a is variable according to the change in the orientation of the link but maintains invariable in Fig. 2.16b. For this reason, the ball-and-socket joint in Fig. 2.16a exceeds the tilting limit at the 15°orientation because of $54.2° > 45°$. In Fig. 2.16b, however, the joint has not reached its limit. Therefore, the mobile platform in Fig. 2.16b can yield higher rotational capability than can that in Fig. 2.16a, for which we can conclude that a parallel mechanism with a parallelogram in its leg may obtain higher rotational capability than can the traditional one. Moreover, because the socket maintains its original orientation at every instant, the rotational capability at the point near the original point does not diminish.

The advantages of using parallelograms in the design of parallel mechanism are clear: desired DOF output, considerably higher stiffness, higher rotational capability, and good kinematic and dynamic performance. The disadvantage stems from the complex structure when evaluated against the combination of a revolute joint and one constant link. A parallelogram has four revolute joints and four links. The axes of all the joints should be parallel to one another, and the lengths of each two of the four links should be equal to one another as well. Therefore, the fabrication accuracy should be sufficiently high enough; otherwise, accumulation of errors occurs, eventually resulting in error accumulation in the device. All the joints have single DOFs, but this is not a critical problem. Considering the advantages that the parallelogram offers, the complex structure is acceptable.

Maximizing the advantage of the desired output of a planar four-bar parallelogram, some researchers proposed parallel mechanisms (Liu and Wang 2003, 2005; Liu et al. 2005b).

2.3.1.3 Some Parallel Mechanisms Containing a Planar Parallelogram

Some parallel mechanisms can be designed on the basis of the concept of a parallelogram. In these designs, a parallelogram plays a key role in desired DOF output as well as in improved rotational capability and stiffness.

Two-DOF Parallel Mechanisms

The most planar 2-DOF parallel mechanisms (Figs. 3.1 and 3.2) are the five-bar mechanisms with prismatic or revolute actuators. The mechanism with revolute actuators comprises five revolute pairs, and the two joints fixed to the base are actuated (Fig. 3.1f). The output of the mechanism is the translational motion of a point on the end-effector. Figure 3.1f shows that the locus of any point on the

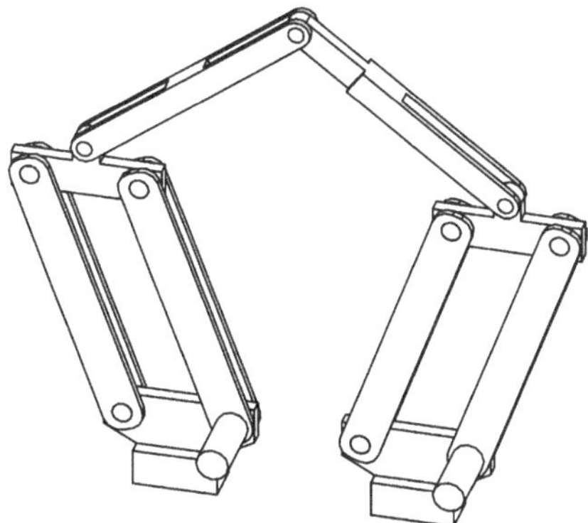

Fig. 2.19 Two-DOF parallel mechanism with a (Pa)RRR(Pa) chain

proximal link, i.e., the actuated link, is a circle, which is the same as that of a point on the output link of a parallelogram. This feature reminds us to replace the proximal link with a parallelogram. The new parallel mechanism is shown in Fig. 2.19. In the same manner, the RRRPR and RRRRP mechanisms shown in Fig. 1.12b and e can be replaced by (Pa)RRPR and (Pa)RRRP, respectively. Figure 2.20a shows another planar 2-DOF parallel mechanism with a (Pa)RR(Pa)P chain, in which one revolute joint in the first (Pa) and the prismatic joint are active. The mobile platform also has two planar DOFs. In these designs, the parallelogram is a substitute for the combination of a revolute joint and a length-fixed link in each leg. This substitution improves the stiffness of the leg. The design shown in Fig. 2.20b differs from those in Figs. 2.19 and 2.20a. It consists of two R(Pa)R chains. In this design, if revolute actuator M_1 is locked, the mechanism is equivalent to a planar four-bar mechanism with actuator M_2. If M_1 is active but M_2 is locked, the mechanism undergoes translation along the x-axis. At the same time, rotation about the x-axis occurs; this rotation is an associated movement. Thus, the output is the translation along the x-axis and another translation in the O-yz plane or a rotation about the x-axis.

As previously mentioned, the output of most planar 2-DOF parallel mechanisms is the planar motion of a point, while orientation changes instantly. In some applications, an object should be transferred with fixed orientation. In such a case, the translational motion of a rigid body with fixed orientation is needed, and the above-mentioned 2-DOF parallel mechanisms are unavailable. The design based on a parallelogram can be used to address this problem. Such a parallel mechanism with a 2-P(Pa) chain is shown in Fig. 2.21a; the two translational DOFs can be achieved if the prismatic joints are active. To obtain the two DOFs of a rigid body in this system, the P(Pa)&PRR kinematic chain shown in Fig. 2.21b is sufficient for guaranteeing the mechanism with two translational DOFs. Two planar four-bar parallelograms are used to increase the stiffness of the system and make the system symmetrical. The parallel mechanism has been applied to the development

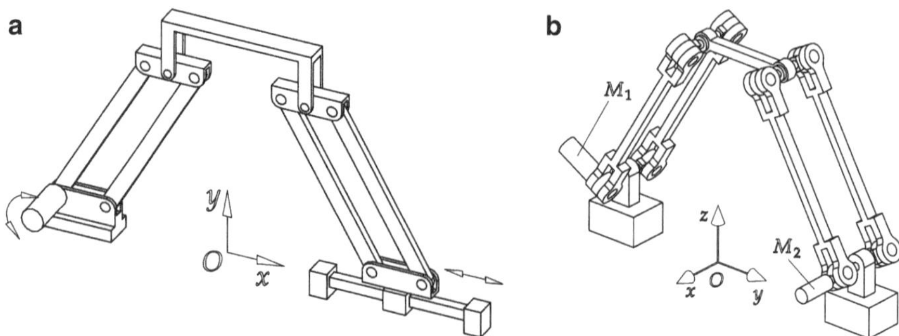

Fig. 2.20 Two kinds of planar 2-DOF parallel mechanisms: (**a**) with a (Pa)RR(Pa)P chain and (**b**) with a 2-R(Pa)R chain

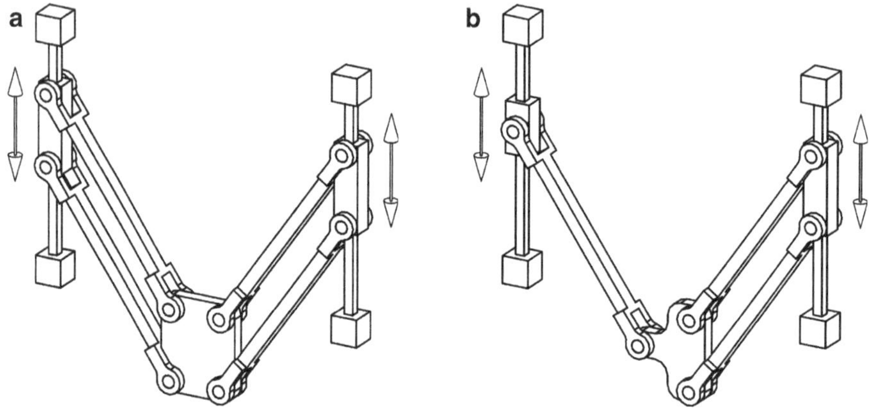

Fig. 2.21 Parallel mechanisms with two pure translational DOFs: (**a**) with symmetrical architecture and (**b**) with asymmetrical architecture

of a five-axis hybrid machine tool (Liu 2001; Liu et al. 2005c), which is now being used to machine a kind of impeller. In the mechanisms shown in Fig. 2.21, some P joints can be replaced by four-bar planar parallelograms. For example, the parallel mechanisms can be equipped with the 4(Pa) and 2(Pa)&3R chains. The latter is illustrated in Fig. 2.22.

Some two-DOF parallel mechanisms based on the above-mentioned design concept are listed in Table 2.1.

The output link of a planar parallelogram has one DOF, and its orientation remains constant. The output link of a double-parallelogram mechanism, denoted as the (Pa)(Pa) chain, should have two DOFs, and its orientation is also constant. Therefore, we can design a 2-DOF parallel mechanism that consists of three legs, with the third one being a passive (Pa)(Pa) chain and the two others active. Each of the actuated legs can be a 3-, 4-, 5- or 6-DOF chain. At the least, it must be a 3-DOF chain, as in the PRR, RRR, RPR, or PPR chain. As an example, the parallel

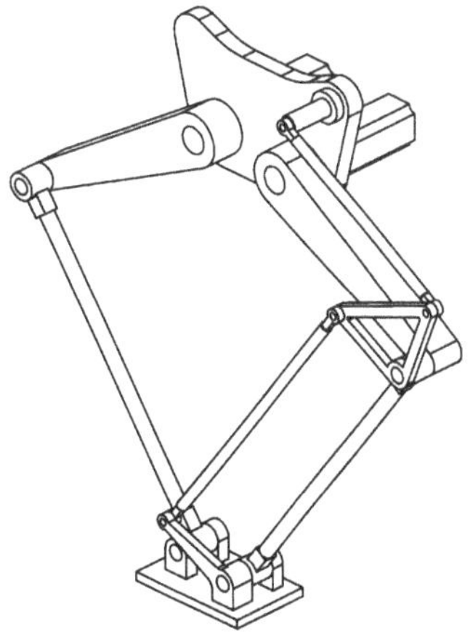

Fig. 2.22 Two-DOF translational parallel mechanism with a 2(Pa)&3R chain

Table 2.1 Some 2-DOF parallel mechanisms

No.	Leg chains: First leg	Second leg	Mechanism chains	Remarks
1	(Pa)RR	(Pa)RR	(Pa)RRR(Pa)	As shown in Fig. 2.17
2	(Pa)RR	RPR	(Pa)RRPR	Comes from the RRRPR mechanism
3	(Pa)RR	PRR	(Pa)RRRP	Comes from the RRRRP mechanism
4	(Pa)R	P(Pa)R	(Pa)R&P(Pa)R	Fig. 2.18a
5	R(Pa)R	R(Pa)R	2-R(Pa)R	Fig. 2.18b
6	P(Pa)	P(Pa) or PRR	P(Pa)(Pa)P(or P(Pa)PRR)	There are two cases: the actuated P joint can be vertical or horizontal. The vertical case is shown in Fig. 2.19
7	(Pa)(Pa)	(Pa)(Pa) or RRR	4(Pa) (or 2(Pa)&3R)	In each leg, a revolute joint in the input link of the parallelogram that attached to the base is actuated. The 2(Pa)&3R mechanism is shown in Fig. 2.20

mechanisms with 2-PRR&1-(Pa)(Pa) and 2-RPR&1-(Pa)(Pa) chains are shown in Fig. 2.23a, b, respectively. Because of the double-parallelogram mechanism, both mechanisms have two pure translational DOFs. A prismatic joint cannot change the slider orientation; thus, the double-parallelogram chain can be replaced by a prismatic parallelogram, referred to as the P(Pa) chain, shown in Fig. 2.24. The possible 2-DOF translational parallel mechanisms with a passive leg are listed in Table 2.2, and a typical instance is illustrated in Fig. 2.25.

Three-DOF Parallel Mechanisms

As introduced in Sect. 3.1.2, many kinds of parallel mechanisms with three DOFs are available. An example is the planar 3-RRR parallel mechanism (Fig. 1.15). The mobile platform has three planar DOFs, which are two translations in the *O*-*xy* plane

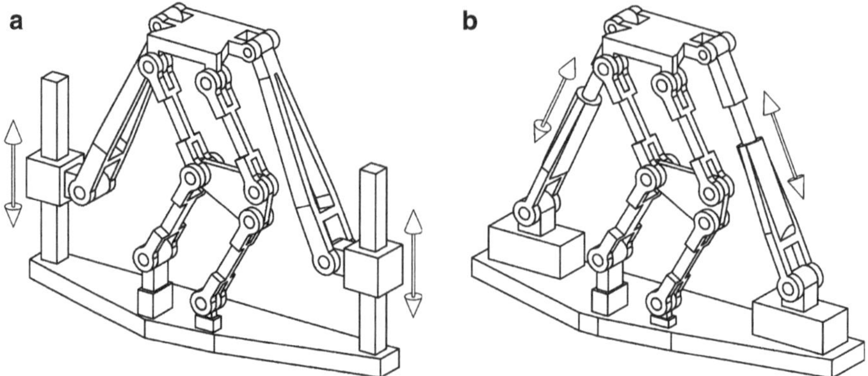

Fig. 2.23 Two 2-DOF translational parallel mechanisms: (**a**) with two active PRR chains and one passive (Pa)(Pa) chain and (**b**) with two active RPR chains and one passive (Pa)(Pa) chain

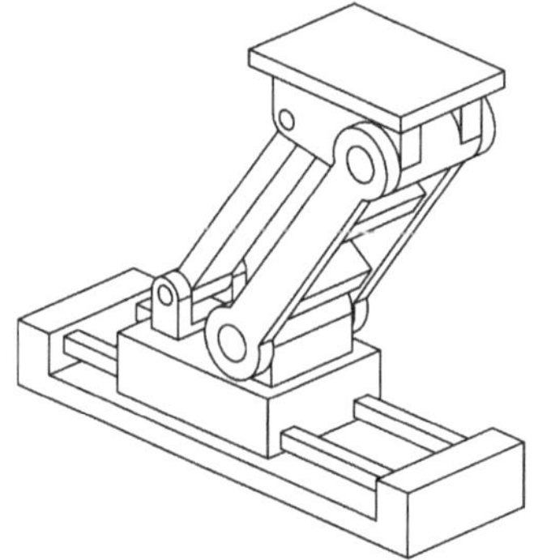

Fig. 2.24 P(Pa) kinematic chain

and a rotation around the *Z*-axis. To increase the stiffness of each leg, a 3-DOF parallel mechanism with parallelograms is employed (Fig. 2.26). This mechanism is kinematically equivalent to the architecture in Fig. 1.15. Similarly, two types of such mechanisms with 3-P(Pa)R chains, which are equivalent to those with 3-PRR chains, are shown in Fig. 2.27. Figure 2.28 shows four types of spatial 3-DOF 3-[PP]S parallel mechanisms, which are the modified versions of the 3-RPS and 3-PRS parallel mechanisms. Comparatively, they possess higher rotational capability and stiffness. In the mechanism shown in Fig. 2.28a, either the P joint or one R joint of the parallelogram attached to the base in each leg can be active. In the three others, the P joints are actuated.

Figure 2.29 shows another spatial 3-DOF parallel mechanism with a 1-R(Pa)R&2-RR(Pa)R chain. If actuator M_1 is locked, the mechanism is equivalent to a planar 2-DOF mechanism with actuators M_2 and M_3. Thus, the output is the translation along the *z*-axis and two planar motions in the *O-xy* plane. Its kinematic chain can also be 1-R(Pa)R&2-PR(Pa)R.

Table 2.2 Some 2-DOF translational parallel mechanisms with a passive leg

No.	Leg chains: First and second legs	Leg chains: Passive leg	Mechanism chains	Remarks
1	RRR	(Pa) (Pa) or P(Pa)	2-RRR&1-(Pa)(Pa) or 2-RRR&1-P(Pa)	
2	PRR		2-PRR&1-(Pa)(Pa) or 2-PRR&1-P(Pa)	The 2-PRR&1-(Pa)(Pa) mechanism is shown in Fig. 2.23a
3	RPR		2-RPR&1-(Pa)(Pa) or 2-RPR&1-P(Pa)	The 2-RPR&1-(Pa)(Pa) mechanism is shown in Fig. 2.23b
4	RRU		2-RRU&1-(Pa)(Pa) or 2-RRU&1-P(Pa)	
5	PRU		2-PRU&1-(Pa)(Pa) or 2-PRU&1-P(Pa)	
6	RPU		2-RPU&1-(Pa)(Pa) or 2-RPU&1-P(Pa)	
7	RRS		2-RRS&1-(Pa)(Pa) or 2-RRS&1-P(Pa)	
8	PRS		2-PRS&1-(Pa)(Pa) or 2-PRS&1-P(Pa)	The 2-PRS&1-P(Pa) mechanism is shown in Fig. 2.25
9	RPS		2-RPS&1-(Pa)(Pa) or 2-RPS&1-P(Pa)	
10	UPS		2-UPS&1-(Pa)(Pa) or 2-UPS&1-P(Pa)	
11	PUS		2-PUS&1-(Pa)(Pa) or 2-PUS&1-P(Pa)	
12	URS		2-URS&1-(Pa)(Pa) or 2-URS&1-P(Pa)	
13	RUS		2-RUS&1-(Pa)(Pa) or 2-RUS&1-P(Pa)	

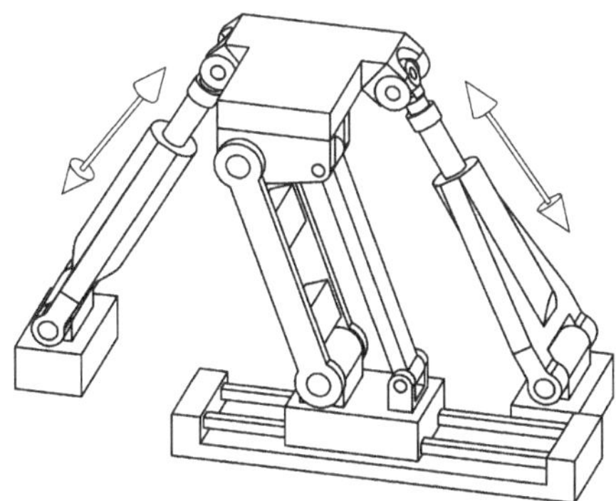

Fig. 2.25 Two-DOF translational parallel mechanism with a 2-PRS&1-P(Pa) chain

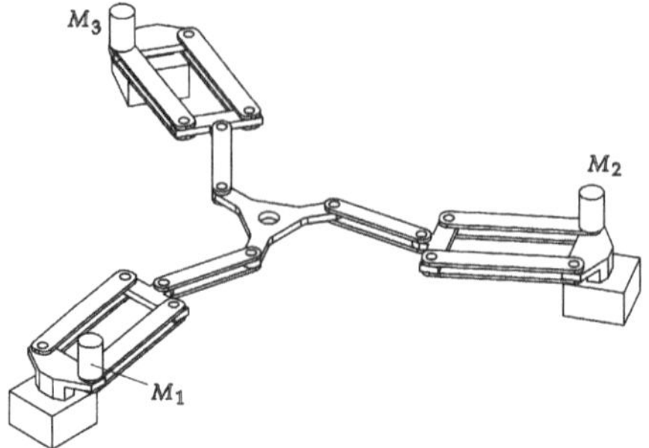

Fig. 2.26 Planar 3-(Pa)RR parallel mechanism

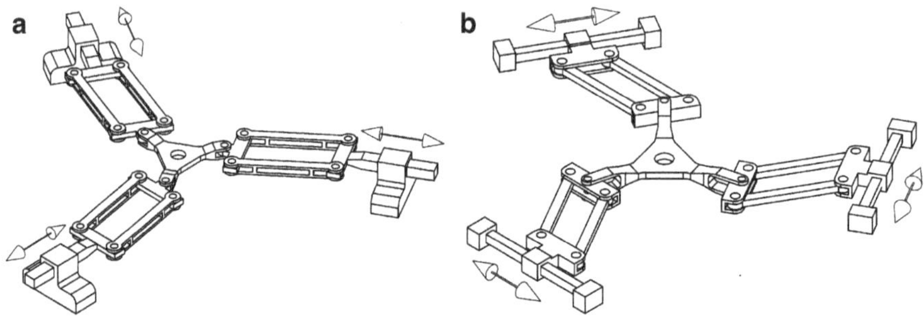

Fig. 2.27 Two kinds of planar 3-P(Pa)R parallel mechanisms

To solve the problem of the UU chain, an R(Pa)R chain was proposed (Fig. 2.11). Accordingly, the 2-PRS&1-PUU mechanism was revised as the 2-PRS&1-PR(Pa)R mechanism (see Fig. 2.15). The observation of the revised mechanism shows that because the PR(Pa)R chain constrains two rotations, a redundant DOF exists in the two spherical joints. Thus, the two PRS chains can be replaced by two PRU chains. The modified mechanism is shown in Fig. 2.30. The topological mechanism with revolute actuators is illustrated in Fig. 2.31. In the two mechanisms, the first and second legs should be in the same plane. Moreover, in these two legs, two axes for the revolute joints in the U joints that are connected to the mobile platform should be collinear. The axis of the revolute joint in the third leg linked to the mobile platform should be parallel to these two axes. Therefore, the two mechanisms have two translations in the *O*-*yz* plane and one rotation about the collinear line. The mechanisms shown in Figs. 2.15 and 2.30 have very high rotational capability and are analyzed in the subsequent chapters.

The collinear axes for the two revolute joints of the mechanism in Fig. 2.30 motivate the redesign of the first and second legs (Fig. 2.32). In these two designs, the two revolute joints are combined into one revolute joint. Figure 2.32a shows that the first and second legs are connected to the mobile platform through one

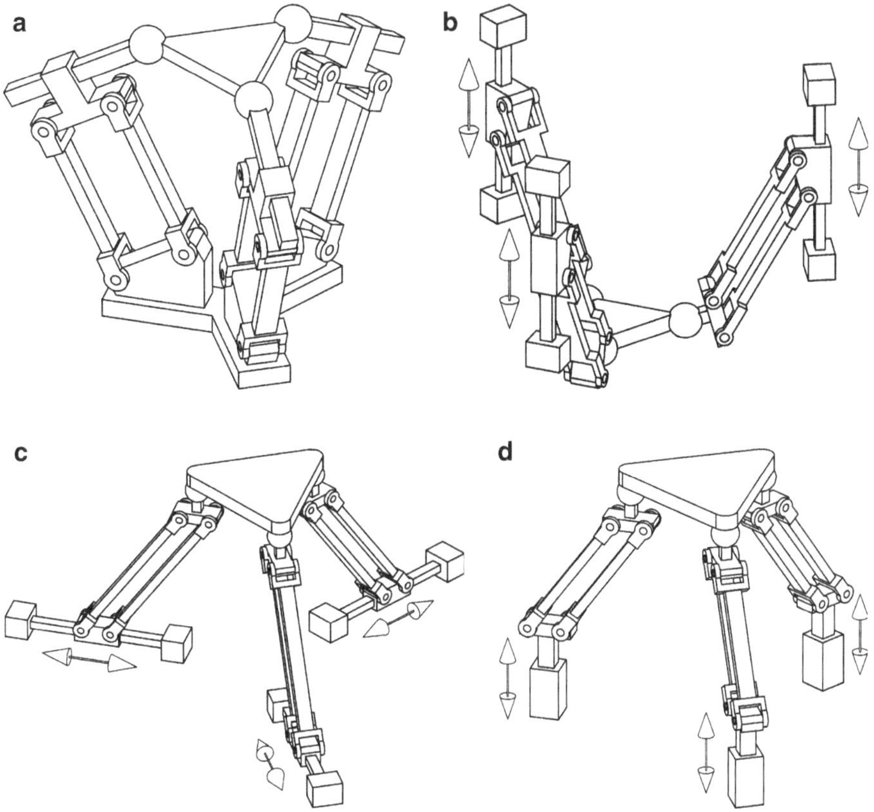

Fig. 2.28 Four kinds of 3-[PP]S parallel mechanisms

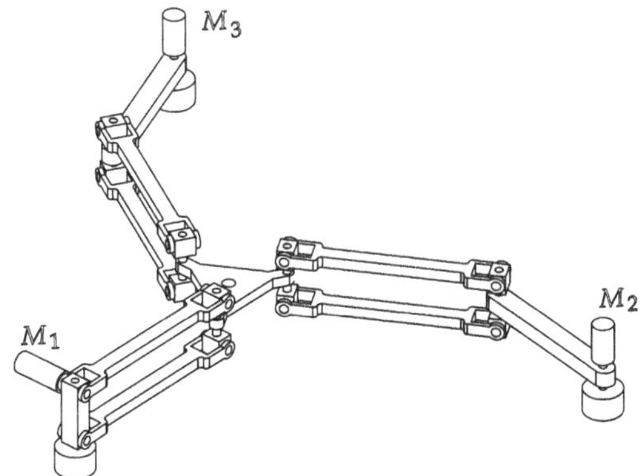

Fig. 2.29 Spatial 3-DOF parallel mechanism with 1-R(Pa)R&2-RR(Pa)R

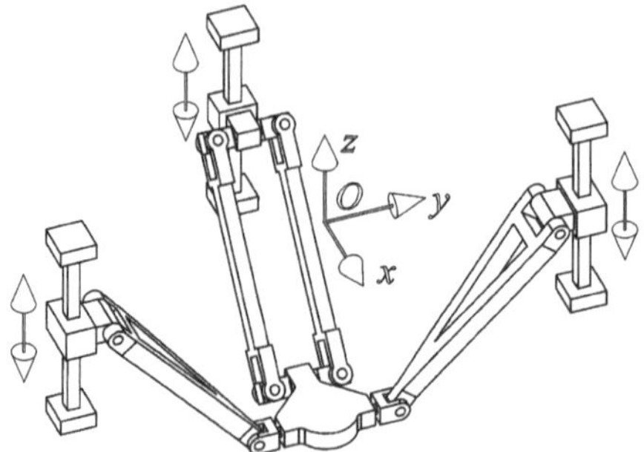

Fig. 2.30 Three-DOF parallel mechanism with a 2-PRU&1-PR(Pa)R chain

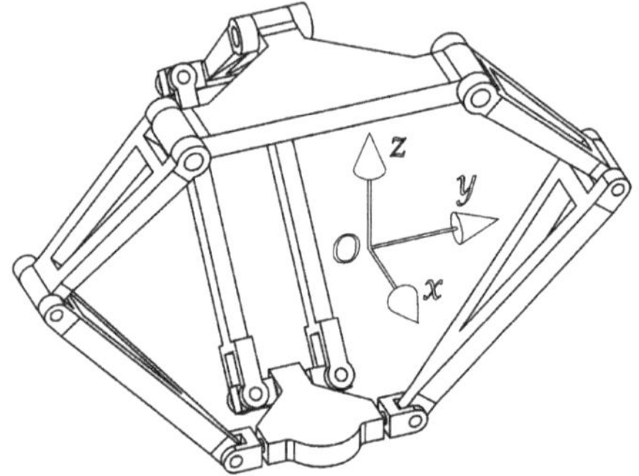

Fig. 2.31 Three-DOF parallel mechanism with a 2-RRU&1-RR(Pa)R chain

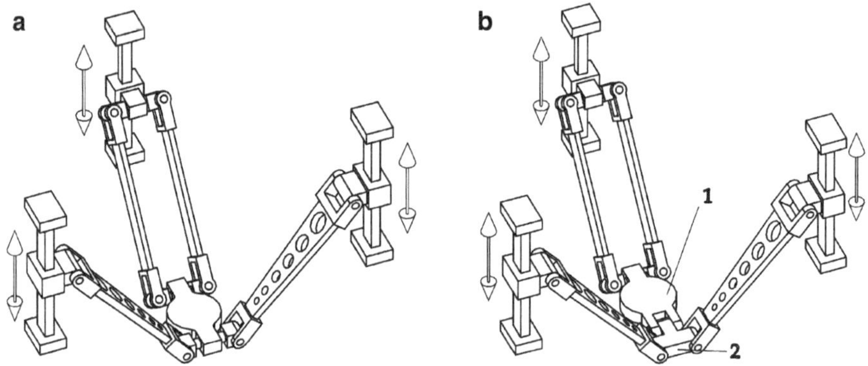

Fig. 2.32 Two topological mechanisms of the mechanism shown in Fig. 2.30

revolute joint. As shown in Fig. 2.32b, the first and second legs with PRR chains are connected to a constant orientation bar 2, which is linked to mobile platform 1 by a revolute joint. If the kinematic chain for the mechanism shown in Fig. 2.30 is 2-PRU&1-PR(Pa)R, the chain for the two designs will be $(PRR)_2R$&PR(Pa)R.

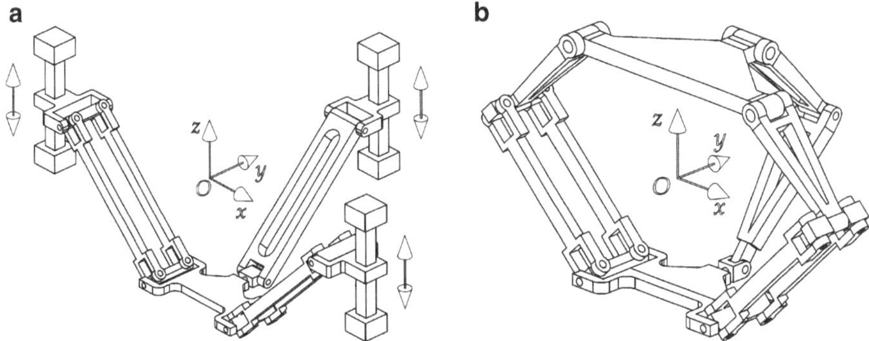

Fig. 2.33 Two spatial 3-DOF parallel mechanisms: (**a**) with a 2-PR(Pa)R&1-PRU chain and (**b**) with a 2-RR(Pa)R&1-RRU chain

Table 2.3 Constraint and DOFs of the parallel mechanism in Fig. 2.33a

Single leg			Combination of three legs	
No.	Chain type	Constraints	Constraints	Remained DOFs
1	PR(Pa)R	$\{RO_x, RO_z\}$	$\{T_x, RO_x, RO_z\}$	$\{T_y, T_z, RO_y\}$
2	PR(Pa)R	$\{RO_x, RO_z\}$		
3	PRU	$\{T_x, RO_z\}$		

This modification, which has no negative influence on the rotational capability of the mechanism, can also be extended to the mechanism (Fig. 2.31) with revolute actuators.

The kinematic chains in the mechanism in Fig. 2.30 shows that the end-effector of a PR(Pa)R chain has three translations and one rotation, and that of a PRU chain has two translations and two rotations. A mechanism with two PRU chains and one PR(Pa)R chain has two translational DOFs and one rotational DOF. *What about a mechanism with two PR(Pa)R chains and one PRU chain?* Such a mechanism (Liu et al. 2005a) is shown in Fig. 2.33a, in which the three R joints attached to the mobile platform are parallel with the others. When the three P joints are active, the mechanism should have three DOFs. The mechanism with revolute actuators is shown in Fig. 2.33b. Given the arrangement of links and joints of the mechanism, the combination of the three legs constrains the rotation of the mobile platform with respect to the x- and z-axes as well as with the translation along the x-axis. This leaves the two mechanisms with two translational DOFs in the O-yz plane and one rotational DOF about the y-axis. Table 2.3 shows the capability description of the mechanism with prismatic actuators. The mechanisms shown in Figs. 2.30 and 2.33a have the same output, but there is an outstanding difference between them in terms of the rotational DOF: the latter is actuation redundant for the DOF but the former is not. That is, the rotational DOF of the latter is determined by the combination of the first and second legs with PR(Pa)R chains. This situation is similar to the mechanisms shown in Figs. 2.31 and 2.33b.

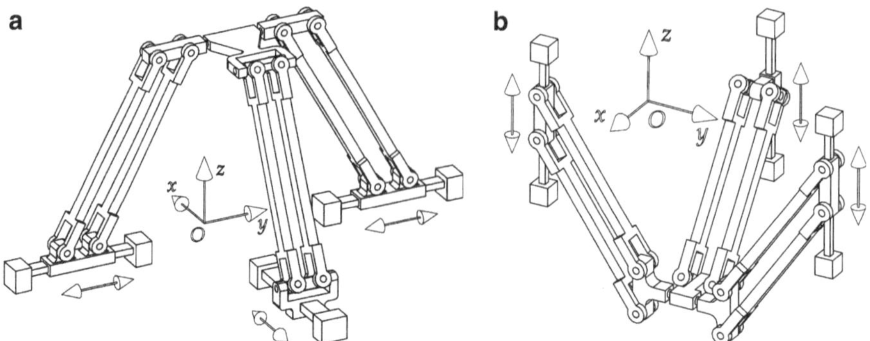

Fig. 2.34 Kinematic architecture of the parallel mechanism with a 2-P(Pa)R&1-PR(Pa)R chain

Table 2.4 Constraint and DOFs of the parallel mechanism in Fig. 2.34

Single leg			Combination of three legs	
No.	Chain type	Constraints	Constraints	Remained DOFs
1	P(Pa)R	$\{RO_x, RO_z, T_x\}$	$\{T_x, RO_x, RO_z\}$	$\{T_y, T_z, RO_y\}$
2	P (Pa)R	$\{RO_x, RO_z, T_x\}$		
3	PR(Pa)R	$\{RO_x, RO_z\}$		

Let us consider another spatial 3-DOF parallel mechanism (Fig. 2.34). The mechanism has two P(Pa)R chains and one PR(Pa)R chain. Table 2.4 presents the capability of the mechanism in detail, showing that the first leg itself can constrain the mobile platform with the translation along the *x*-axis and rotations about the *z*- and *x*-axes. The second leg can be identical to or different from the first leg (e.g., the second leg can be equipped with a PRU chain). The combination of the first and second legs can also impose the same constraints on the mechanism, i.e., the translation along the *x*-axis and rotations about the *z*-and *x*-axes. Moreover, the third leg can be a 4-, 5-, or 6-DOF chain (e.g., a PUU or PUS). It can also be a traditional PSS chain. The possible chains for the three legs and mechanism are shown in Table 2.5. The table demonstrates that six types of mechanisms have different legs. For example, the mechanisms with P(Pa)R-PRU-PR(Pa)R and P(Pa)R-P(Pa)R-PUU chains are shown in Fig. 2.35a, b, respectively. In the PUU chain, the axis of the revolute joint in the first U joint that is attached to the P joint and the axis of the revolute joint in the second U joint that is connected to the mobile platform must both be parallel to the *y*-axis. Then, the mechanisms can also have three DOFs, which are two translational DOFs in the *O*-*yz* plane and one rotational DOF about the *y*-axis. For the PUS chain in Table 2.5, the axis of the revolute joint in the first U joint that is attached to the P joint should also be parallel to the *y*-axis. Many topological architectures based on such a concept have been proposed, but these are not comprehensively described in this chapter.

We compare the mechanism shown in Fig. 2.34 and its topological mechanisms with the mechanism in Fig. 2.30. They have the same output, but there are some

Table 2.5 Possible legs and topological mechanisms of the parallel mechanism in Fig. 2.34

Leg chain			
First leg	Second leg	Third leg	Mechanism chains
P(Pa)R	P(Pa)R	PR(Pa)R	P(Pa)R-P(Pa)R-PR(Pa)R
			P(Pa)R-PRU-PR(Pa)R
	PRU	PUU	P(Pa)R-P(Pa)R-PUU
			P(Pa)R-PRU-PUU
		PUS (or PSS)	P(Pa)R-P(Pa)R-PUS
			P(Pa)R-PRU-PUS

Where *P* prismatic joint, *R* revolute joint, *U* universal joint, *(Pa)* parallelogram, *S* spherical joint

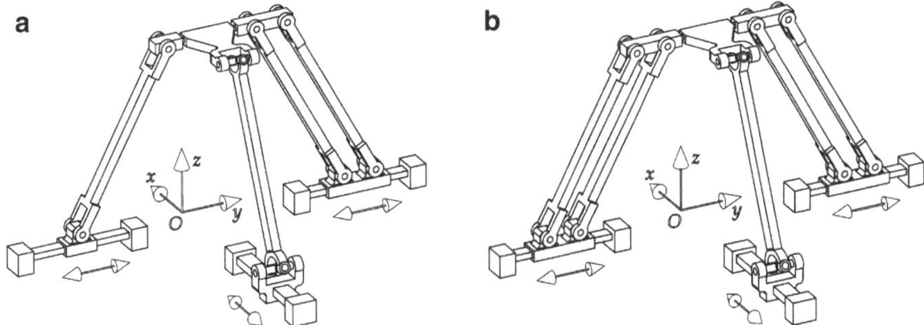

Fig. 2.35 Two kinds of topological architectures of the mechanism in Fig. 2.34a

differences between them. In the mechanism in Fig. 2.30, the first and second legs are PRU chains, which can constrain the mobile platform with the translation along the x-axis and the rotation about the z-axis. The combination of the two legs causes the mobile platform to rotate freely about the line parallel to the x-axis. Because it is the undesirable output, the rotational DOF should be constrained by the third leg. In such a situation, the stiffness of the third leg should be sufficiently high for addressing the inner torque of the mobile platform. Conversely, in the mechanism shown in Fig. 2.34, the first and second legs themselves can constrain the rotation about the x-axis; no additional requirement on the stiffness of the third leg is necessary. As a result, the system stiffness of the mechanism is relatively higher.

Similar to a 2-DOF parallel mechanism, the planar four-bar parallelogram can also be used in the design of a 3-DOF parallel mechanism with a passive leg. As an example, Fig. 2.36 shows a spatial 3-DOF parallel mechanism with three active UPS chains and one passive leg with a (Pa)U chain, which leads to one degree of translational freedom and two degrees of rotational freedom of the mechanism.

The possible 3-DOF parallel mechanisms with a planar four-bar parallelogram are listed in Table 2.6.

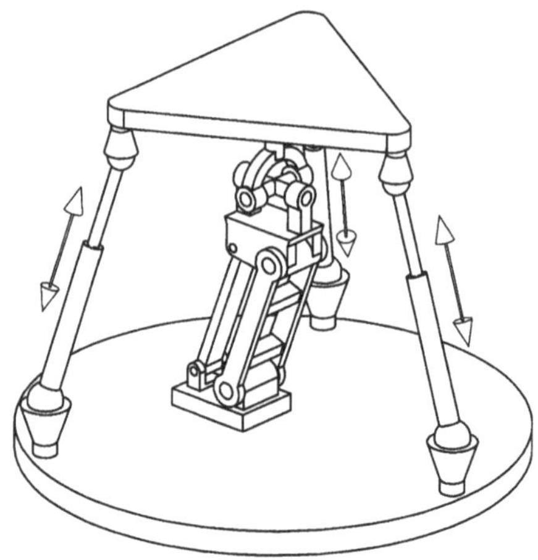

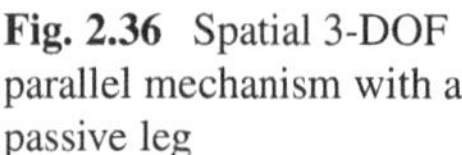

Fig. 2.36 Spatial 3-DOF parallel mechanism with a passive leg

Four-DOF Parallel Mechanisms

Several parallel mechanisms with four DOFs can be designed on the basis of the concept of a four-bar parallelogram. Figure 2.37a shows a 4-DOF parallel mechanism with a 2-PR(Pa)U& 2-PR(Pa)R chain, and in Fig. 2.37b, the mechanism is equipped with a 2-PUU&2-PR(Pa)R chain. The mobile platforms for both these mechanisms have four DOFs, which are three translations and one rotation about the y-axis with respect to the base. If the quadrangle of the mobile platform is similar to that of the base, the mechanisms will be in their singular configurations at the original position.

Figure 2.38 shows two kinds of 4-DOF parallel mechanisms with a passive leg. Each of these designs is equipped with four legs having UPS (or SPS) chains, in which the P joint is actuated; the fifth (passive) leg in the mechanism in Fig. 2.38a has a (Pa)PU chain; and the mechanism in Fig. 2.38b has a (Pa)S chain.

The other parallel mechanisms with four DOFs are listed in Table 2.7.

Five-DOF Parallel Mechanisms

As mentioned earlier, designing a 5-DOF symmetrical parallel mechanism is difficult. Such a mechanism with fully symmetric architecture is also difficult to design using a parallelogram. Two kinds of 5-DOF parallel mechanisms with a parallelogram in one of their legs are illustrated in Fig. 2.39. In these designs, the fifth leg is actuated and also serves as the leading leg, i.e., the output of the mechanism is dependent on the leg. The other four legs are equipped with UPS or SPS chains, in which the P joints are also actuated. In Fig. 2.39a, the kinematic chain of the fifth leg is the P(Pa)S, and that in Fig. 2.39b is the P(Pa)PU, in which the prismatic joints attached to the base are actuated. Thus, the mechanisms have two translations and three rotations, as well as three translations and two rotations, respectively.

The other parallel mechanisms with five DOFs are enumerated in Table 2.8.

Table 2.6 Some three-DOF parallel mechanisms with a planar four-bar parallelogram

No.	Leg chains			Mechanism chains	Remarks
	First leg	Second leg	Third leg		
1	(Pa)RR	(Pa)RR	(Pa)RR	3-(Pa)RR	As shown in Fig. 2.26
2	P(Pa)R	P(Pa)R	P(Pa)R	3- P(Pa)R	Two examples are shown in Fig. 2.27
3	(Pa)PS	(Pa)PS	(Pa)PS	3- (Pa)PS	As shown in Fig. 2.28a
4	P(Pa)S	P(Pa)S	P(Pa)S	3- P(Pa)S	As shown in Fig. 2.28b-d
5	R(Pa)R	RR(Pa)R	RR(Pa)R	1-R(Pa)R&2-RR(Pa)R	As shown in Fig. 2.29
6	PRU(S)	PRU(S)	PR(Pa)R	2-PRU(S)&1-PR(Pa)R	As shown in Fig. 2.30
7	RRU(S)	RRU(S)	RR(Pa)R	2-RRU(S) &1-RR(Pa)R	As shown in Fig. 2.31
8	PR(Pa)R	PR(Pa)R	PRU	2-PR(Pa)R&1-PRU	As shown in Fig. 2.33a
9	RR(Pa)R	RR(Pa)R	RRU	2-RR(Pa)R&1-RRU	As shown in Fig. 2.33b
10	P(Pa)R	P(Pa)R	PR(Pa)R	2-P(Pa)R&1-PR(Pa)R	Fig. 2.34, please see Table 2.5 for other topology architectures
	The first, second and third legs	The fourth (passive) leg			
11	UPS (PUS or RUS)	(Pa)U		3-UPS&1-(Pa)U	As shown in Fig. 2.36
12		(Pa)PR		3-UPS&1-(Pa)PR	
13		(Pa)RP		3-UPS&1-(Pa)RP	

Where *P* prismatic joint, *R* revolute joint, *U* universal joint, *(Pa)* parallelogram, *S* spherical joint

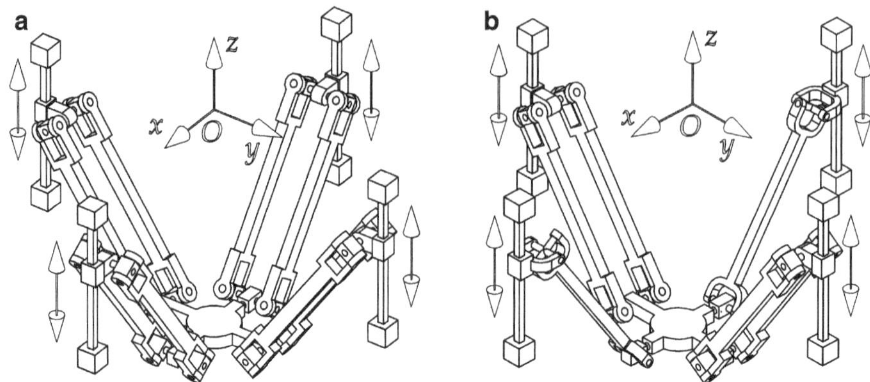

Fig. 2.37 Two 4-DOF parallel mechanisms: (**a**) with a 2-PR(Pa)U& 2-PR(Pa)R chain and (**b**) with 2-PUU&2-PR(Pa)R chain

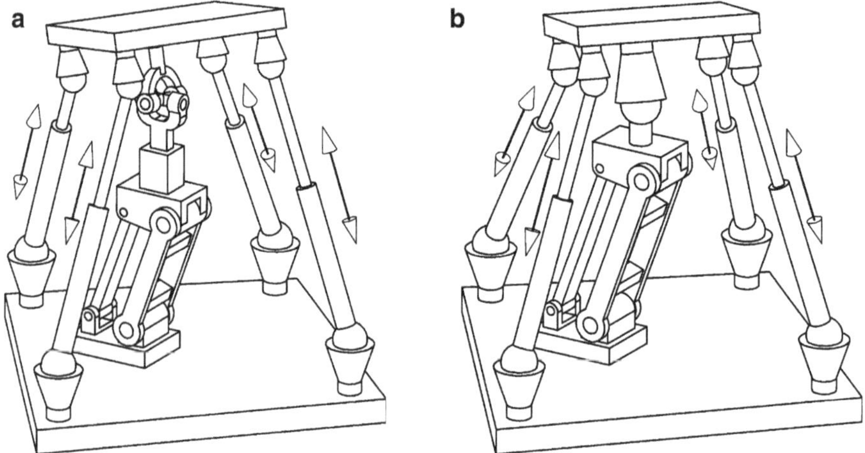

Fig. 2.38 Two kinds of 4-DOF parallel mechanisms with a passive leg: (**a**) with a 4-SPS&1-(Pa)PU chain and (**b**) with a 4-SPS&1-(Pa)S chain

Six-DOF Parallel Mechanisms

In the past two decades, 6-DOF parallel mechanisms have been the most frequently studied mechanisms. These mechanisms can provide all of the DOFs that a rigid body can have in 3-D space. This kind of parallel mechanism has been eliciting increasing attention in the field. Generally, each leg of this mechanism is a typical serial chain with six DOFs. The fully parallel mechanism is usually one that includes at least six such legs. Some 6-DOF parallel mechanisms have three legs. Two actuators are attached to each leg. Compared with a 6-DOF mechanism with six legs, a fully parallel mechanism has a considerably larger workspace, simpler forward and inverse kinematic solutions, and fewer moving parts and joints. The

Table 2.7 Some four-DOF parallel mechanisms with a four-bar parallelogram

No.	Leg chains: First leg	Second leg	Third leg	Fourth leg	Mechanism chains	Remarks
1	PR(Pa)U	PR(Pa)R	PR(Pa)U	PR(Pa)R	2-PR(Pa)U&2-PR(Pa)R	As shown in Fig. 2.37a
2	PUU	PR(Pa)R	PUU	PR(Pa)R	2-PUU&2-PR(Pa)R	As shown in Fig. 2.37b
	Four active legs		The passive leg			
3	UPS (PUS or RUS)		(Pa)PU		4-UPS&1-(Pa)PU	As shown in Fig. 2.38a
4			(Pa)RU		4-UPS&1-(Pa)RU	
5			(Pa)S		4-UPS&1-(Pa)S	As shown in Fig. 2.38b
6			P(Pa)U		4-UPS&1-P(Pa)U	
7			(Pa)UP		4-UPS&1-(Pa)UP	
8			R(Pa)U		4-UPS&1-R(Pa)R	

Where *P* prismatic joint, *R* revolute joint, *U* universal joint, *(Pa)* parallelogram, *S* spherical joint

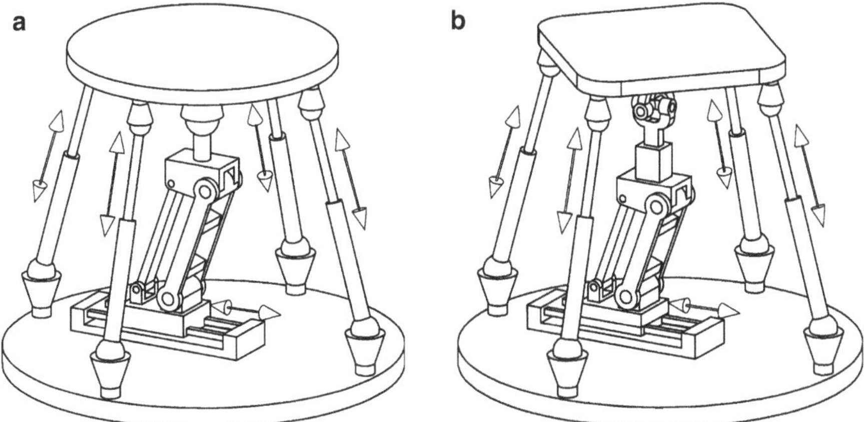

Fig. 2.39 Two kinds of new 5-DOF parallel mechanisms: (**a**) with a 4-SPS&1-P(Pa)S chain and (**b**) with a 4-SPS&1-P(Pa)PU chain

parallelogram concept is also used in the design of such mechanisms. Application examples are TURIN (Sorli et al. 1997) with a 3-(Pa)(Pa)PS chain and a mechanism with 3-R(Pa)S chain proposed by Ebert-Uphoff (1998).

In the present chapter, three types of 6-DOF parallel mechanisms with parallelograms are discussed (Fig. 2.40). The mechanism in Fig. 2.40a consists of three identical legs, which are PP(Pa)S chains. In each of the three legs, two prismatic joints are actuated. For each of the two mechanisms shown in Fig. 2.40b, c, the mobile platform is connected to the base through the three legs as well; the mechanisms are equipped with 2-DOF planar actuators. Compared with the 3-PPRS mechanism and others with planar actuators, these parallel mechanisms have a much higher tilting capability. Table 2.9 lists some 6-DOF parallel mechanisms.

Table 2.8 Some five-DOF parallel mechanisms with a parallelogram in their legs

No.	Leg chains: Legs with identical kinematic chains	Leg chains: The fifth leg	Mechanism chains	Remarks
1	UPS (PUS or RUS)	P(Pa)S	4-UPS&1-P(Pa)S	There are two cases: the actuated P joint can be vertical or horizontal. The vertical case is shown in Fig. 2.39a
2		P(Pa)PU	4-UPS&1-P(Pa)PU	There are two cases: the actuated P joint can be vertical or horizontal. The vertical case is shown in Fig. 2.39b
3		P(Pa)RU	4-UPS&1-P(Pa)RU	In the fifth leg, the P joint is actuated
4		P(Pa)UP	4-UPS&1-P(Pa)UP	In the fifth leg, the first P joint is actuated
5		(Pa)PS	4-UPS&1-P(Pa)S	In the fifth leg, the P joint is actuated
6		(Pa)RS	4-UPS&1-(Pa)RS	In the fifth leg, one of the four revolute joints in the parallelogram is actuated
7		R(Pa)S	4-UPS&1-R(Pa)S	In the fifth leg, the R joint is actuated

2.3.2 Type Synthesis Based on the Evolution Method

Many parallel mechanisms can be proposed using different methods. To a certain extent, these parallel mechanisms can provide inspiration for the design of potential new mechanisms. The modified mechanism may have relative advantages and good industrial applicability. In this section, we introduce some novel *3-DOF parallel mechanisms* that evolved from others.

Some spatial 3-DOF fully parallel mechanisms were introduced in section "3-DOF parallel mechanisms" (Figs. 2.30, 2.31, 2.32, 2.33, 2.34, and 2.35). In these mechanisms, at least one leg consists of a parallelogram. The parallelogram eventually presents difficulties in manufacturing and assembly as well as affects the accuracy and application of the mechanisms. These issues motivate us to identify a solution to overcome these problems.

As discussed in Sect. 4.1, the use of a parallelogram in each mechanism shown in Figs. 2.30, 2.31, 2.32, 2.33, 2.34, and 2.35 guarantees the unique rotational DOF of the mobile platform. For example, the parallelogram in a mechanism can restrict the rotations about the z- and x-axes (Fig. 2.30). The translation in the O-yz plane of the mobile platform is implemented by actuating the sliders of the two legs (denoted as the first and second legs) with identical kinematic chains. The two legs are in the same plane, i.e., the O-yz plane. Therefore, the leg with a parallelogram (referred to as the third leg) provides the mobile platform with the active rotation about the axis parallel to the y-axis and the passive translations along the x-, y-,

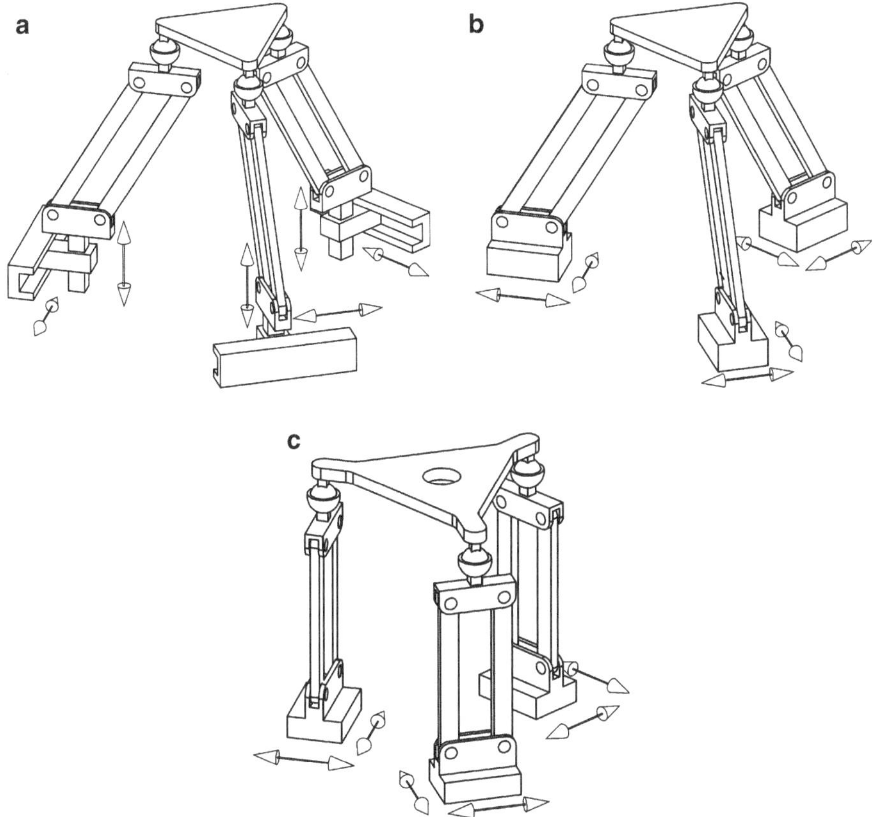

Fig. 2.40 Three new types of 6-DOF parallel mechanisms

and z-axes. The translation along the x-axis is a parasitic motion. Observing the mechanism shows that at any position (y, z) of the mobile platform, the movement of the mobile platform and third leg is actually that of a slider-crank mechanism. If the z-coordinate of the mobile platform is specified, the change in shape of the parallelogram in the third leg conforms to the translation along the y-axis. Thus, replacing the third leg of the parallel mechanism with a PRC kinematic chain is feasible. The modified parallel mechanism is shown in Fig. 2.41a. In the new mechanism, legs 1 and 2 have identical kinematic chains, i.e., PRU (U-universal) chains. A U joint is usually composed of two R joints. In the two U joints, the axes of the two R joints that are connected to the mobile platform should be collinear. Otherwise, the mobile platform loses one DOF. Because of the PRC chain of leg 3, the two U joints can be replaced by two S (spherical joint) joints. Undoubtedly, the new mechanism also has the same DOFs as that of the mechanism shown in Fig. 2.30, i.e., the translation in the O-yz plane and the rotation about the axis parallel to the y-axis if the P joints are actuated.

Table 2.9 Some six-DOF parallel mechanisms with a four-bar parallelogram

No.	Leg chains			Mechanism chains	Remarks
	First leg	Second leg	Third leg		
1	PP(Pa)S	PP(Pa)S	PP(Pa)S	3-PP(Pa)S	As shown in Fig. 2.40a
2	$(PA)_2$(Pa)S	$(PA)_2$(Pa)S	$(PA)_2$(Pa)S	3-$(PA)_2$(Pa)S	As shown in Fig. 2.40b
3	$(PA)_2$(Pa)S	$(PA)_2$(Pa)S	$(PA)_2$(Pa)S	3-$(PA)_2$(Pa)S	As shown in Fig. 2.40c

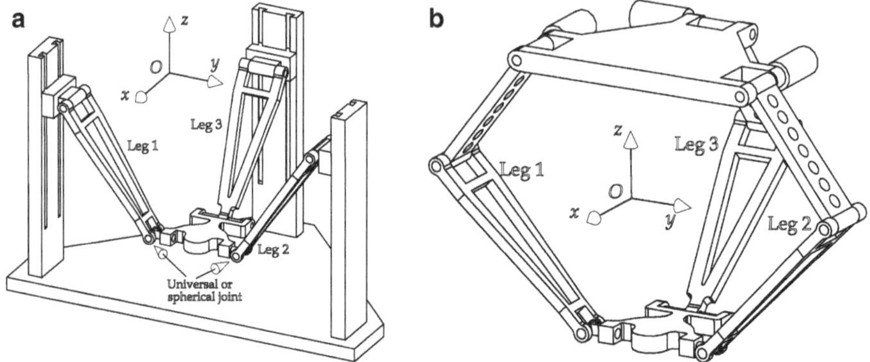

Fig. 2.41 Modified parallel mechanisms of the mechanism shown in Fig. 2.30: (**a**) with linear actuators and (**b**) with revolute actuators

Although they have the same output, a huge difference in the kinematic motion of the third leg exists between the old and modified mechanisms. When the z-coordinate of the mobile platform is specified but the platform translates along the y-axis, the slider in the third leg of the old mechanism must be active accordingly to maintain the orientation of the platform. On the other hand, the slider in the new mechanism must be locked, indicating that the translation along the y-axis and rotation of the modified mechanism are decoupled. This also means that the mechanism is energy saving. Input error must exist in the actuator. Therefore, the active input in the mechanism in Fig. 2.30 may lead to the rotational error of the mobile platform. Thus, the difference enables the mechanism in Fig. 2.41a to accuracy that is better than that of the old mechanism. Additionally, the kinematic problem of the new mechanism is accordingly simpler (see the example in Sect. 3.8).

The modified parallel mechanism is clearly more complex than the old mechanism because the parallelogram is not used in the latter. In a planar parallelogram, every two links should be parallel to each other. This parallelism requires manufacturing accuracy and increases the difficulty of link machining and assembly, as well as costs. No parallelogram is used in the third leg; thus, the architecture of the new mechanism is simpler. The manufacturing is also easier, and the cost accordingly reduced. As shown in Fig. 2.41a, self-calibration can be implemented by attaching the sensors to the two revolute joints and cylinder joints that are connected to the mobile platform. Hereby, the accuracy of the modified parallel mechanism can be

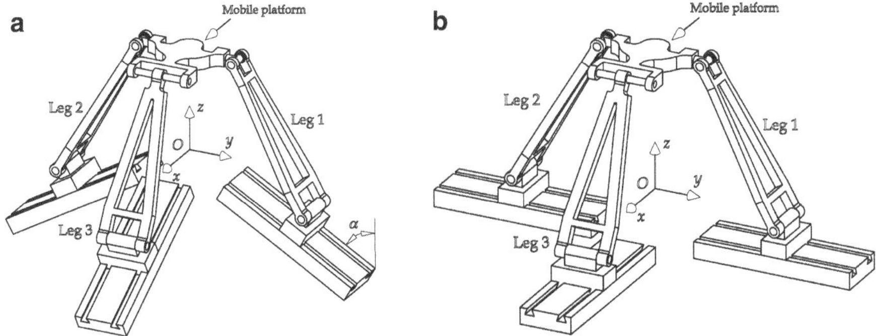

Fig. 2.42 Modified parallel mechanisms: (**a**) with inclined angle α and (**b**) with horizontal actuators

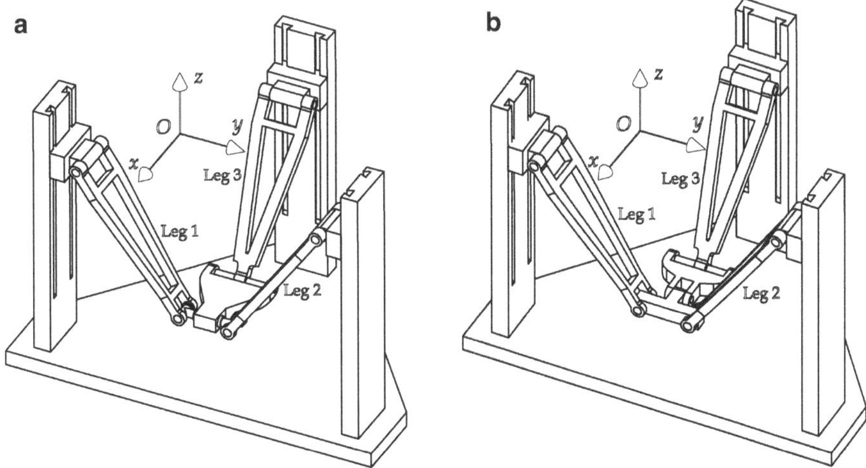

Fig. 2.43 Modified parallel mechanism with a $(PRR)_2$R-PRC chain

improved. Compared with the old version, therefore, the new mechanism will be more popular in practical applications.

Figure 2.41b shows the modified parallel mechanism with revolute actuators, wherein the R joints fixed to the base are active. Notably, the actuation direction of all the sliders in the parallel mechanism shown in Fig. 2.41a with prismatic actuators may be inclined at an α angle with respect to the vertical line (Fig. 2.42a). Figure 2.42b illustrates a typical example at a horizontal actuation direction.

Accordingly, kinematic chain $(PRR)_2$R&PR(Pa)R of the parallel mechanisms shown in Fig. 2.32 may be replaced by the $(PRR)_2$R-PRC chain (see Fig. 2.43a). This modification, which has no negative influence on the kinematics and rotational capability of the mechanism, can be also extended to the parallel mechanism with revolute actuators (Fig. 2.43b). In the modified parallel mechanisms, the universal joints connected to the mobile platform can be replaced by spherical joints.

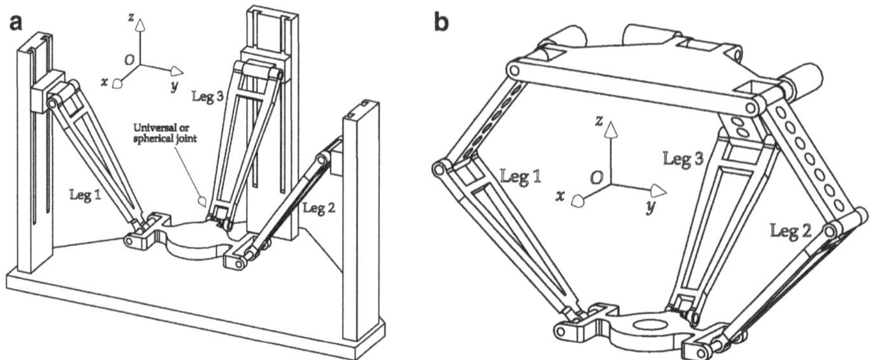

Fig. 2.44 Modified parallel mechanisms of those shown in Fig. 2.31: (**a**) with linear actuators and (**b**) with revolute actuators

In the mechanism in Fig. 2.33a, two legs (the first and second legs) consist of a parallelogram. The kinematic chain of each of the two legs is the same as that of the third leg in the parallel mechanism shown in Fig. 2.30. Therefore, the two legs can also be replaced by the PRC chain. The new version of the parallel mechanism is illustrated in Fig. 2.44a. The new mechanism with revolute actuators is shown in Fig. 2.44b, in which the R joints fixed to the base platform are active.

As shown in Fig. 2.44a, the mobile platform of the mechanism is connected to the base by a PRU or PRS chain (leg 3) and two PRC chains (legs 1 and 2). The axes of the two C joints in legs 1 and 2 must be parallel to each other. If leg 3 uses the PRU chain, the axes must also be parallel to the R joint of the U joint that is connected to the mobile platform. Given the arrangement of the links and joints of the mechanism, the combination of the three legs constrains the rotation of the mobile platform with respect to the y- and z-axes and the translation along the y-axis, leaving the mechanism with two translational DOFs in the O-xz plane and one rotational DOF about the axis parallel to the x-axis.

Similarly, the actuation direction of all the sliders in the parallel mechanism shown in Fig. 2.44a may be inclined at an α angle with respect to the vertical line (Fig. 2.45a). Figure 2.45b illustrates a typical example at a horizontal actuation direction. They have the same mobility as the mechanism shown in Fig. 2.44a.

Compared with the mechanisms shown in Fig. 2.33, the parallel mechanisms introduced here also present advantages in terms of kinematics, architecture, manufacturing, energy cost, accuracy, and assembly for similar reasons.

No planar parallelogram is used in each mechanism of the new family; thus, every leg can be designed as a telescopic link. The parallel mechanisms with such links are shown in Fig. 2.46.

Although the modified mechanisms have some comparative advantages over the old parallel mechanisms, they are still characterized by disadvantages. For example, the adoption of the C joint may cause operational failure because it is a passive

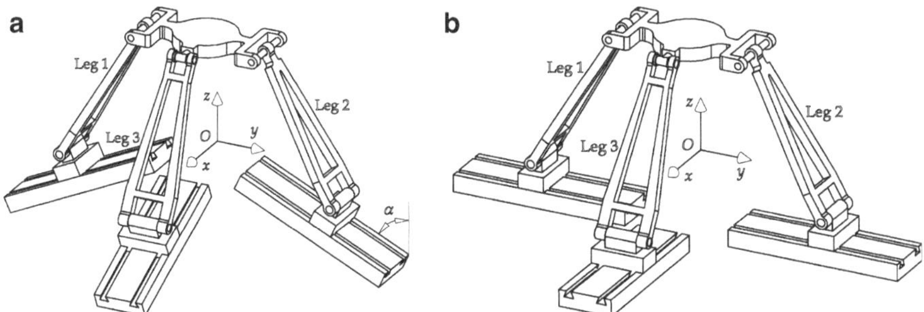

Fig. 2.45 Modified parallel mechanisms: (**a**) with inclined angle α and (**b**) with horizontal actuators

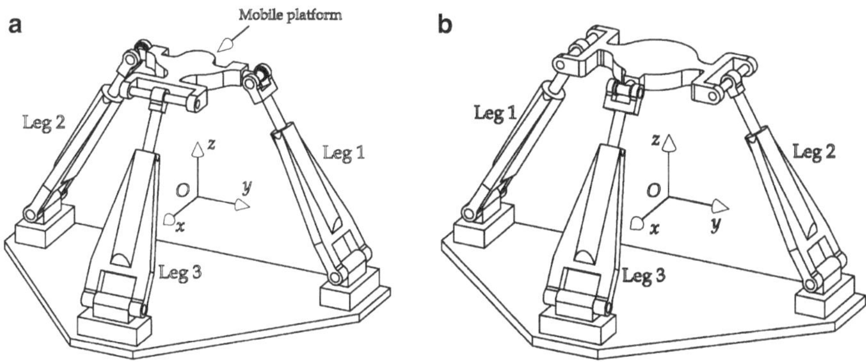

Fig. 2.46 Modified parallel mechanisms with telescopic legs

joint. To avoid this problem, we can apply the passive DOF to the mechanisms. In the mechanisms shown in Fig. 2.41, for example, the joints connected to the mobile platform in the first and second legs can be spherical joints. In each of the spherical joints, one redundant DOF exists. A redundant R joint may also be added to the third leg of the mechanism in Fig. 2.41. On the other hand, to decrease the operational failure, the demand for parallelism of the C axes in the first and second legs should be very high.

The mechanism shown in Fig. 2.41a shows that with the C joint connected to the mobile platform, the translation along the y-axis is limited. If the C joint is excessively long, it will exhibit poor stiffness. Additionally, enhancing the accuracy of a C joint is usually a difficult task. Therefore, we can split the C joint into two joints, i.e., one R joint and one P joint, and move the P to the position near the base. The modified mechanism with a 2-PRRR&1-PPRR chain is shown in Fig. 2.47. With this change, both the stiffness and accuracy increase. Such a change can be realized in other mechanisms.

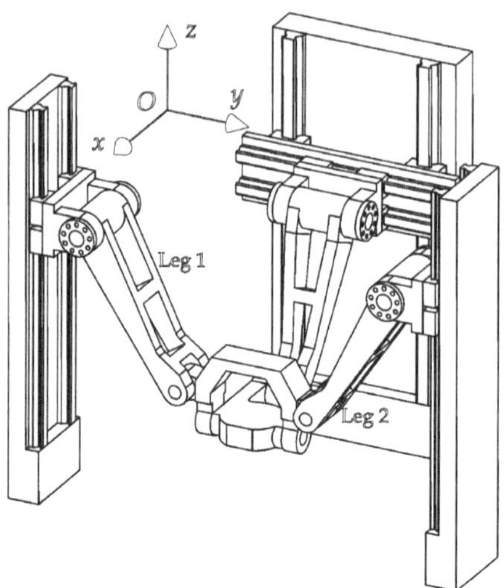

Fig. 2.47 2-PRRR&1-PPRR mechanism

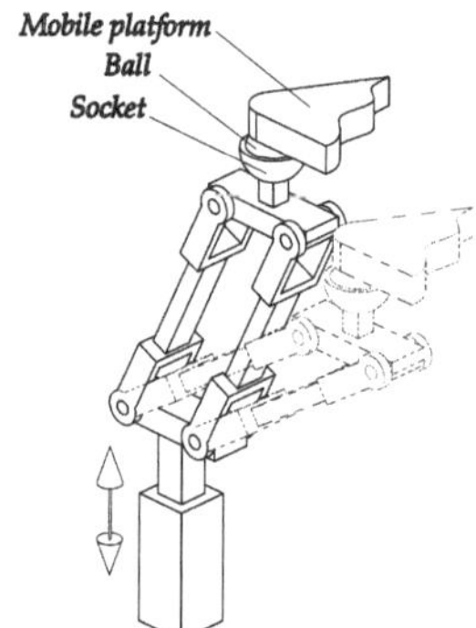

Fig. 2.48 One leg of the 3-[PP]S parallel mechanism shown in Fig. 2.26d, in which the leg changes its configuration and the socket of the spherical joint maintains its orientation

The 3-[PP]S parallel mechanism shown in Fig. 2.28d illustrates that in each leg, the center point of the spherical joint moves in a plane and the socket link maintains its orientation when the leg changes its configuration because of the use of the parallelogram (Fig. 2.48). Consequently, the tilting capability of the mobile platform improves. The output link of a prismatic joint can retain its orientation. We consider replacing the parallelogram with a combined link and P joint. Such a mechanism is called a 3-PPS parallel mechanism. Figure 2.49a shows a general architecture for this type of mechanism, in which the angle from the horizontal line to the direction of the first P joint is α', α is the angle between the directions of the two P joints, and $|\alpha - \alpha'| \neq 90°$. We assume that in each leg, the prismatic joint nearest the base would be actuated. Such a joint is denoted as the first P joint; meanwhile, another prismatic joint is referred to as the second P joint. Given that a 3-PPS mechanism can move vertically and the second prismatic joint is not actuated, the direction of the joint cannot be vertical. The 3-PPS mechanism can be modified into a 3-PCU mechanism because a cylinder joint is kinematically identical to the combined prismatic and revolute joint. For example, the 3-PPS mechanism at

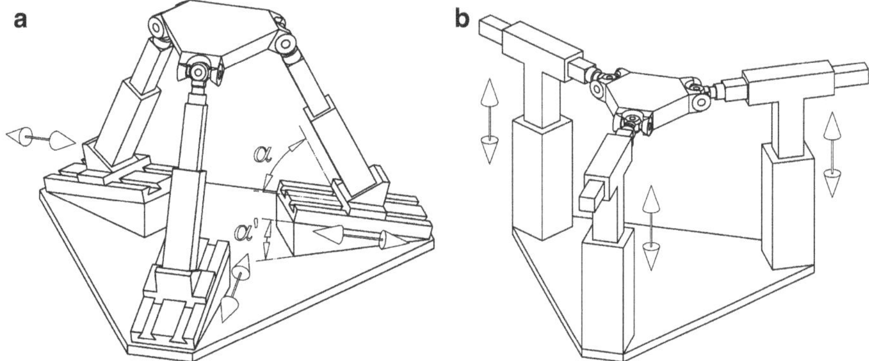

Fig. 2.49 3-PPS parallel mechanisms: (**a**) in a general architecture (**b**) in a special case, $\alpha = \alpha' = 90°$

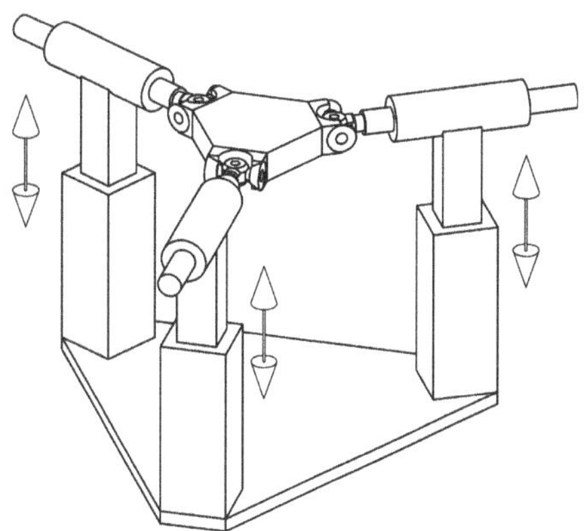

Fig. 2.50 3-PCU parallel mechanism

$\alpha = \alpha' = 90°$ (Fig. 2.49b) is identical to the 3-PCU mechanism shown in Fig. 2.50. The mechanism was proposed and analyzed in Liu et al. (2004). As analyzed in Liu and Bonev (2008), compared with the 3-PRS mechanism shown in Fig. 1.21, the 3-PPS mechanism in Fig. 2.49b exhibits a comparatively stable performance within its entire workspace. Generally, the parallel mechanism whose actuation directions are identical has an advantage in terms of direction, i.e., its performance is the same along that direction (Liu et al. 2006). Therefore, the 3-PPS and 3-PCU mechanisms in Figs. 2.49b and 2.50 are usually desirable.

References

Ball RS (1990) A treatise on the theory of screws. Cambridge University Press, Cambridge

Carricato M, Parenti-Castelli V (2003) A family of 3-DOF translational parallel manipulators. ASME J Mech Des 125(2):302–307

Ceccarelli M (1997) A new 3 D.O.F. spatial parallel mechanism. Mech Mach Theory 32(8): 895–902

Chung Y-H, Lee J-W (2001) Design of a new 2 DOF parallel mechanism. In: Proceedings of IEEE/ASME international conference on advanced intelligent mechatronics, IEEE press, Piscataway, N.J., Como, pp 129–134
Dafaoui E-M, Amirat Y, Pontnau J, Francois C (1998) Analysis and design of a six-DOF parallel mechanism, modeling, singular configurations, and workspace. IEEE Trans Robot Autom 14(1):78–92
Dai JS, Huang Z, Lipkin H (2006) Mobility of overconstrained parallel mechanisms. ASME J Mech Des 128:220–229
Ebert-Uphoff I, Gosselin CM (1998) Kinematic study of a new type of spatial parallel platform mechanism. In: Proceedings of DETC ASME design engineering technical conferences, Atlanta, Georgia, DETC/MECH-5962, ASME press, New York
Fang Y, Tsai L-W (2002) Structure synthesis of a class of 4-DOF and 5-DOF parallel manipulators with identical limb structures. Int J Robot Res 21(9):799–810
Frisoli A, Salsedo F, Bergamasco M (1999) Design of a new tendon driven haptic interface with six degrees of freedom. In: Proceedings of IEEE international workshop on robot and human interaction, IEEE press, Piscataway, N.J., Pisa, pp 303–308
Gogu G (2005) Mobility of mechanisms: a critical review. Mech Mach Theory 40:1068–1097
Hervé JM (1999) The Lie group of rigid body displacements, a fundamental tool for mechanism design. Mech Mach Theory 34(5):719–730
Honegger M, Brega R, Schweitzer G (2000) Application of a nonlinear adaptive controller to a 6 dof parallel manipulator. In: Proceedings of IEEE international conference on robotics & automation, IEEE press, Piscataway, N.J., San Francisco, CA, pp 1930–1935
Huang T, Li M, Zhao XM et al (2005a) Conceptual design and dimensional synthesis for a 3-DOF Module of the TriVariant—a novel 5-DOF reconfigurable hybrid robot. IEEE Trans Robot 21(3):449–456
Huang T, Wang ZX, Li M, et al (2005b) A 3-DOF pure translational parallel mechanism with asymmetrical architecture. China Patent, No. 03144282.X
Huang T, Zhao X, Whitehouse DJ (2002) Stiffness estimation of a tripod-based parallel kinematic machine. IEEE Trans Robot Autom 18(1):50–58
Huang Z, Li QC (2002) General methodology for the type synthesis of lower-mobility symmetrical parallel mechanisms and several novel mechanisms. Int J Robot Res 21(2):131–145
Huang Z, Li QC (2003) Type synthesis of symmetrical lower mobility parallel mechanisms using the constraint-synthesis method. Int J Robot Res 22(1):59–79
Hudgens J, Tesar D (1988) A fully-parallel six degree-of-freedom micromanipulator: kinematics analysis and dynamic model. In: Proceedings of 20th Biennial ASME mechanisms conference, ASME press, Kissimmee, Florida, New York, pp 29–38
Hunt KH (1978) Kinematic geometry of mechanisms. Clarendon Press, Oxford
Kim W-K, Lee J-Y, Yi BJ (1997) Analysis for a planar 3 degree-of-freedom parallel mechanism with actively adjustable stiffness characteristics. In: Proceedings of IEEE international conference on robotics and automation, IEEE press, Piscataway, N.J., New Mexico, pp 2663–2670
Kong X, Gosselin CM (2004a) Type synthesis of 3-DOF translational parallel manipulators based on screw theory. ASME J Mech Des 126(1):83–92
Kong X, Gosselin CM (2004b) Type synthesis of 3T1R 4-DOF parallel manipulators based on screw theory. IEEE Trans Robot Autom 20(2):181–190
Kutzbach K (1933) Einzelfragen Aus Dem Gebiet Der Maschinenteile. Zeitschrift der Verein Deutscher Ingenieur 77:1168–1169
Lerbet J (1987) Mecanique des systemes de solides rigides comportant des boucles fermees. Ph.D thesis, Paris VI, Paris
Liu X-J (2001) Mechanical and kinematics design of parallel robotic mechanisms with less than six degrees of freedom. Post-Doctoral research report, Tsinghua University, Beijing (in Chinese)
Liu X-J, Wang J, Gao F, Wang L-P (2001a) On the analysis of a new spatial three degrees of freedom parallel manipulator. IEEE Trans Robot Autom 17(6):959–968

Liu X-J, Wang J, Wang L-P, Gao F (2001b) On the design of 6-DOF parallel micro-motion manipulators. In: Proceedings of the IEEE/RSJ international conference on intelligent robots and systems, IEEE press, Piscataway, N.J., Maui, Hawaii, pp 343–348

Liu X-J, Bonev IA (2008) Orientation capability, error analysis, and dimensional optimization of two articulated tool heads with parallel kinematics. ASME J Manuf Sci Eng 130(1), Article Number: 011015

Liu X-J, Pruschek P, Pritschow G (2004) A new 3-DOF parallel mechanism with full symmetrical structure and parasitic motions. In: Proceedings of international conference on intelligent manipulation and grasping, IEEE press, Piscataway, N.J., Genoa, pp 389–394

Liu X-J, Tang X, Wang J (2005a) HANA: a novel spatial parallel manipulator with one rotational and two translational degrees of freedom. Robotica 23(2):257–270

Liu X-J, Wang J (2003) Some new parallel mechanisms containing the planar four-bar parallelogram. Int J Robot Res 22(9):717–732

Liu X-J, Wang J (2005) Some new spatial 3-DOF fully-parallel robotic mechanisms with high or improved rotational capability. In: Liu JX (ed) Robots manipulators: new research. Nova Science Publishers, New York, pp 33–64

Liu X-J, Wang J, Pritschow G (2005b) A new family of spatial 3-DOF fully-parallel manipulators with high rotational capability. Mech Mach Theory 40(4):475–494

Liu X-J, Wang Q-M, Wang J (2005c) Kinematics, dynamics and dimensional synthesis of a novel 2-DOF translational manipulator. J Intell Robot Syst 41(4):205–224

Liu X-J, Wang J, Kim J (2006) Determination of the link lengths for a spatial 3-DOF parallel manipulator. J Mech Des 128:365–373

Meng J, Liu G, Li Z (2007) A geometric theory for analysis and synthesis of sub-6 DOF parallel manipulators. IEEE Trans Robot 23(4):625–649

Merlet J-P (2000) Parallel robots. Kluwer Academic Publishers, Dordrecht, pp 15–49

Ohya Y, Arai T, Mae Y et al (1999) Development of 3-DOF finger module for micro manipulation. In: Proceedings of the 1999 IEEE/RSJ international conference on intelligent robots and systems, IEEE press, Piscataway, N.J., Kyongju, pp 894–899

Pernette E, Clavel R (1996) Parallel robots and microrobotics. In: ISRAM'96, Montpellier, ASME press, New York, pp 535–542

Pierrot F, Dauchez P, Fournier A (1991) Towards a fully-parallel 6 d.o.f. robot for high speed applications. In: Proceedings of IEEE international conference on robotics & automation, Sacramento, IEEE press, Piscataway, N.J., CA, pp 1288–1293

Rico JM, Aguilera LD, Gallardo J et al (2006) A more general mobility criterion for parallel platforms. ASME J Mech Des 128:207–219

Sorli M, Ferraresi C, Kolarski M et al (1997) Mechanics of TURIN parallel robot. Mech Mach Theory 32(1):51–57

Tahmasebi F (1992) Kinematic synthesis and analysis of a novel class of six-DOF parallel minimanipulators. Ph.D thesis, University of Maryland, College Park

Wang J, Gosselin CM (1999) Static balancing of spatial three-degree-of-freedom parallel mechanisms. Mech Mach Theory 34:437–452

Yang T-L (2004) Topology structure design of robot mechanisms (in Chinese). China Machine Press, Beijing

Zhao TS (2000) Some theoretical issues on analysis and synthesis for spatial imperfect-DOF parallel robots (in Chinese), Ph.D thesis, Yanshan University, Qinhuangdao

Zhang D, Gosselin C M (2002) Kinetostatic analysis and design optimization of the Tricept machine tool family. ASME J Manuf Sci Eng 124(3):725–733

Part II
Kinematic Analysis

Chapter 3
Position Analysis of Parallel Mechanisms

Abstract This chapter shows both inverse and direct position analyses of several typical parallel mechanisms including a planar 5R type, a 2-DOF spherical type, two 3-DOF types, and a 6-RUS type. Since for the parallel mechanism, it is more convenient to employ the geometric method to solve its kinematics, the position analysis established upon writing geometric constraint equations for each leg is introduced. In particular, the position analysis of 3-[PP]S type of parallel mechanism in terms of a special orientation description is presented to simplify its kinematics.

Keywords Position analysis • Kinematics • 3-[PP]S type • Parallel mechanism • Orientation description

The position analysis involves two problems, i.e., inverse and direct kinematics. The inverse kinematics involves mapping a known pose (position and orientation) of the mobile platform of the mechanism to a set of input joint variables that will achieve that pose. The direct kinematics is the problem of finding the pose of the mobile platform for specified inputs. For a parallel mechanism, its inverse kinematics is easier than the direct kinematics.

For the serial mechanism, the Denavit-Hartenberg method (Denavit and Hartenberg 1955) is a general used method. However, this method is complicated by the existence of multiple closed loops. For parallel mechanisms, it is often more convenient to employ the geometric method. Generally, there are two methods. One is writing a vector-loop equation for each leg, and the passive joint variables are eliminated among these equations. Another one is writing a geometric constraint equation for each leg. The equations usually describe the relationship between input and output vectors. In the following sections, we analyze the kinematics of several parallel manipulators to illustrate the later method.

X.-J. Liu and J. Wang, *Parallel Kinematics: Type, Kinematics, and Optimal Design*, Springer Tracts in Mechanical Engineering, DOI 10.1007/978-3-642-36929-2_3,

3.1 Position Analysis of the 5R Parallel Mechanism

The planar 5R parallel mechanism, as shown in Fig. 1.12f, is such a mechanism that the output point is connected to the base by two legs, each of which consists of three revolute joints and two links. The two legs are connected to a common point with the common revolute joint at the end of each leg. In each of the two legs, the revolute joint connected to the base is actuated. Such a mechanism can position a point freely in a plane.

A kinematics model of the mechanism is developed as shown in Fig. 3.1. Each actuated joint is denoted as A_i $(i = 1,\ 2)$, the other end of each actuated link is denoted as B_i, and the common joint of the two legs is denoted as P, which is also the output point. A fixed global reference system O-xy is located at the center of A_1A_2 with the y-axis normal to A_1A_2 and the x-axis directed along A_1A_2. For the structure symmetry, there are $OA_1 = OA_2$, $OA_1 = OA_2$, and $B_1P = B_2P$. The length of the actuated link for each leg is denoted as $A_iB_i = R_1(r_1)$. Additionally, $B_iP = R_2(r_2)$ and $OA_i = R_3(r_3)$. Here, R_j $(j = 1, 2, 3)$ are link lengths with dimension and r_j $(j = 1, 2, 3)$ are those with non-dimension.

3.1.1 Inverse Kinematics

The position of the output point P in the reference system O-xy can be described by the position vector $\boldsymbol{p}$, and there is

$$\boldsymbol{p} = \begin{pmatrix} x & y \end{pmatrix}^{\mathrm{T}}. \tag{3.1}$$

In the reference frame O-xy, the position vectors $\boldsymbol{b}_i$ of points B_i $(i = 1,\ 2)$ can be written as

$$\boldsymbol{b}_1 = \begin{pmatrix} r_1\cos\theta_1 - r_3 & r_1\sin\theta_1 \end{pmatrix}^{\mathrm{T}} \quad \text{and} \quad \boldsymbol{b}_2 = \begin{pmatrix} r_1\cos\theta_2 + r_3 & r_1\sin\theta_2 \end{pmatrix}^{\mathrm{T}} \tag{3.2}$$

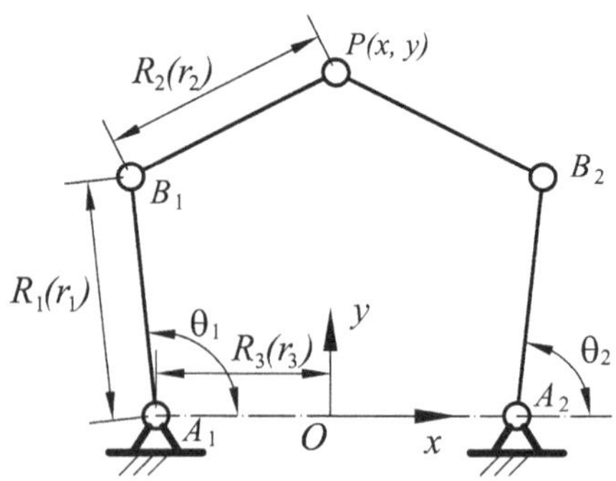

Fig. 3.1 The planar 5R parallel mechanism

where θ_1 and θ_2 are the input angles of the two legs. Then, the inverse kinematic problem can be solved by writing the following constraint equation:

$$|\boldsymbol{p} - \boldsymbol{b}_i| = r_2, \quad i = 1, 2 \tag{3.3}$$

in another form

$$(x - r_1 \cos\theta_1 + r_3)^2 + (y - r_1 \sin\theta_1)^2 = r_2^2 \tag{3.4}$$

$$(x - r_1 \cos\theta_2 - r_3)^2 + (y - r_1 \sin\theta_2)^2 = r_2^2 \tag{3.5}$$

from which, if the position of output point P is known, the inputs to reach the position can be obtained as

$$\theta_i = 2\tan^{-1}(z_i), \quad i = 1, 2 \tag{3.6}$$

where

$$z_i = \frac{-b_i + \sigma_i\sqrt{b_i^2 - 4a_i c_i}}{2a_i}, \quad i = 1, 2 \tag{3.7}$$

in which

$$\begin{aligned}
\sigma_i &= 1 \text{ or } -1\\
a_1 &= r_1^2 + y^2 + (x + r_3)^2 - r_2^2 + 2(x + r_3) r_1\\
b_1 &= -4yr_1\\
c_1 &= r_1^2 + y^2 + (x + r_3)^2 - r_2^2 - 2(x + r_3) r_1\\
a_2 &= r_1^2 + y^2 + (x - r_3)^2 - r_2^2 + 2(x - r_3) r_1\\
b_2 &= b_1 = -4yr_1\\
c_2 &= r_1^2 + y^2 + (x - r_3)^2 - r_2^2 - 2(x - r_3) r_1.
\end{aligned}$$

From Eq. (3.7), one may see that there are four solutions for the inverse kinematic problem of the 5R mechanism. The configuration shown in Fig. 3.1 can be obtained if $\sigma_1 = 1$ and $\sigma_2 = -1$. Such a configuration is denoted as the "+−" model. Then there are three others, which are "−+", "−−", and "++" models, respectively. These four inverse kinematics models correspond to four types of working modes.

3.1.2 Direct Kinematics

The direct kinematic problem is to obtain the output with respect to a set of given inputs. From Eqs. (3.4) and (3.5), one obtains

$$x^2 + y^2 - 2(r_1\cos\theta_1 - r_3)x - 2r_1\sin\theta_1 y - 2r_1 r_3\cos\theta_1 + r_3^2 + r_1^2 - r_2^2 = 0 \tag{3.8}$$

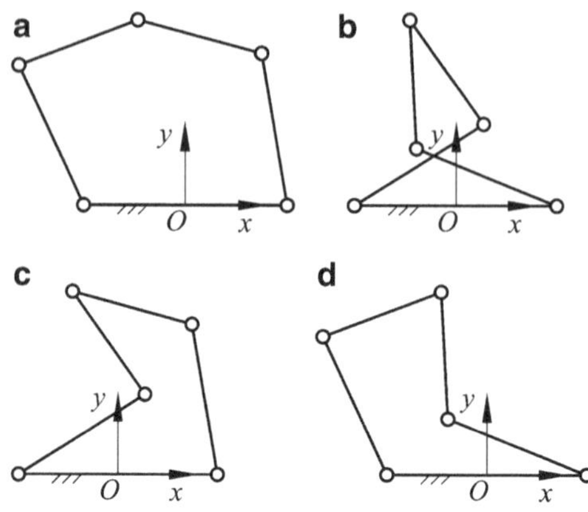

Fig. 3.2 Four inverse kinematic models: (**a**) "+−" model; (**b**) "−+" model; (**c**) "− −" model; (**d**) "++" model

$$x^2 + y^2 - 2(r_1 \cos\theta_2 + r_3)x - 2r_1 \sin\theta_2 y + 2r_1 r_3 \cos\theta_2 + r_3^2 + r_1^2 - r_2^2 = 0. \tag{3.9}$$

Equations (3.8) and (3.9) yields

$$x = ey + f \tag{3.10}$$

in which $e = \frac{r_1(\sin\theta_1 - \sin\theta_2)}{2r_3 + r_1\cos\theta_2 - r_1\cos\theta_1}$ and $f = \frac{r_1 r_3(\cos\theta_2 + \cos\theta_1)}{2r_3 + r_1\cos\theta_2 - r_1\cos\theta_1}$. Substituting Eq. (3.10) to Eq. (3.8) yields

$$dy^2 + gy + h = 0 \tag{3.11}$$

where

$$d = 1 + e^2$$
$$g = 2(ey - er_1\cos\theta_1 + er_3 - r_1\sin\theta_1)$$
$$h = f^2 - 2f(r_1\cos\theta_1 - r_3) - 2r_1 r_3 \cos\theta_1 + r_3^2 + r_1^2 - r_2^2$$

Then, y can be obtained from Eq. (3.11) as

$$y = \frac{-g + \sigma\sqrt{g^2 - 4dh}}{2d} \tag{3.12}$$

in which $\sigma = 1$ or -1. From Eqs. (3.10) and (3.12), one sees that there are two solutions for the direct kinematic problem of the mechanism. They correspond to types of assembly modes. The corresponding configurations are called the *up*- and *down-configurations*. The *up-configuration* can be achieved when $\sigma = 1$.

For example, if $r_1 = 1.2$, $r_2 = 1.0$, and $r_3 = 0.8$, the four inverse kinematic models of the mechanism are shown in Fig. 3.2, where the specified position of the

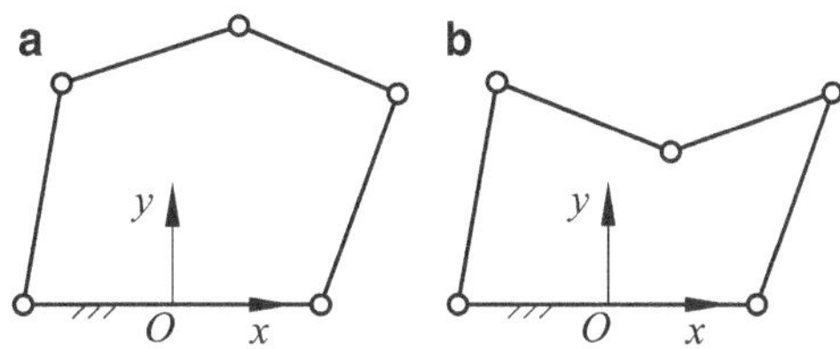

Fig. 3.3 Two direct kinematic models: (**a**) the up-configuration (**b**) the down-configuration

output point is ($x = -0.37,\ \ y = 1.44$). And the two direct kinematic models are shown in Fig. 3.3, where the inputs are given as $\theta_1 = 4\pi/9$ and $\theta_2 = 7\pi/18$.

In this chapter, what we are concerned about is the mechanism with the "+−" model and, at the same time, the *up-configuration*.

3.2 Position Analysis of a 2-DOF Translational Parallel Mechanism

3.2.1 Two-P(Pa) Parallel Mechanism

In this chapter, we consider the kinematics of the parallel mechanism shown in Fig. 2.19. Its schematic diagram is illustrated in Fig. 3.4, where a reference frame $\mathfrak{R}$: O-xy is fixed to the base and a moving reference frame $\mathfrak{R}'$: O'-$x'y'$ is attached to the mobile platform, where O' is the reference point on the mobile platform. For the characteristic of a planar four-bar parallelogram, we may consider chains P_1B_1 and P_2B_2 to resolve the kinematics of the manipulator. And vectors $\boldsymbol{p}_{i\mathfrak{R}'}$ and $\boldsymbol{p}_{i\mathfrak{R}}$ ($i = 1,\ 2$) will be defined as the position vectors of points P_i in frames $\mathfrak{R}'$ and $\mathfrak{R}$, respectively, vectors $\boldsymbol{b}_{i\mathfrak{R}}$ ($i = 1,\ 2$) as the position vectors of points B_i in frame $\mathfrak{R}$. The geometric parameters of the manipulator are $P_iB_i = L$ ($i = 1,\ 2$), $P_1P_2 = 2r$, and the distance between two guideways $2R$. The position of point O' in the fixed frame $\mathfrak{R}$ is denoted as vector

$$\boldsymbol{c}_{\mathfrak{R}} = \left(x,\ y\right)^{\mathrm{T}} \tag{3.13}$$

The vectors of $\boldsymbol{b}_{i\mathfrak{R}}$ in the fixed frame $\mathfrak{R}$ can be written as

$$\boldsymbol{b}_{1\mathfrak{R}} = \left(R\ y_1\right)^{\mathrm{T}},\quad \boldsymbol{b}_{2\mathfrak{R}} = \left(-R\ y_2\right)^{\mathrm{T}} \tag{3.14}$$

where y_i are the actuated inputs for the two legs. And vectors $\boldsymbol{p}_{i\mathfrak{R}}$ in the fixed frame $\mathfrak{R}$ can be written as

$$\boldsymbol{p}_{i\mathfrak{R}} = \boldsymbol{p}_{i\mathfrak{R}'} + \boldsymbol{c}_{\mathfrak{R}},\quad i = 1,\ 2 \tag{3.15}$$

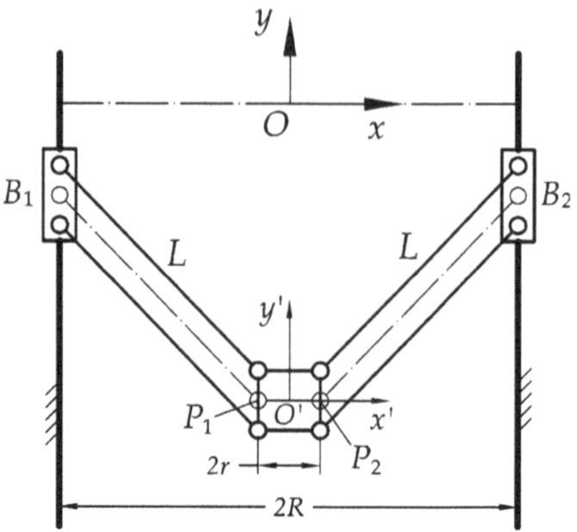

Fig. 3.4 The schematic diagram of the 2-P(Pa) parallel mechanism

where

$$\boldsymbol{p}_{1\Re'} = \left(r \ 0\right)^{\mathrm{T}}, \quad \boldsymbol{p}_{2\Re'} = \left(-r \ 0\right)^{\mathrm{T}}. \tag{3.16}$$

Then the inverse kinematics problem of the manipulator can be solved by writing the following constraint equation:

$$\|\boldsymbol{p}_{i\Re} - \boldsymbol{b}_{i\Re}\| = L \quad i = 1, 2 \tag{3.17}$$

that is,

$$y_1 = \pm\sqrt{L^2 - (r + x - R)^2} + y \tag{3.18}$$

$$y_2 = \pm\sqrt{L^2 - (x - r + R)^2} + y \tag{3.19}$$

from which we maysee that there are four solutions for the inverse kinematics of the manipulator. Hence, for a given manipulator and for prescribed values of the position of the mobile platform, the required actuated inputs can be directly computed from Eqs. (3.18) and (3.19). To obtain the configuration as shown in Fig. 3.4, the "±" in Eqs. (3.18) and (3.19) should be "+".

From Eqs. (3.18) and (3.19), the direct kinematics of the manipulator can be described as

$$x = ay + b \tag{3.20}$$

where

$$a = \frac{y_2 - y_1}{2\,(R - r)}, \quad b = \frac{y_1^2 - y_2^2}{4\,(R - r)} \tag{3.21}$$

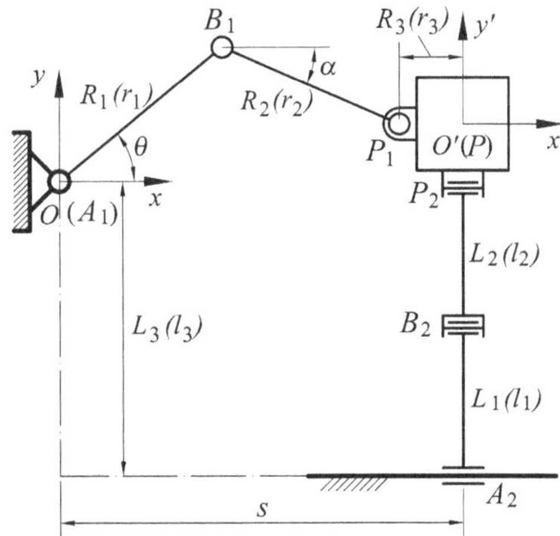

Fig. 3.5 The schematic of the RRR&PRRR parallel mechanism

and

$$y = \frac{-f \pm \sqrt{f^2 - 4eg}}{2e} \tag{3.22}$$

in which

$$e = a^2 + 1, \quad f = 2a(r + b - R) - 2y_1, \quad g = (r + b - R)^2 + y_1^2 - L^2. \tag{3.23}$$

Hence, the solutions of the direct kinematics of the manipulator can reach two. To obtain the direct configuration as shown in Fig. 3.4, the "$\pm$" in Eq. (3.22) should be "$-$".

From above equations, we may see that the inverse and direct kinematics problems of the mechanism are very easy and can be described as closed forms.

3.2.2 The RRR&PRRR Parallel Mechanism

The two translational parallel mechanisms considered here are shown in Fig. 1.13c. As illustrated in Fig. 3.5, a reference frame $\Re$:O-xy is fixed to the base at the joint point A_1 and a moving reference frame $\Re'$:O'-$x'y'$ is attached to the end-effector, where O' is the reference point on the end-effector. Vectors $\boldsymbol{p}_{i\Re}$ $(i = 1, 2)$ are defined as the position vectors of points P_i in the frame $\Re$, and vectors $\boldsymbol{b}_{i\Re}$ $(i = 1, 2)$ as the position vectors of points B_i in frame $\Re$. The geometric parameters of the mechanism are $A_1B_1 = R_1(r_1)$, $B_1P_1 = R_2(r_2)$, $PP_i = R_3(r_3)$, $A_2B_2 = L_1(l_1)$, and $B_2P_2 = L_2(l_2)$, and the distance from the point A_1 to the guideway is $L_3(l_3)$, where R_n and L_n $(n = 1, 2, 3)$ are dimensional parameters and

r_n and l_n non-dimensional parameters. The position of point O' in the fixed frame $\Re$ is denoted as vector

$$\boldsymbol{c}_{\Re} = \left(x,\, y\right)^{\mathrm{T}}. \tag{3.24}$$

The vectors of $\boldsymbol{b}_{1\Re}$ in the fixed frame $\Re$ can be written as

$$\boldsymbol{b}_{1\Re} = \left(R_1 \cos\theta \;\; R_1 \sin\theta\right)^{\mathrm{T}} \tag{3.25}$$

where θ is the actuated input for the RRR leg. Vector $\boldsymbol{p}_{1\Re}$ in the fixed frame $\Re$ can be written as

$$\boldsymbol{p}_{1\Re} = (-R_3 \;\; 0)^{\mathrm{T}} + \boldsymbol{c}_{\Re} = (x - R_3 \;\; y)^{\mathrm{T}}. \tag{3.26}$$

The inverse kinematics problem of the RRR leg can be solved by writing the following constraint equation:

$$\|\boldsymbol{p}_{1\Re} - \boldsymbol{b}_{1\Re}\| = R_2 \tag{3.27}$$

that is,

$$(x - R_3 - R_1 \cos\theta)^2 + (y - R_1 \sin\theta)^2 = R_2^2. \tag{3.28}$$

Then, there is

$$\theta = 2\tan^{-1}(m) \tag{3.29}$$

where

$$m = \frac{-b + \sigma\sqrt{b^2 - 4ac}}{2a} \tag{3.30}$$

$\sigma = 1 \;\text{ or} - 1$
$va = (x - R_3)^2 + y^2 + R_1^2 - R_2^2 + 2\,(x - R_3)\,R_1$
$b = -4yR_1$
$c = (x - R_3)^2 + y^2 + R_1^2 - R_2^2 - 2\,(x - R_3)\,R_1.$

For the PRRR leg, it is obvious that

$$s = x \tag{3.31}$$

in which s is the input of the PRRR leg. From Eqs. (3.30) and (3.31), we may see that there are two solutions for the inverse kinematics of the mechanism. Hence, for a given mechanism and for prescribed values of the position of the end-effector, the required actuated inputs can be directly computed from Eqs. (3.29) and (3.31). To obtain the configuration as shown in Fig. 3.5, parameter σ in Eq. (3.30) should be 1. This configuration is called the "+" working mode. When $\sigma = -1$, the corresponding configuration is referred to as the "—" working mode.

The direct kinematic problem is to obtain the output with respect to a set of given inputs. From Eqs. (3.28) and (3.31), one obtains

$$y = e + \sigma\sqrt{f} \tag{3.32}$$

and

$$x = s \tag{3.33}$$

where $e = R_1 \sin\theta$ and $f = R_2^2 - (s - R_3 - R_1 \cos\theta)^2$. Therefore, there are also two direct kinematic solutions for the mechanism. The parameter $\sigma = -1$ corresponds to the configuration shown in Fig. 3.5, which is denoted as the down-configuration. When $\sigma = 1$, the configuration is referred to as the up-configuration. These two kinds of configurations correspond to two kinds of assembly modes of the mechanism.

3.3 Position Analysis of a 2-DOF Spherical Parallel Mechanism

A spherical 2-DOF parallel mechanism with 5R kinematic chain is shown in Fig. 3.6, where the mechanism is symmetric in kinematic structure. The five revolute joints intersect at one common point O, which is referred to as the motion center. Points A_i, B_i ($i = 1, 2$) and P are the center points of the revolute joints. The angle between lines OA_i and OB_i is α_1. α_0 and α_2 are the angles between lines OA_1 and OA_2 and between lines PB_1 and PB_2, respectively. A global reference frame O-xyz is fixed to the base with the z-axis perpendicular to the plane defined by points O, A_1 and A_2, and the x-axis passing through the center of A_1A_2. Local frames O-$x_iy_iz_i$ ($i = 1, 2$) are defined as that the x_i-axis and z_i-axis are coincident with OA_i and the z-axis, respectively.

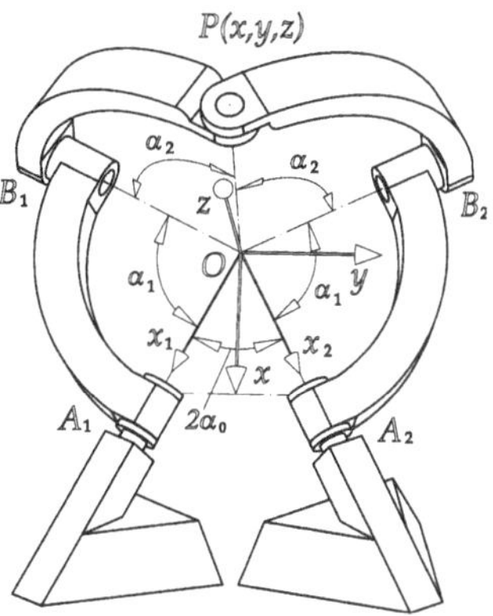

Fig. 3.6 A spherical 2-DOF 5R parallel mechanism

Then, the orientation of the local frame $O\text{-}x_iy_iz_i$ with respect to the fixed frame can be expressed by a matrix $\boldsymbol{R}$, i.e.,

$$\boldsymbol{R}_z(\varepsilon_i) = \begin{bmatrix} \cos\varepsilon_i & -\sin\varepsilon_i & 0 \\ \sin\varepsilon_i & \cos\varepsilon_i & 0 \\ 0 & 0 & 1 \end{bmatrix} \quad i = 1, 2 \tag{3.34}$$

in which $\varepsilon_1 = -\alpha_0$ and $\varepsilon_2 = \alpha_0$.

The vectors of revolute joints B_i in the frame $O\text{-}x_iy_iz_i$ can be written as

$$\boldsymbol{b}_{ii} = \left[c_{\alpha_1} \; s_{\alpha_1} c_{\theta_i} \; s_{\alpha_1} s_{\theta_i} \right]^{\mathrm{T}} \tag{3.35}$$

where $s_{\alpha_1} = \sin\alpha_1$, $c_{\alpha_1} = \cos\alpha_1$, and θ_i are the inputs. Then, in the frame $O\text{-}xyz$, the vectors of B_i can be expressed as

$$\boldsymbol{b}_{\mathrm{i}} = \boldsymbol{R}_z(\varepsilon_i)\,\boldsymbol{b}_{ii} = \left[l_i \; m_i \; n_i \right]^{\mathrm{T}} \tag{3.36}$$

in which $l_i = c_{\alpha_1} c_{\varepsilon_i} - s_{\alpha_1} c_{\theta_i} s_{\varepsilon_i}$, $m_i = c_{\alpha_1} s_{\varepsilon_i} + s_{\alpha_1} c_{\theta_i} c_{\varepsilon_i}$, and $n_i = s_{\alpha_1} s_{\theta_i}$.

The vector of point P in the frame $O\text{-}xyz$ is defined as $\boldsymbol{p}$, then

$$\boldsymbol{p} = \left[x \; y \; z \right]^{\mathrm{T}}. \tag{3.37}$$

Supposing that the locus of point P is the surface of a unit sphere, then $x^2 + y^2 + z^2 = 1$. Letting the angle between the vector $\boldsymbol{p}$ and the z-axis be ϕ, and the angle between the projecting line of the vector $\boldsymbol{p}$ in the plane $O\text{-}xy$ and the x-axis be φ, then there is

$$\begin{cases} x = \sin\phi\cos\varphi \\ y = \sin\phi\sin\varphi \\ z = \cos\phi \end{cases}. \tag{3.38}$$

As shown in Fig. 3.6, there is the constraint equation like

$$\boldsymbol{b}_{\mathrm{i}} \cdot \boldsymbol{p} = \cos\alpha_2, \quad i = 1, 2 \tag{3.39}$$

3.3.1 Inverse Kinematics

For this mechanism, the inverse kinematic problem is finding the solution of input (θ_1, θ_2) for a given output $(x, \; y, \; z)$. From Eq. (3.39), there is

$$\cos\alpha_2 = (c_{\alpha_1} c_{\varepsilon_i} - s_{\alpha_1} c_{\theta_i} s_{\varepsilon_i}) \cdot x + (c_{\alpha_1} s_{\varepsilon_i} + s_{\alpha_1} c_{\theta_i} c_{\varepsilon_i}) \cdot y + (s_{\alpha_1} s_{\theta_i}) \cdot z \tag{3.40}$$

which can be rewritten as

$$A_i \sin\theta_i + B_i \cos\theta_i = C_i, \quad i = 1,2 \tag{3.41}$$

where $A_i = s_{\alpha_1}\cdot z$, $B_i = -s_{\alpha_1}s_{\varepsilon_i}\cdot x + s_{\alpha_1}c_{\varepsilon_i}\cdot y$, and $C_i = \cos\alpha_2 - c_{\alpha_1}c_{\varepsilon_i}\cdot x - c_{\alpha_1}s_{\varepsilon_i}\cdot y$. Letting

$$\tan\frac{\theta_i}{2} = \eta_i \tag{3.42}$$

there are $\sin\theta_i = \frac{2\eta_i}{1+\eta_i^2}$ and $\cos\theta_i = \frac{1-\eta_i^2}{1+\eta_i^2}$. Then, Eq. (3.41) can be rewritten as

$$(B_i + C_i)\,\eta_i^2 - 2A_i\eta_i + (C_i - B_i) = 0. \tag{3.43}$$

Therefore, the inverse kinematic solution of the mechanism can be obtained as

$$\theta_i = 2\tan^{-1}(\eta_i) = 2\tan^{-1}\left(\frac{A_i \pm \sqrt{A_i^2 + B_i^2 - C_i^2}}{B_i + C_i}\right), \quad i = 1,2 \tag{3.44}$$

from which one may see that there are four solutions for the inverse kinematics of the spherical 5R parallel mechanism. The four solutions correspond to four working modes of the mechanism.

3.3.2 Direct Kinematics

The direct kinematic problem is finding the solution of output $(x,\ y,\ z)$ or the orientation $(\phi,\ \varphi)$ for a given input $(\theta_1,\ \theta_2)$. From Eq. (3.39), there is

$$\begin{cases} l_1\cdot x + m_1\cdot y + n_1\cdot z = \cos\alpha_2 \\ l_2\cdot x + m_2\cdot y + n_2\cdot z = \cos\alpha_2 \\ x^2 + y^2 + z^2 = 1 \end{cases} . \tag{3.45}$$

Then,

$$Az^2 + Bz + C = 0 \tag{3.46}$$

where

$$\begin{aligned} A &= (l_1m_2 - l_2m_1)^2 + (l_1n_2 - l_2n_1)^2 + (m_1n_2 - m_2n_1)^2 \\ B &= -2q\left[(l_1n_2 - l_2n_1)(l_1 - l_2) + (m_1n_2 - m_2n_1)(m_1 - m_2)\right] \\ C &= (l_1q - l_2q)^2 - (l_1m_2 - l_2m_1)^2 + (m_1q - m_2q)^2 \\ q &= \cos\alpha_2 \end{aligned}$$

When $B^2 - 4AC > 0$, there are two solutions, i.e.,

$$z = \frac{-B \pm \sqrt{B^2 - 4AC}}{2B}. \tag{3.47}$$

Substituting Eq. (3.47) into Eq. (4.38), the orientation (ϕ, φ) can be obtained. Then, there are two solutions for the direct kinematic problem of the spherical 5R parallel mechanism. They correspond to two kinds of assembling modes.

3.4 Position Analysis of a 3-DOF Translational Parallel Mechanism

3.4.1 DELTA

A 3-DOF parallel robot, called DELTA, was proposed by Clavel and others in Switzerland at EPFL (Clavel 1988). The DELTA, shown in Fig. 1.22, was then investigated as an example of high-speed applications (e.g., pick-and-place). Since a DELTA is symmetric, we may only consider one of three legs of a DELTA as shown in Fig. 3.7. The coordinate systems can be established as follows:

O-xyz: the reference frame system denoted $\Re$, where O is the center of equilateral triangle defined by B_i $(i = 1, 2, 3)$, x is along OB_1, and z is orthogonal to the base plane determined by OB_i $(i = 1, 2, 3)$

O-$x_i y_i z$: the coordinate system denoted $\Re_i$, where x_i is along OB_i

O'-$x'y'z'$: the moving coordinate system noted as $\Re'$, where O' is the center of equilateral triangle defined by P_i $(i = 1, 2, 3)$, x' is along $O'P_1$, and z' is orthogonal to the base plane determined by $O'P_i$ $(i = 1, 2, 3)$

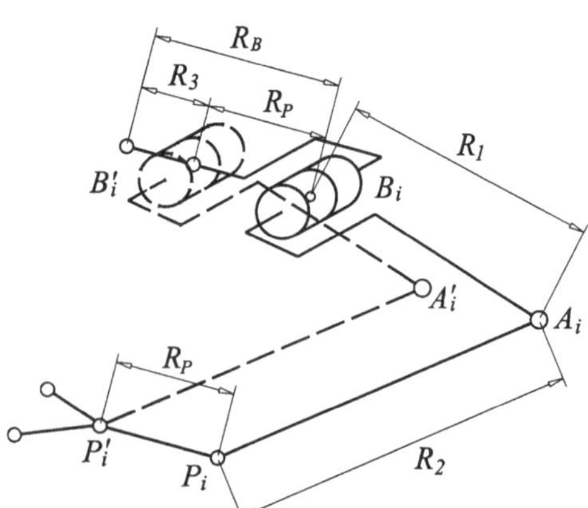

Fig. 3.7 Geometric parameters of a DELTA

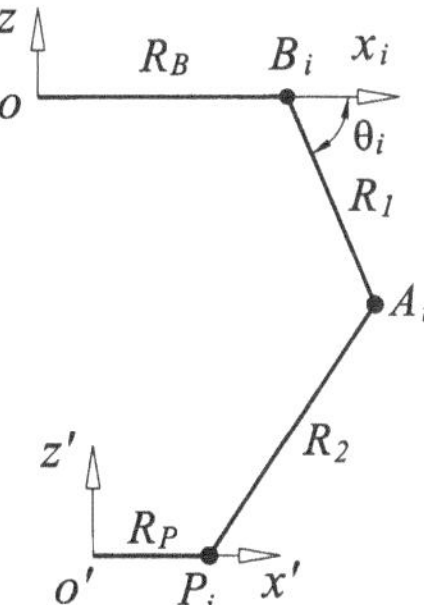

Fig. 3.8 One leg in DELTA

The point $\boldsymbol{A}_i$ can be written in frame $\Re$ as

$$(\boldsymbol{A}_i)_{\Re} = \boldsymbol{Q}(\boldsymbol{A}_i)_{\Re_i} = \begin{pmatrix} (R_1\cos\theta_i + R_B)\cos\alpha_i \\ (R_1\cos\theta_i + R_B)\sin\alpha_i \\ R_1 \ \sin\theta_i \end{pmatrix}, \quad i = 1, 2, 3 \tag{3.48}$$

where θ_i is the input parameter, and

$$\alpha_1 = 0, \ \alpha_2 = \frac{2\pi}{3}, \ \alpha_3 = -\frac{2\pi}{3} \tag{3.49}$$

and

$$\boldsymbol{Q} = \begin{bmatrix} \cos\alpha_i & -\sin\alpha_i & 0 \\ \sin\alpha_i & \cos\alpha_i & 0 \\ 0 & 0 & 1 \end{bmatrix}. \tag{3.50}$$

$\boldsymbol{Q}$ is the transformation matrix from O-xyz to O-$x_i y_i z$. The coordinate of point P_i in the frame $\Re$ can be written as

$$(\boldsymbol{P}_i)_{\Re} = (\boldsymbol{P}_i)_{\Re'} + (\boldsymbol{R})_{\Re} = \begin{pmatrix} R_P\cos\alpha_i + x \\ R_P\sin\alpha_i + y \\ z \end{pmatrix} \tag{3.51}$$

where

$$(\boldsymbol{R})_{\Re} = \begin{pmatrix} x \ y \ z \end{pmatrix}^T \tag{3.52}$$

which is the position vector of point O' and can be the position of the end-effector for the DELTA. From Fig. 3.8, the inverse kinematics problem of the robots can be derived by writing the constraint equation of link $A_i P_i$ as follows:

$$\|(\boldsymbol{P}_i - \boldsymbol{A}_i)_{\Re}\| = R_2, \quad i = 1, 2, 3 \tag{3.53}$$

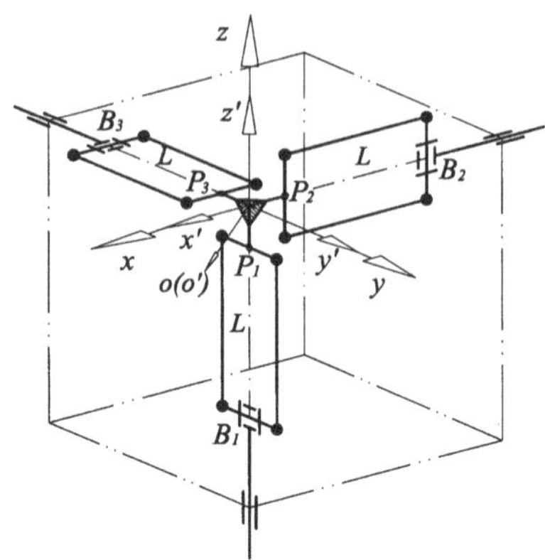

Fig. 3.9 A kinematics model of the parallel cube-manipulator

3.4.2 A Parallel Cube-Manipulator

A parallel cube-manipulator is shown in Fig. 1.26. A kinematics model of the manipulator can be developed as shown in Fig. 3.9. The center of the link in each of the three legs connected to the base by prismatic joints is denoted as B_i $(i = 1, 2, 3)$. And the center of the interval between the two spherical joints connecting the mobile platform for each chain is denoted as P_i $(i = 1, 2, 3)$, which is in one of three sides of the cubic mobile platform, respectively. B_i and P_i are the centers of the revolute joints in each leg for the design of Fig. 1.26b. A fixed global reference system: $\Re : o\text{-}xyz$ is located at the intersection point of three axes of actuated prismatic joints with the z-axis and the y-axis directed along the first and third actuation directions, respectively, as shown in Fig. 3.9. Another reference system $\Re' : o'\text{-}x'y'z'$ is located at the center of the cubic mobile platform. The z'-axis and y'-axis directed along P_1o' and P_3o', respectively, as shown in Fig. 3.9. Related geometric parameters are $o'P_i = r$ and $B_iP_i = L$, where $i = 1, 2, 3$.

3.4.2.1 Inverse Kinematics Problem

The objective of the inverse kinematics is to find the inputs of the manipulator with the given position of reference point o'. The position vector $\boldsymbol{c}_{\Re}$ of point o' in frame $\Re$ can be written as

$$\boldsymbol{c}_{\Re} = \begin{bmatrix} x & y & z \end{bmatrix}^{\mathrm{T}} \tag{3.54}$$

As shown in Fig. 3.9, the coordinate of the point P_i in the frame $\Re'$ can be described by the vector $\boldsymbol{p}_{i\Re'}$ $(i = 1, 2, 3)$, which can be expressed as

$$\boldsymbol{p}_{1\Re'} = \begin{bmatrix} 0 \\ 0 \\ -r \end{bmatrix}, \quad \boldsymbol{p}_{2\Re'} = \begin{bmatrix} -r \\ 0 \\ 0 \end{bmatrix}, \quad \boldsymbol{p}_{3\Re'} = \begin{bmatrix} 0 \\ -r \\ 0 \end{bmatrix}. \tag{3.55}$$

Then vectors $\boldsymbol{p}_{i\Re}$ $(i = 1,\ 2,\ 3)$ in frame $O\text{-}xyz$ can be written as

$$\boldsymbol{p}_{i\Re} = \boldsymbol{p}_{i\Re'} + \boldsymbol{c}_{\Re}. \tag{3.56}$$

As shown in Fig. 3.9, vector $\boldsymbol{b}_{i\Re}$ $(i = 1,\ 2,\ 3)$ is defined as the position vector of point B_i in frame $\Re$, and

$$\boldsymbol{b}_{1\Re} = \begin{bmatrix} 0 \\ 0 \\ z_1 \end{bmatrix}, \quad \boldsymbol{b}_{2\Re} = \begin{bmatrix} x_2 \\ 0 \\ 0 \end{bmatrix}, \quad \boldsymbol{b}_{3\Re} = \begin{bmatrix} 0 \\ y_3 \\ 0 \end{bmatrix} \tag{3.57}$$

where z_1, x_2, and y_3 are the inputs of the manipulator, which will be denoted as ρ_1, ρ_2, and ρ_3, respectively. Then the inverse kinematics of the parallel cube-manipulator can be solved by writing the following constraint equation:

$$\left\| \boldsymbol{p}_{i\Re} - \boldsymbol{b}_{i\Re} \right\| = L, \quad i = 1,2,3 \tag{3.58}$$

that is,

$$x^2 + y^2 + (z - \rho_1 - r)^2 = L^2 \tag{3.59}$$

$$(x - \rho_2 - r)^2 + y^2 + z^2 = L^2 \tag{3.60}$$

$$x^2 + (y - \rho_3 - r)^2 + z^2 = L^2 \tag{3.61}$$

from which, if the position of the mobile platform is specified, the inputs of the manipulator can be obtained as

$$\rho_1 = \pm\sqrt{L^2 - x^2 - y^2} + z - r \tag{3.62}$$

$$\rho_2 = \pm\sqrt{L^2 - y^2 - z^2} + x - r \tag{3.63}$$

$$\rho_3 = \pm\sqrt{L^2 - x^2 - z^2} + y - r \tag{3.64}$$

from which one may see that there are eight inverse kinematics solutions for the manipulator. To get the inverse configuration as shown in Fig. 3.9, the sign "±" in each of the Eqs. (3.62), (3.63), and (3.64) should be "–". Actually, any other assembly modes except for the one as shown in Fig. 3.9 will undoubtedly result in the interference between the mobile platform and the legs. Therefore, the cube-manipulator can only be assembled as the configuration shown in Fig. 3.9, which means that there is only one inverse kinematics solution for the manipulator.

3.4.2.2 Direct Kinematics Problem

From Eqs. (3.59), (3.60), and (3.61), we may see that, in each equation, the maximum degree of the polynomial is two, the coefficient of which is 1, and, what is more, there is only one variable among x, y, and z with one degree. This allows us to express the two of three variables x, y, and z with the third one very easily, e.g., x and y can be described as the function of z, that is,

$$x = \frac{\rho_1 + r}{\rho_2 + r} z - \frac{(\rho_1 + r)^2}{2(\rho_2 + r)} + \frac{\rho_2 + r}{2} \tag{3.65}$$

$$y = \frac{\rho_1 + r}{\rho_3 + r} z - \frac{(\rho_1 + r)^2}{2(\rho_3 + r)} + \frac{\rho_3 + r}{2}. \tag{3.66}$$

Substituting Eqs. (3.65) and (3.66) to Eq. (3.59) and rearranging it will lead to

$$az^2 + bz + c = 0. \tag{3.67}$$

Then there is

$$z = \frac{-b \pm \sqrt{b^2 - 4ac}}{2a} \tag{3.68}$$

in which

$$a = 1 + \left(\frac{\rho_1 + r}{\rho_2 + r}\right)^2 + \left(\frac{\rho_1 + r}{\rho_3 + r}\right)^2$$

$$b = -(\rho_1 + r)^3 \left[\frac{1}{(\rho_2 + r)^2} + \frac{1}{(\rho_3 + r)^2}\right]$$

$$c = \frac{(\rho_2 + r)^2}{4} + \frac{(\rho_3 + r)^2}{4} + \frac{(\rho_1 + r)^4}{4} \left[\frac{1}{(\rho_2 + r)^2} + \frac{1}{(\rho_3 + r)^2}\right] - L^2.$$

x and y will be reached substituting Eq. (3.68) to Eqs. (3.65) and (3.66), respectively, if inputs ρ_1, ρ_2, and ρ_3 are given. And from Eqs. (3.65), (3.66), and (3.68), one may see that there are only two direct kinematics solutions for the manipulator. To obtain the direct configuration as shown in Fig. 3.9, the sign "±" in Eq. (3.68) should be "+". The second direct kinematics solution, in which the sign "±" in Eq. (3.68) is "−", will lead to the interference between the mobile platform and legs.

3.5 Position Analysis of a 3-DOF Spherical Parallel Mechanism

A 3-DOF spherical parallel mechanism is shown in Figs. 1.18 and 3.10. It is such a mechanism that the axes of all nine revolute joints intersect at one common point, which will henceforth be called the center of the mechanism. All the moving bodies are in pure rotation with respect to this point. The three actuated joints of the mechanism are fixed on the frame.

Here we introduce the inverse kinematic problem of the mechanism. The Cartesian coordinates are established as follows: the coordinate systems $O\text{-}X_{i1}Y_{i1}Z_{i1}$, $O\text{-}X_{i2}Y_{i2}Z_{i2}$, and $O\text{-}X_{i3}Y_{i3}Z_{i3}$ can be defined according to the Denavit-Hartenberg notation (Denavit and Hartenberg 1955); the coordinate system $O\text{-}XYZ$ is defined as that the Z-axis is directed along the HO, and that the X-axis is orthogonal to the Z-axis and is in the plane defined by the Z and Z_{11} axes; and the coordinate system $O\text{-}X_0Y_0Z_0 - Z_0$ is directed along the OH', and the X_0-axis is orthogonal to the Z_0-axis and is in the plane defined by the Z_0 and Z_{13} axes. In this mechanism, there are four kinematic angular parameters: α_1, α_2, β_1, and β_2. Define vectors as the following: $\boldsymbol{v}_i$ $(i = 1, 2, 3)$ along the Z_{i3}-axis, $\boldsymbol{\mu}_i$ $(i = 1, 2, 3)$ along the Z_{i1}-axis, and $\boldsymbol{w}_i$ $(i = 1, 2, 3)$ along the Z_{i2}-axis.

For the mechanism studied here, the Cartesian coordinates in fact represent the orientation of the end-effector, i.e., the orientation of frame $O\text{-}X_0Y_0Z_0$ with respect to frame $O\text{-}XYZ$. The orientation is expressed by using three Euler angles, which are noted ϕ_1, ϕ_2 and ϕ_3, respectively. Its inputs are given by three angles: θ_1, θ_2, and θ_3. When ϕ_1, ϕ_2, and ϕ_3 are given, vectors $\boldsymbol{v}_i\left(\boldsymbol{v}_{ix}, \boldsymbol{v}_{iy}, \boldsymbol{v}_{iz}\right)$ can be obtained easily. The solution to the inverse kinematic problem is obtained by writing the closure equations as follows:

$$\boldsymbol{\omega}_i \cdot \boldsymbol{v}_i = \cos\alpha_2, \quad i = 1, 2, 3 \tag{3.69}$$

which, for each leg, leads to a quadratic equation of the form

$$A_i x_i^2 + 2B_i x_i + C_i = 0 \tag{3.70}$$

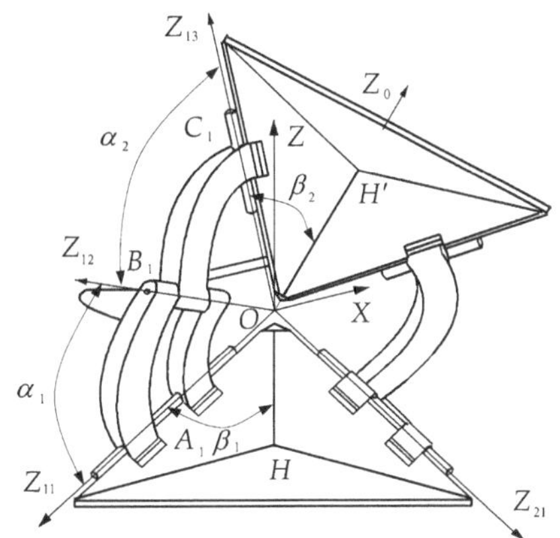

Fig. 3.10 A 3-DOF spherical parallel mechanism

where

$$x_i = \tan(\theta_i/2) \tag{3.71}$$

and

$$\begin{aligned}
A_i &= \cos\eta_i \sin\alpha_1 \nu_{iy} - \cos\alpha_1 \cos\eta_i \sin\beta_1 \nu_{ix} - \sin\alpha_1 \sin\eta_i \nu_{ix} \\
&\quad - \cos\alpha_1 \sin\beta_1 \sin\eta_i \nu_{iy} - \cos\alpha_1 \cos\beta_1 \nu_{iz} - \cos\alpha_2 \\
B_i &= -\cos\beta_1 \cos\eta_i \sin\alpha_1 \nu_{ix} - \cos\beta_1 \sin\alpha_1 \sin\eta_i \nu_{iy} + \sin\alpha_1 \sin\beta_1 \nu_{iz} \\
C_i &= \sin\alpha_1 \sin\eta_i \nu_{ix} - \cos\alpha_1 \cos\eta_i \sin\beta_1 \nu_{ix} - \cos\eta_i \sin\alpha_1 \nu_{iy} \\
&\quad - \cos\alpha_1 \sin\beta_1 \sin\eta_i \nu_{iy} - \cos\alpha_1 \cos\beta_1 \nu_{iz} - \cos\alpha_2
\end{aligned}$$

where

$$\begin{aligned}
\nu_{ix} &= -q_{11} \cos\eta_i \sin\beta_2 - q_{12} \sin\eta_i \sin\beta_2 + q_{13} \cos\beta_2 \\
\nu_{iy} &= -q_{21} \cos\eta_i \sin\beta_2 - q_{22} \sin\eta_i \sin\beta_2 + q_{23} \cos\beta_2 \\
\nu_{iz} &= -q_{31} \cos\eta_i \sin\beta_2 - q_{32} \sin\eta_i \sin\beta_2 + q_{33} \cos\beta_2
\end{aligned}$$

further

$$\begin{aligned}
q_{11} &= -\sin\phi_1 \sin\phi_3 + \cos\phi_1 \cos\phi_3 \cos\phi_2 \\
q_{12} &= -\sin\phi_1 \cos\phi_3 - \cos\phi_1 \sin\phi_3 \cos\phi_2 \\
q_{13} &= \cos\phi_1 \sin\phi_2 \\
q_{21} &= \cos\phi_1 \sin\phi_3 + \sin\phi_1 \cos\phi_3 \cos\phi_2 \\
q_{22} &= \cos\phi_1 \cos\phi_3 - \sin\phi_1 \sin\phi_3 \cos\phi_2 \\
q_{23} &= \sin\phi_1 \sin\phi_2 \\
q_{31} &= -\cos\phi_3 \sin\phi_2 \\
q_{32} &= \sin\phi_3 \sin\phi_2 \\
q_{33} &= \cos\phi_2
\end{aligned}$$

and

$$\eta_1 = 0, \quad \eta_2 = \frac{2}{3}\pi, \quad \eta_3 = \frac{4}{3}\pi.$$

Then, if x_i is obtained from Eq. (3.70) as

$$x_i = \frac{-B_i \pm \sqrt{B_i^2 - A_i C_i}}{A_i} \tag{3.72}$$

the inputs can be reached by

$$\theta_i = 2\tan^{-1}(x_i) \quad i = 1, 2, 3 \tag{3.73}$$

from which one may see that there are eight solutions for the inverse kinematic problem of the mechanism.

The direct kinematic problem of the mechanism was discussed in Gosselin et al. (1994) and Huang and Yao (1999).

3.6 Position Analysis of the 2-PRU&1-PR(Pa)R Parallel Mechanism

3.6.1 Inverse Kinematics

The 2-PRU&1-PR(Pa)R parallel mechanism is shown in Fig. 2.28. A kinematics model of the mechanism is developed as shown in Fig. 3.11. Vertices of the output platform are denoted as platform joints P_i $(i = 1, 2, 3)$, and vertices of the base platform are denoted as b_i $(i = 1, 2, 3)$. A fixed global reference system $\Re : O\text{-}xyz$ is located at the center of the side b_1b_2 with the z-axis normal to the base plate and the y-axis directed along b_1b_2. Another reference frame, called the top frame $\Re' : O'\text{-}x'y'z'$, is located at the center of the side P_1P_2. The z'-axis is perpendicular to the output platform and y'-axis directed along P_1P_2. The length of link for each leg is denoted as L, where $P_iB_i = L, i = 1, 2, 3$. What we should note that, in some case, the length of the link P_3B_3 can be different from that of P_1B_1 and P_2B_2.

For inverse kinematic analysis, the pose of the mobile platform is considered known, and the position is given by the position vector $(\boldsymbol{O}')_{\Re}$ and the orientation is given by a matrix $\boldsymbol{Q}$. And there are

$$(\boldsymbol{O}')_{\Re} = (x\ y\ z)^{\mathrm{T}} \tag{3.74}$$

where $x = 0$, and

$$\boldsymbol{Q} = \begin{bmatrix} \cos\phi & 0 & \sin\phi \\ 0 & 1 & 0 \\ -\sin\phi & 0 & \cos\phi \end{bmatrix} \tag{3.75}$$

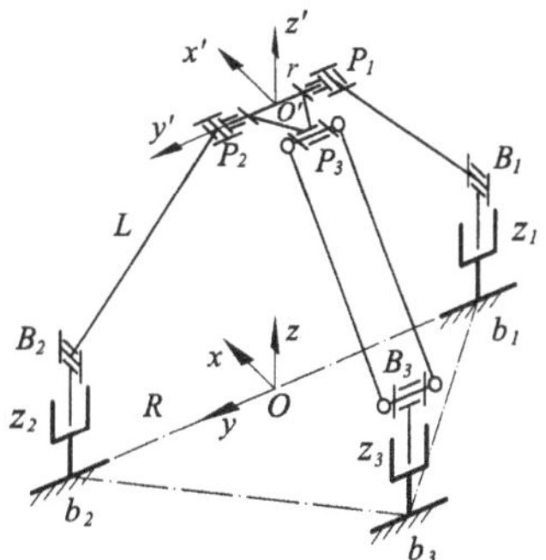

Fig. 3.11 Geometric parameters of the spatial 2-PRU&1-PR(Pa)R parallel manipulator

where the angle ϕ is the rotational degree of freedom of the output platform with respect to y-axis. The coordinate of the point P_i in the frame $\Re'$ can be described by the vector $(\boldsymbol{p}_i)_{\Re'}$ $(i = 1,\ 2,\ 3)$, and

$$(\boldsymbol{p}_1)_{\Re'} = \begin{pmatrix} 0 \\ -r \\ 0 \end{pmatrix}, \quad (\boldsymbol{p}_2)_{\Re'} = \begin{pmatrix} 0 \\ r \\ 0 \end{pmatrix}, \quad (\boldsymbol{p}_3)_{\Re'} = \begin{pmatrix} -r \\ 0 \\ 0 \end{pmatrix}. \tag{3.76}$$

Vectors $(\boldsymbol{b}_i)_{\Re}$ $(i = 1,\ 2,\ 3)$ will be defined as the position vectors of base joints in frame $\Re$, and

$$(\boldsymbol{b}_1)_{\Re} = \begin{pmatrix} 0 \\ -R \\ z_1 \end{pmatrix}, \quad (\boldsymbol{b}_2)_{\Re} = \begin{pmatrix} 0 \\ R \\ z_2 \end{pmatrix}, \quad (\boldsymbol{b}_3)_{\Re} = \begin{pmatrix} -R \\ 0 \\ z_3 \end{pmatrix}. \tag{3.77}$$

The vector $(\boldsymbol{p}_i)_{\Re}$ $(i = 1,\ 2,\ 3)$ in frame O-xyz can be written as

$$(\boldsymbol{p}_i)_{\Re} = \boldsymbol{Q}(\boldsymbol{p}_i)_{\Re'} + (\boldsymbol{O}')_{\Re} \tag{3.78}$$

Then the inverse kinematics of the parallel mechanism can be solved by writing the following constraint equation:

$$\left\| [\boldsymbol{p}_i - \boldsymbol{b}_i]_{\Re} \right\| = L \quad i = 1, 2, 3. \tag{3.79}$$

Hence, for a given mechanism and for prescribed values of the position and orientation of the platform, the required actuator inputs can be directly computed from Eq. (3.79), that is,

$$z_1 = \pm\sqrt{L^2 - (-r + y + R)^2} + z \tag{3.80}$$

$$z_2 = \pm\sqrt{L^2 - (r + y - R)^2} + z \tag{3.81}$$

$$z_3 = \pm\sqrt{L^2 - (-r\cos\phi + R)^2 - y^2} + r\sin\phi + z. \tag{3.82}$$

From Eqs. (3.80), (3.81), and (3.82), we may see that there are eight inverse kinematics solutions for a given pose of the parallel mechanism. To obtain the inverse configuration as shown in Fig. 3.11, each one of the signs "$\pm$" in Eqs. (3.80), (3.81), and (3.82) should be "$+$".

3.6.2 Direct Kinematics

The objective of the direct kinematics solution is to define a mapping from the known set of the actuated inputs to the unknown pose of the output platform. For the architecture with prismatic actuators, the inputs that are considered known are z_1, z_2, and z_3. The unknown pose of the output platform is described by the position vector $(\boldsymbol{O}')_{\Re}$ and the angle ϕ. From Eqs. (3.80) and (3.81), we may obtain

$$y = m\,z - n \tag{3.83}$$

where $m = \frac{z_1 - z_2}{2(R-r)}$, $n = \frac{z_1^2 - z_2^2}{4(R-r)}$. Substituting Eq. (3.83) into Eq. (3.80) leads to

$$e\,z^2 + f\,z + g = 0 \tag{3.84}$$

where

$e = m^2 + 1$
$f = 2m\,(R - r - n) - 2z_1$
$g = (R - r - n)^2 + z_1^2 - L^2.$

The solution of z can be written, from Eq. (4.84), as

$$z = \frac{-f \pm \sqrt{f^2 - 4eg}}{2e}. \tag{3.85}$$

If z and y are obtained from Eqs. (3.85) and (3.83), the direct solutions of angle ϕ can be reached as

$$\phi = 2\tan^{-1}(t) \tag{3.86}$$

where

$$t = \frac{A \pm \sqrt{A^2 + B^2 - C^2}}{C + B} \tag{3.87}$$

with

$A = 2r\,(z - z_3)$
$B = -2R\,r$
$C = L^2 - y^2 - r^2 - (z - z_3)^2 - R^2.$

From Eqs. (3.85), (3.86), (3.87), and (3.83), we may see that for the given values of z_1, z_2, and z_3, there are two solutions for y and z, respectively, and four solutions for ϕ. Therefore, the solution of the direct kinematics of the architecture with linear actuators can reach four. To obtain the direct configuration as shown in Fig. 3.11, the sign "±" in Eq. (3.85) should be "−" and that of Eq. (3.87) should be "+".

From the above analysis, we may reach the results that the solution of inverse kinematics for the spatial 3-DOF parallel mechanism can reach eight and the solution of direct kinematics can reach four, and all the solutions can be described in closed forms.

3.7 Position Analysis of the 2-RRU&1-RR(Pa)R Parallel Mechanism

3.7.1 *Inverse Kinematics*

The 2-RRU&1-RR(Pa)R parallel mechanism is shown in Fig. 2.29. A kinematics model of the mechanism is developed as shown in Fig. 3.12. Vertices of the output platform are denoted as P_i ($i = 1, 2, 3$), and vertices of the base platform are denoted as B_i ($i = 1, 2, 3$). A fixed global reference system $\Re$:O-xyz is located at the center of the side B_1B_2 with the z-axis normal to the base plate and the y-axis directed along B_1B_2. Another reference frame, called the top frame $\Re'$:O'-$x'y'z'$, is located at the center of the side P_1P_2. Let R and r be the characteristic lengths of OB_i and $O'P_i$, respectively. The z'-axis is perpendicular to the output platform and y'-axis directed along P_1P_2. Connecting joints between the upper and lower links are denoted as E_i. Lengths of upper and lower links for each leg are denoted as L_a and L_b, where $L_a = P_iE_i$ and $L_b = B_iE_i$ ($i = 1, 2, 3$). What should be noted that, in some case, lengths of the links P_3E_3 and B_3E_3 can be different from that of P_1E_1 (P_2E_2) and B_1E_1 (B_2E_2).

For inverse kinematic analysis of the mechanism, the pose of the mobile platform is considered known, and the position is given by the position vector $(\boldsymbol{c})_{\Re} = \begin{pmatrix} 0 & y & z \end{pmatrix}^{\mathrm{T}}$ and the orientation is given by a matrix $\boldsymbol{Q}$ of Eq. (3.75). The coordinate of the point P_i in the frame $\Re'$ can be described by the vector $(\boldsymbol{p}_i)_{\Re'}$ ($i = 1, 2, 3$), and

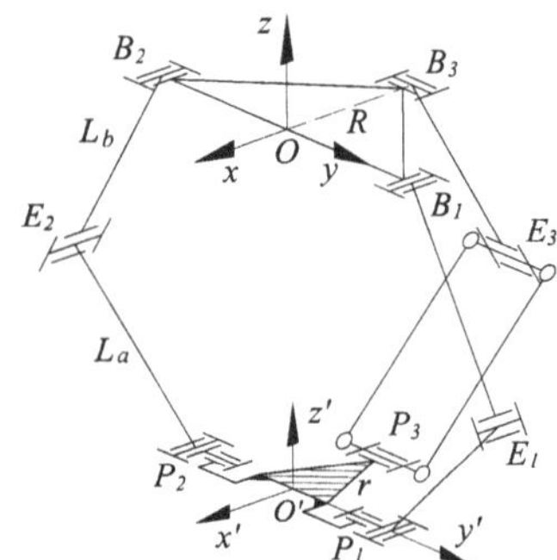

Fig. 3.12 Geometric parameters of the 2-RRU&1-RR(Pa)R parallel mechanism

$$(\boldsymbol{p}_1)_{\Re'} = \begin{pmatrix} 0 \\ r \\ 0 \end{pmatrix}, \quad (\boldsymbol{p}_2)_{\Re'} = \begin{pmatrix} 0 \\ -r \\ 0 \end{pmatrix}, \quad (\boldsymbol{p}_3)_{\Re'} = \begin{pmatrix} -r \\ 0 \\ 0 \end{pmatrix}. \tag{3.88}$$

Then vectors $(\boldsymbol{p}_i)_{\Re}$ $(i = 1,\ 2,\ 3)$ in frame $O\text{-}xyz$ can be written as

$$(\boldsymbol{p}_i)_{\Re} = \boldsymbol{Q}(\boldsymbol{p}_i)_{\Re'} + (\boldsymbol{c})_{\Re}. \tag{3.89}$$

Vectors $(\boldsymbol{e}_i)_{\Re}$ $(i = 1,\ 2,\ 3)$ will be defined as the position vectors of connecting joints E_i in frame $\Re$, and

$$[\boldsymbol{e}_1]_{\Re} = \begin{pmatrix} 0 \\ R + L_b \cos\theta_1 \\ L_b \sin\theta_1 \end{pmatrix}, \quad [\boldsymbol{e}_2]_{\Re} = \begin{pmatrix} 0 \\ -R + L_b \cos\theta_2 \\ L_b \sin\theta_2 \end{pmatrix},$$

$$[\boldsymbol{e}_3]_{\Re} = \begin{pmatrix} -R + L_b \cos\theta_3 \\ 0 \\ L_b \sin\theta_3 \end{pmatrix}. \tag{3.90}$$

Then the inverse kinematics of the parallel mechanism can be solved by writing the following constraint equation:

$$\left\| [\boldsymbol{p}_i - \boldsymbol{e}_i]_{\Re} \right\| = L_a, \quad i = 1, 2, 3 \tag{3.91}$$

from which one obtains

$$(r + y - R - L_b \cos\theta_1)^2 + (z - L_b \sin\theta_1)^2 = L_a^2 \tag{3.92}$$

$$(-r + y + R - L_b \cos\theta_2)^2 + (z - L_b \sin\theta_2)^2 = L_a^2 \tag{3.93}$$

$$(r \cos\phi - R + L_b \cos\theta_3)^2 + y^2 + (r \sin\phi + z - L_b \sin\theta_3)^2 = L_a^2 \tag{3.94}$$

and

$$A_i s_i^2 + B_i s_i + C_i = 0 \tag{3.95}$$

in which

$$s_i = \tan(\theta_i / 2) \tag{3.96}$$

$$A_1 = (r + y - R)^2 + z^2 + L_b^2 - L_a^2 + 2(r + y - R) L_b$$

$$B_1 = -4zL_b$$

$$C_1 = (r + y - R)^2 + z^2 + L_b^2 - L_a^2 - 2(r + y - R) L_b$$

$$A_2 = (-r + y + R)^2 + z^2 + L_b^2 - L_a^2 + 2(-r + y + R) L_b$$

$$B_2 = B_1$$

$$C_2 = (-r + y + R)^2 + z^2 + L_b^2 - L_a^2 - 2(-r + y + R) L_b$$

$$A_3 = (r\cos\phi - R)^2 + y^2 + L_b^2 + (z + r\sin\phi)^2 - L_a^2 - 2(r\cos\phi - R) L_b$$

$$B_3 = -4(z + r\sin\phi) L_b$$

$$C_3 = (r\cos\phi - R)^2 + y^2 + L_b^2 + (z + r\sin\phi)^2 - L_a^2 + 2(r\cos\phi - R) L_b.$$

Hence, for a given mechanism and prescribed values of the position and orientation of the mobile platform, the required actuator inputs can be directly computed as

$$\theta_i = 2\tan^{-1}(s_i) \tag{3.97}$$

where

$$s_i = \left(-B_i \pm \sqrt{B_i^2 - 4A_iC_i}\right) \Big/ (2A_i). \tag{3.98}$$

From Eq. (3.98), we may see that there are eight inverse kinematics solutions for a given pose of the parallel mechanism. To obtain the inverse configuration as shown in Fig. 3.12, the sign "±" in Eq. (3.98) should be "+" for $i = 1$ and "−" for $i = 2, 3$.

3.7.2 Direct Kinematics

The objective of the direct kinematics solution is to define a mapping from the known set of the actuated inputs to the unknown pose of the output platform. For the architecture with prismatic actuators, the inputs that are considered known are θ_1, θ_2, and θ_3. The unknown pose of the output platform is described by the position vector $(\boldsymbol{c})_{\Re}$ and the angle ϕ. From Eqs. (3.92) and (3.93), we may obtain

$$y = m\,z - n \tag{3.99}$$

where

$$m = \frac{L_b(\sin\theta_1 - \sin\theta_2)}{2r - 2R - L_b(\cos\theta_1 - \cos\theta_2)}, \quad n = \frac{L_b(R-r)(\cos\theta_1 + \cos\theta_2)}{2r - 2R - L_b(\cos\theta_1 - \cos\theta_2)}. \tag{3.100}$$

Substituting Eq. (3.99) into Eq. (3.92) leads to

$$e z^2 + f z + g = 0. \tag{3.101}$$

The solution of z can be written, from Eq. (3.101), as

$$z = \left(-f \pm \sqrt{f^2 - 4eg}\right) \Big/ (2e) \tag{3.102}$$

where

$$\begin{aligned} e &= m^2 + 1 \\ f &= 2m(r - R - n - L_b\cos\theta_1) - 2L_b\sin\theta_1 \\ g &= (r - R - n - L_b\cos\theta_1)^2 + (L_b\sin\theta_1)^2 - L_a^2. \end{aligned}$$

If z and y are obtained from Eqs. (3.102) and (3.99), the direct solutions of angle ϕ can be reached as

$$\phi = 2\tan^{-1}(t) \tag{3.103}$$

where

$$t = \left(A \pm \sqrt{A^2 + B^2 - C^2}\right) \Big/ (C + B) \tag{3.104}$$

with

$$\begin{aligned} A &= 2r(z - L_b\sin\theta_3) \\ B &= 2r(L_b\cos\theta_3 - R) \\ C &= L_a^2 - y^2 - r^2 - (z - L_b\sin\theta_3)^2 - (L_b\cos\theta_3 - R)^2. \end{aligned}$$

From Eqs. (3.99), (3.102), and (3.104), we may see that for the given values of θ_1, θ_2, and θ_3, there are two solutions for y and z, respectively, and four solutions for ϕ. Therefore, the solution of the direct kinematics of the architecture with linear actuators can reach four. To obtain the direct configuration as shown in Fig. 3.12, the sign "±" in Eq. (3.102) should be "−" and that of Eq. (3.104) should be "+".

From the above analysis, we may reach the results that the solution of inverse kinematics for the spatial 3-DOF parallel mechanism can reach eight and the solution of direct kinematics can reach four, and all the solutions can be described in closed forms.

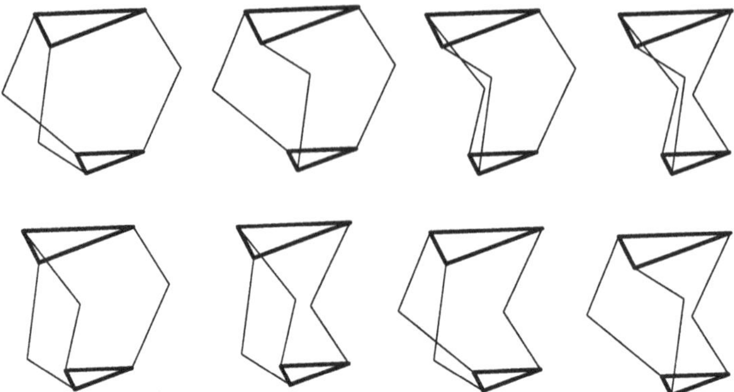

Fig. 3.13 The eight solutions of the inverse kinematics problem

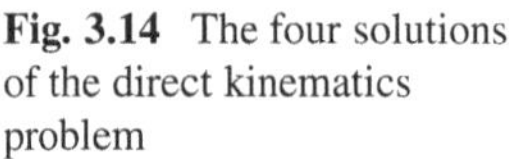

Fig. 3.14 The four solutions of the direct kinematics problem

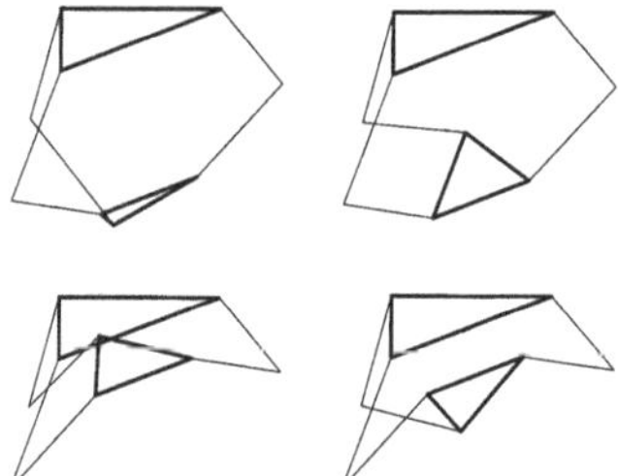

For an example, we here consider a mechanism with $R = 1.0$, $r = 0.6$, and $L_a = L_b = 1.2$. Its inverse and direct configurations are shown in Figs. 3.13 and 3.14, respectively. The given pose of the mechanism is $y = 0.37$, $z = -2.0$, and $\phi = 0.0$ for inverse kinematics. The inputs for the direct kinematics problem are $\theta_1 = -5\pi/18$, $\theta_2 = -2\pi/3$, and $\theta_3 = -0.6\pi$.

3.8 Position Analysis of a 2-PRU&1-PRC Parallel Mechanism

The 2-PRU&1-PRC parallel mechanism is illustrated in Fig. 2.39a. As mentioned in Sect. 2.3.2, the kinematics of the mechanism will be simpler than that of the 2-PRU&1-PR(Pa)R parallel mechanism. We here give the inverse kinematic problem of the mechanism.

The kinematic schematic of the mechanism is shown in Fig. 3.15. Vertices of the mobile platform are denoted as platform joints P_i $(i = 1,2,3)$; and central points of the three revolute joints attached to the sliders are denoted as B_i (i = 1,2,3). A fixed global reference frame O-xyz is located at the center point of the line segment ab with the z-axis normal to the plane abc and the y-axis directed along ab. Another reference frame, called the moving frame, O'-$x'y'z'$, is located at the

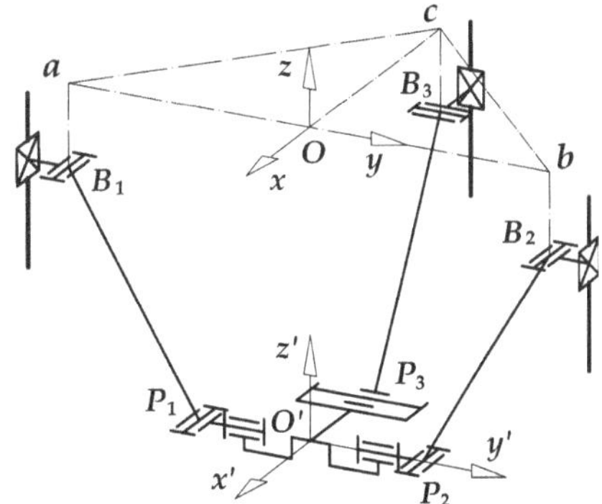

Fig. 3.15 Kinematic schematic of the 2-PRU&1-PRC parallel mechanism

center of the side P_1P_2. The z'-axis is perpendicular to the mobile platform and the y'-axis directed along P_1P_2. Since the first and second legs are the same in kinematic chains and the third leg is different, the geometric parameters for the first and second legs can be the same, but different from those of the third leg. Then, the geometric parameters will be $O'P_1 = O'P_2 = r$, $B_1P_1 = B_2P_2 = R_2$, $O'P_3 = L_1$, $P_3B_3 = L_2$, the normal distance L_3 from the point O to the straight-line path of the joint point B_3, i.e., $Oc = L_3$, and $Oa = Ob = R$.

The inverse kinematic problem of this mechanism is somewhat similar to that of the 2-PRU&1-PR(Pa)R mechanism. Vectors $\boldsymbol{b}_i$ $(i = 1,2,3)$ are defined as the position vectors of points B_i in the reference frame O-xyz and can be written as

$$\boldsymbol{b}_1 = (0 - R \;\; z_1)^{\mathrm{T}}, \quad \boldsymbol{b}_2 = (0 \;\; R \;\; z_2)^{\mathrm{T}}, \quad \boldsymbol{b}_3 = (-L_3 \;\; 0 \;\; z_3)^{\mathrm{T}}. \tag{3.105}$$

In the reference frame O-xyz, position vectors $\boldsymbol{p}_i$ $(i = 1,2,3)$ of points P_i can be written as

$$\boldsymbol{p}_1 = (0 \;\; y - r \;\; z)^{\mathrm{T}}, \quad \boldsymbol{p}_1 = (0 \;\; y + r \;\; z)^{\mathrm{T}}, \quad \boldsymbol{p}_3 = (-L_1\cos\phi \;\; 0 \;\; z + L_1\sin\phi)^{\mathrm{T}} \tag{3.106}$$

where $(y, \;\; z, \;\; \phi)$ is the pose of the mechanism and ϕ is the rotating angle of the mobile platform about the y'-axis. The kinematic problem of the mechanism can be solved by writing

$$|\boldsymbol{b}_i\boldsymbol{p}_i| = B_iP_i. \tag{3.107}$$

Then, there are

$$(R - r + y)^2 + (z_1 - z)^2 = R_2^2 \tag{3.108}$$

$$(R - r - y)^2 + (z_2 - z)^2 = R_2^2 \tag{3.109}$$

$$(z_3 - z - L_1 \sin\phi)^2 + (L_3 - L_1 \cos\phi)^2 = L_2^2. \tag{3.110}$$

For a given pose $(y,\ z,\ \phi)$, the inputs y_i ($i = 1,2,3$) can be obtained as

$$z_1 = \pm\sqrt{R_2^2 - (R - r + y)^2} + z \tag{3.111a}$$

$$z_2 = \pm\sqrt{R_2^2 - (R - r - y)^2} + z \tag{3.111b}$$

$$z_3 = \pm\sqrt{L_2^2 - (L_3 - L_1 \cos\phi)^2} + z + L_1 \sin\phi. \tag{3.111c}$$

Therefore, there are eight inverse kinematic solutions for the mechanism. The configuration shown in Fig. 3.15 corresponds to the solution when the "±" signs in Eqs. (3.111a), (3.111b), and (3.111c) are all "+". Observing the kinematic equations of the 2-PRU&1-PRC and 2-PRU&1-PR(Pa)R parallel mechanisms, one may see that the kinematics of the 2-PRU&1-PRC mechanism is relatively simpler.

Letting $R - r$ in Eqs. (3.111a) and (3.111b) be R_1, i.e., $R_1 = R - r$, there are

$$z_1 = \pm\sqrt{R_2^2 - (R_1 + y)^2} + z \tag{3.112}$$

$$z_2 = \pm\sqrt{R_2^2 - (R_1 - y)^2} + z \tag{3.113}$$

from which we may see that the inverse kinematic problem of the first and second legs is actually that of the PRRRP symmetrical parallel mechanism (Liu et al. 2006), which is kinematically a planar parallel mechanism. If the position of point O' is specified, the kinematic equation of Eq. (3.111c) is actually that of a slider-crank mechanism (Söylemez 2002).

3.9 Position Analysis of the 3-[PP]S Type of Parallel Mechanism

As shown in Fig. 3.16, a [PP]S parallel mechanism is such a mechanism that when the mobile platform moves, the center points P_1, P_2, and P_3 of the three spherical joints are always in the vertical planes Π_1, Π_2, and Π_3, respectively, which are intersecting at a common line at 120° angles and are defined as the restricting planes. That means any one of the three spherical joints acts a planar motion. For such a reason, this type of parallel mechanism is referred to as a [PP]S parallel mechanism. The typical mechanism is the 3-PRS or 3-RPS (P, prismatic joint; R, revolute joint; and S, spherical joint) parallel mechanism. As analyzed in Sect. 2.1.3, the mobile platform of a [PP]S mechanism has three output degrees of freedom (DOFs), which

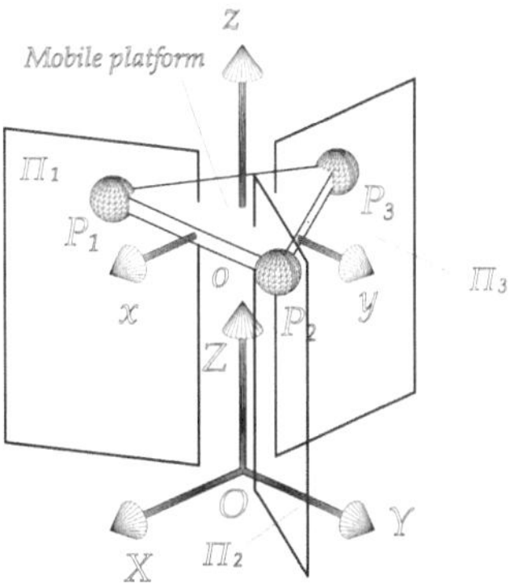

Fig. 3.16 Kinematic geometry of a general 3-[PP]S parallel mechanism

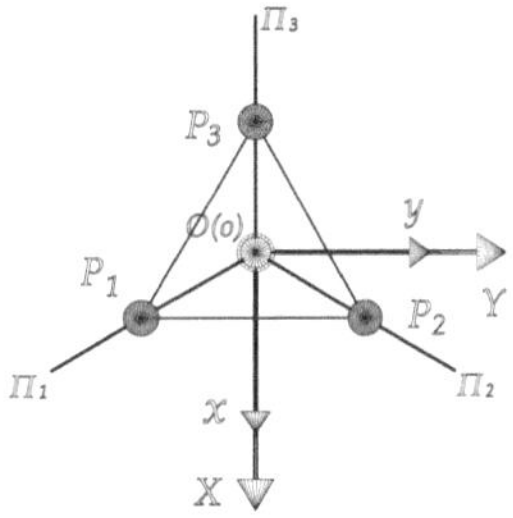

Fig. 3.17 Configuration when the mobile platform is at the original orientation

are two rotations and one translation. It is well noteworthy that the rotating axis of the rotation must be the arbitrary axis in the mobile platform plane, which is defined by the center points of three S joints in the platform and is usually a regular triangle. There exists no rotation about the axis normal to the mobile platform plane. Figure 3.17 shows the mobile platform when it is at the original orientation 0. The moving frame $\Re'$: o-xyz is attached the mobile platform with the original point coincident with the center of the mobile platform plane. The reference frame $\Re$: O-XYZ is such a frame that is fixed to the base and cannot rotate with the mobile platform.

3.9.1 Orientation Description

For such a mechanism, the question to be solved first is how to describe the orientation of the mobile platform. Most researchers used the Euler angles to represent the orientation (see Sect. 2.1.3). The method involves three angles, which makes the question complex. Different from this method, there is a representation introduced in (Bonev 2002; Liu and Bonev 2008), where only two angular parameters, referred to as the *azimuth* and *tilting* angles in this chapter, are involved. Actually, the *azimuth* angle, denoted as ϕ, defines any a-axis passing through the original point o; and the *tilting* angle, denoted as θ, describes the swing angle about the axis. If we use the two angles to express the orientation of a [PP]S mechanism, will the three points P_1,

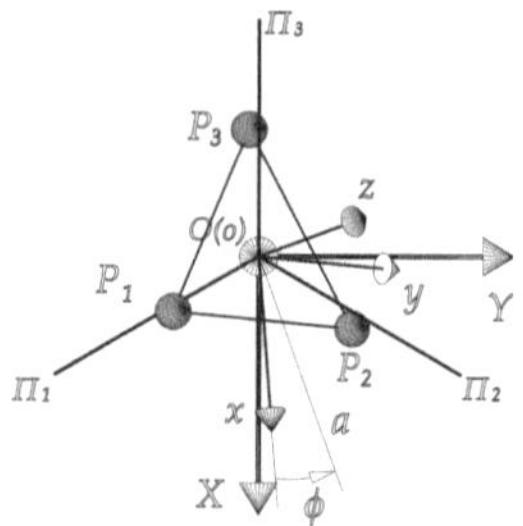

Fig. 3.18 Configuration when the mobile platform rotates an angle θ about an arbitrary axis a when the points P_i are released from their restricting plane

P_2, and P_3 still stay in their restricting planes by only translating the platform along X and Y axes? In other words, after the mobile platform rotates an angle about an arbitrary a-axis in the mobile platform plane, is there a parasitic rotation about the Z-axis, together with necessary translations along X and Y axes, to make sure that the three points are restricted in their planes where they should be?

As shown in Fig. 3.18, supposing the mobile platform rotates an angle θ about the a-axis, the rotation matrix of $\boldsymbol{R}_a(\theta)$ can be written as

$$\boldsymbol{R}_a(\theta) = \begin{bmatrix} \cos^2\phi\,(1-\cos\theta)+\cos\theta & \sin\phi\cos\phi\,(1-\cos\theta) & \sin\phi\sin\theta \\ \sin\phi\cos\phi\,(1-\cos\theta) & \sin^2\phi\,(1-\cos\theta)+\cos\theta & -\cos\phi\sin\theta \\ -\sin\phi\sin\theta & \cos\phi\sin\theta & \cos\theta \end{bmatrix} \tag{3.114}$$

where the angle ϕ, which is from the x-axis (NOT X-axis) to the a-axis, is used to define the a-axis that passes through the original point o. Since the line defined by ϕ is the same as that defined by $180° + \phi$, here, we are supposing $-90° \le \phi \le 90°$.

In the moving frame o-xyz, vectors $\boldsymbol{p}'_i$ ($i = 1,2,3$) that are defined as the position vectors of points P_i can be written as

$$\boldsymbol{p}'_i = \begin{bmatrix} r\cos\psi_i & r\sin\psi_i & 0 \end{bmatrix}^{\mathrm{T}}, \quad i = 1,2,3 \tag{3.115}$$

in which $\psi_i = (2i-3)\pi/3$ and r is the radius of the mobile platform, i.e., the distance from points P_i to the original point o. Then, by releasing the points P_i from their restricting planes, after the rotation of angle θ about an arbitrary axis a, the position vectors $\boldsymbol{p}_i$ of points P_i in the reference frame O-XYZ can be expressed as

$$\boldsymbol{p}_i = \begin{bmatrix} P_{ix} & P_{iy} & P_{iz} \end{bmatrix}^{\mathrm{T}} = \boldsymbol{R}_a(\theta)\,\boldsymbol{p}'_i. \tag{3.116}$$

Then, there are

$$\begin{aligned} P_{1x} &= \tfrac{1}{2}r\left[\cos^2\phi\,(1-\cos\theta)+\cos\theta\right] - \tfrac{\sqrt{3}}{2}r\sin\phi\cos\phi\,(1-\cos\theta) \\ P_{1y} &= \tfrac{1}{2}r\sin\phi\cos\phi\,(1-\cos\theta) - \tfrac{\sqrt{3}}{2}r\left[\sin^2\phi\,(1-\cos\theta)+\cos\theta\right] \\ P_{1z} &= -\tfrac{1}{2}r\sin\phi\sin\theta - \tfrac{\sqrt{3}}{2}r\cos\phi\sin\theta \end{aligned}$$

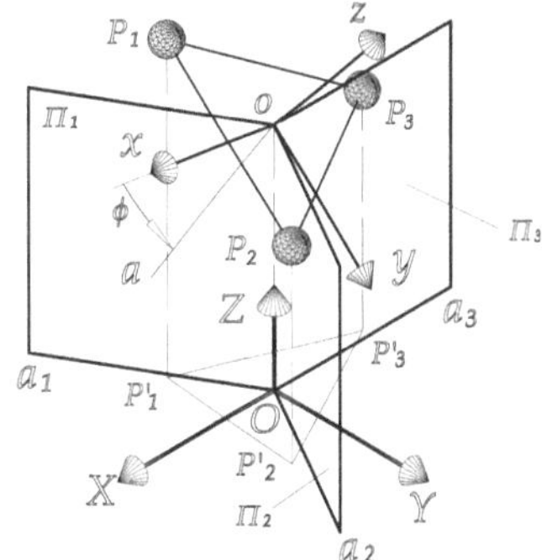

Fig. 3.19 Configuration of the mobile platform after the rotation and its project in the *O-XY* plane

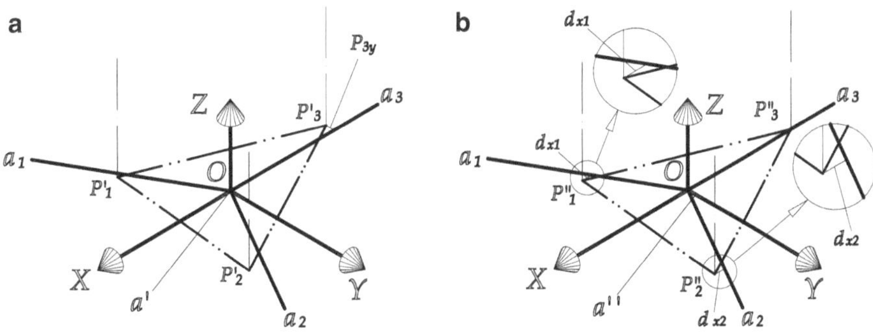

Fig. 3.20 Project of the mobile platform in the *O-XY* plane and its offsets along *X* and *Y* axes

$$
\begin{aligned}
P_{2x} &= \tfrac{1}{2}r\left[\cos^2\phi\,(1-\cos\theta)+\cos\theta\right]+\tfrac{\sqrt{3}}{2}r\sin\phi\cos\phi\,(1-\cos\theta)\\
P_{2y} &= \tfrac{1}{2}r\sin\phi\cos\phi\,(1-\cos\theta)+\tfrac{\sqrt{3}}{2}r\left[\sin^2\phi\,(1-\cos\theta)+\cos\theta\right]\\
P_{2z} &= -\tfrac{1}{2}r\sin\phi\sin\theta+\tfrac{\sqrt{3}}{2}r\cos\phi\sin\theta\\
P_{3x} &= -r\left[\cos^2\phi\,(1-\cos\theta)+\cos\theta\right]\\
P_{3y} &= -r\sin\phi\cos\phi\,(1-\cos\theta)\\
P_{3z} &= r\sin\phi\sin\theta.
\end{aligned}
$$

Therefore, after the rotation, points P_i may move out from their restricting planes Π_i. If we want move them back where they should stay, firstly, we should move the mobile platform with $-P_{3y}$ along the Y-axis, followed by the displacement dx along the X-axis.

As shown in Fig. 3.19, after the rotation about the arbitrary a-axis, the projects of three points P_i in the O-XY plane are denoted as P_i'. We refer to the intersection lines between planes Π_i and the O-XY plane as a_i.

Return to the O-XY plane as shown in Fig. 3.20a, the offset of the point P_3 (or P_3') from the X-axis is actually P_{3y}. Moving the mobile platform with the displacement $-P_{3y}$ along Y-axis leads to the new positions P_i'' of the three points P'_i (see Fig. 3.20b). As a result, the point P_3 is back its restricting plane Π_3. But, the points P_1 and P_2 are still outside their restricting planes Π_1 and Π_2, respectively.

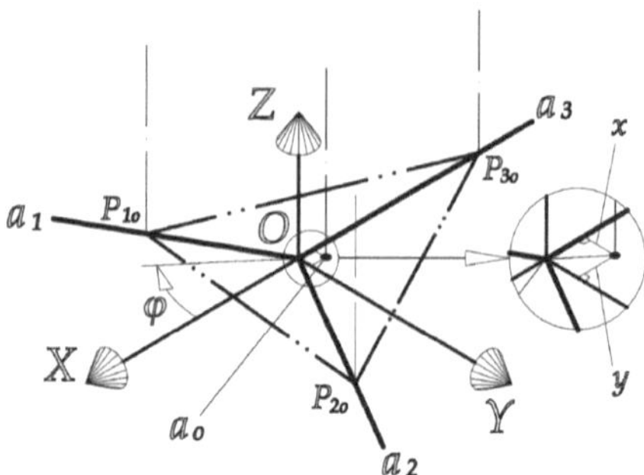

Fig. 3.21 Parasitic motions of the mobile platform along X and Y axes. Points P_{io} are the projecting points in the O-XY plane of points P_i when the mobile platform rotate with respect to the *azimuth* and *tilt* angles of a [PPS] mechanism. The a_o-axis is the project of the a-axis in the O-XY plane.

To put them in the planes possibly and prevent the point P_3 being out the plane Π_3 again, what we may do is just translating the mobile platform along the X-axis. By a displacement dx, will both points P_1 and P_2 be back in their restricting plane simultaneously?

We here denote the displacements of points P_1'' and P_2'' from the lines a_1 and a_2, respectively, along the X-axis as d_{x1} and d_{x2}. It is not difficult to write d_{x1} and d_{x2} as follows:

$$d_{x1} = -\frac{\sqrt{3}}{3}\left(P_{1y} - P_{3y}\right) - P_{1x} \tag{3.117}$$

$$d_{x2} = \frac{\sqrt{3}}{3}\left(P_{2y} - P_{3y}\right) - P_{2x}. \tag{3.118}$$

Then, by substituting corresponding elements of Eq. (3.116) into Eqs. (3.117) and (3.118), we have

$$d_{x1} = \frac{1}{2}r\left(\sin^2\phi - \cos^2\phi\right)(1 - \cos\theta) \tag{3.119}$$

$$d_{x2} = \frac{1}{2}r\left(\sin^2\phi - \cos^2\phi\right)(1 - \cos\theta) \tag{3.120}$$

which indicate obviously that $d_{x1} = d_{x2}$. That means, by translating the mobile platform with the displacement $d_x = d_{x1} = d_{x2}$ along the X-axis, we may definitely let points P_1 and P_2 be back in their restricting plane simultaneously.

This process proves that two angles like ϕ and θ, which are referred to as *azimuth* and *tilting* angles (Bonev 2002), can sure represent the orientation of the mobile platform of a [PP]S mechanism. Unfortunately, there are parasitic motions along X and Y axes, as shown in Fig. 3.21. They are

$$x = d_x = -\frac{1}{2}r\cos 2\phi\,(1 - \cos\theta) \tag{3.121}$$

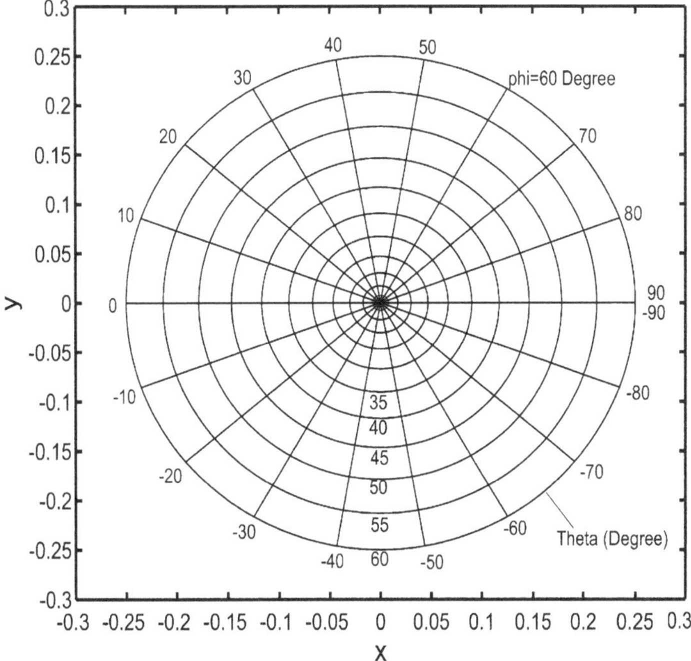

Fig. 3.22 Horizontal offset as function of orientation for a [PP]S parallel mechanism

$$y = -P_{3y} = \frac{1}{2} r \sin 2\phi \, (1 - \cos\theta) \tag{3.122}$$

which indicate that the center of the mobile platform does not always lie on the central Z-axis. It may move away from the Z-axis. The offset, denoted as ν, can be written as

$$\nu = \sqrt{x^2 + y^2} = \frac{1}{2} (1 - \cos\theta) \tag{3.123}$$

with the direction defined by φ

$$\varphi = \tan^{-1}\left(\frac{y}{x}\right) = \begin{cases} \pi - 2\phi, & \pi/2 \geq \phi \geq \pi/4 \\ -2\phi, & \pi/4 > \phi > -\pi/4 \\ -\pi - 2\phi, & -\pi/4 \geq \phi \geq -\pi/2 \end{cases} \tag{3.124}$$

To understand more clearly the coupling between position and orientation, we have shown in Fig. 3.22 the curves for x and y for constant ϕ (phi) or θ (theta).

Now, we may conclude that, by using only five (not six) parameters x, y, z, the *azimuth* angle ϕ, and the *tilting* angle θ, we may completely represent the pose of a [PP]S mechanism. Thanks to the identification of the exact nature of

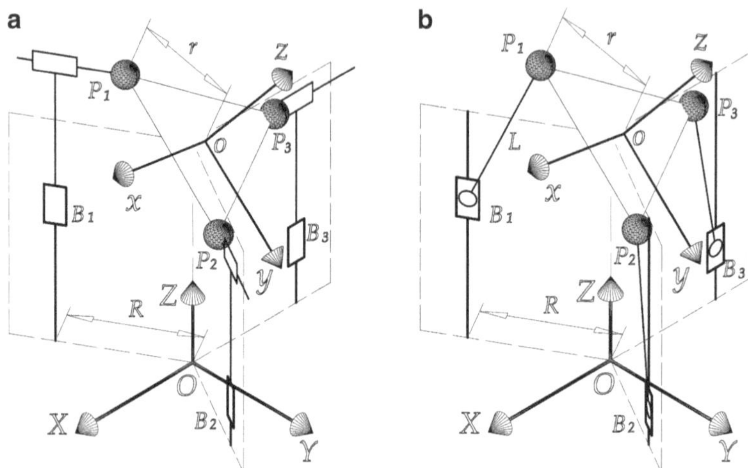

Fig. 3.23 Schematics of the two [PP]S parallel mechanisms: (**a**) with 3-$\underline{P}_V P_H$S parallel kinematics (**b**) with 3-$\underline{P}_V$RS parallel kinematics

the interdependence of the orientation parameters and its geometric significance by using the Tilt-and-Torsion (T&T) angles by Bonev (2002), the computation and representation of the orientation workspace of the [PP]S mechanism will be very simple.

3.9.2 Kinematics of the Two [PP]S Parallel Mechanisms

As an example, the inverse kinematics of two [PP]S parallel mechanisms, such as 3-$\underline{P}_V P_H$S and 3-$\underline{P}_V$RS (P, R, and S standing for prismatic, revolute, and spherical joints, respectively; the subscripts V and H indicating that the direction of the P joint is vertical or horizontal; and the joint symbol with underline means the joint is active) parallel mechanisms, will be considered here. As shown in Fig. 3.23a, the 3-$\underline{P}_V P_H$S parallel mechanism comprises three identical legs, each consisting of one spherical joint and two prismatic joints. In each leg, the prismatic joint attached to the base is vertical and actuated, while the other prismatic joint is horizontal and passive. The 3-$\underline{P}_V$RS (R standing for revolute joint) parallel mechanism shown in Fig. 3.23b comprises three identical legs, each consisting of a spherical joint, a revolute joint, and a prismatic joint. In each leg, the prismatic joint is vertical and actuated. Finally, in each design, the three legs remain in three vertical planes, called the restricting planes, which intersect at a common line at an angle of 120°, and the centers of the spherical joints form an equilateral triangle.

Referring again to Fig. 3.23, let R be the distance from point O to the actuated vertical prismatic joint; r is the radius of the mobile platform, i.e., $r = |oP_i|$

($i = 1, 2, 3$); and L is the leg length of the 3-P$_V$RS mechanism. If the pose $(\phi,\ \theta,\ z)$ of the mobile platform is given, the position vector of each point P_i $(i = 1, 2, 3)$ in the base frame is defined as

$$\boldsymbol{p}_{i\,\Re} = \begin{bmatrix} x_{i\,\Re} & y_{i\,\Re} & z_{i\,\Re} \end{bmatrix}^{\mathrm{T}} = \boldsymbol{R}_a(\theta)\,\boldsymbol{p}'_i + \boldsymbol{c}_{\Re}, \quad i = 1, 2, 3 \tag{3.125}$$

where $\boldsymbol{R}_a(\theta)$ is the rotation matrix defined in Eq. (3.112); $\boldsymbol{p}'_i$ is the position vector of P_i in the mobile frame as expressed in Eq. (3.115); $\boldsymbol{c}_{\Re} = \begin{bmatrix} x & y & z \end{bmatrix}^{\mathrm{T}}$ is the position vector of the platform center o expressed in the base frame; and x and y are the parasitic motions given by Eqs. (3.121) and (3.122).

For the 3-$\underline{\mathrm{P}}_V$P$_H$S mechanism, supposing that the inputs of the legs are ρ_{i_PPS} $(i = 1, 2, 3)$, the inverse kinematic problem of the mechanism can be written as

$$\rho_{i\_\mathrm{PPS}} = z_{i\,\Re} \tag{3.126}$$

where $z_{i\,\Re}$ can be obtained from Eq. (3.125) when the pose $(\phi,\ \theta,\ z)$ of the mobile platform is given. Then, we have

$$\rho_{1\_\mathrm{PPS}} = -\frac{1}{2}r\sin\theta\left(\sin\phi + \sqrt{3}\cos\phi\right) + z \tag{3.127}$$

$$\rho_{2\_\mathrm{PPS}} = -\frac{1}{2}r\sin\theta\left(\sin\phi - \sqrt{3}\cos\phi\right) + z \tag{3.128}$$

$$\rho_{3\_\mathrm{PPS}} = r\sin\phi\sin\theta + z. \tag{3.129}$$

We may see that there is only one geometric parameter, i.e., r, in the kinematics of the mechanism. What is more, the inverse kinematic solution is unique.

For the 3-$\underline{\mathrm{P}}_V$RS mechanism, vectors $\boldsymbol{b}_{i\,\Re}$ $(i = 1, 2, 3)$, which are defined as the position vectors of points B_i, can be expressed in the base frame as

$$\boldsymbol{b}_{i\,\Re} = \begin{bmatrix} R\cos\psi_i & R\sin\psi_i & \rho_{i\_\mathrm{PRS}} \end{bmatrix}^{\mathrm{T}}, \quad i = 1, 2, 3 \tag{3.130}$$

where ρ_{i_PRS} constitutes the Z-coordinates of points B_i. The inverse kinematic problem of the mechanism can be solved by writing the following equations:

$$\left|\boldsymbol{p}_{i\Re}\boldsymbol{b}_{i\Re}\right| = L. \tag{3.131}$$

Substituting Eqs. (3.125) and (3.130) into Eq. (3.131), we have

$$\rho_{i\_\mathrm{PRS}} = \frac{1}{2}\left(-D_i \pm \sqrt{D_i^2 - 4F_i}\right) \tag{3.132}$$

in which

$$
\begin{aligned}
D_1 &= r\sin\theta\left(\sin\phi+\sqrt{3}\cos\phi\right)-2z,\\
F_1 &= \left[r\left(1-\cos\theta\right)\sin\phi\left(\sin\phi-\sqrt{3}\cos\phi\right)+r\cos\theta-R\right]^2\\
&\quad+\left[z-\frac{1}{2}r\sin\theta\left(\sin\phi+\sqrt{3}\cos\phi\right)\right]^2-L^2,\\
D_2 &= r\sin\theta\left(\sin\phi-\sqrt{3}\cos\phi\right)-2z,\\
F_20 &= \left[r\left(1-\cos\theta\right)\sin\phi\left(\sin\phi+\sqrt{3}\cos\phi\right)+r\cos\theta-R\right]^2\\
&\quad+\left[z-\frac{1}{2}r\sin\theta\left(\sin\phi-\sqrt{3}\cos\phi\right)\right]^2-L^2,\\
D_3 &= -2\left(r\sin\phi\sin\theta+z\right),\\
F_3 &= \left[R-\tfrac{1}{2}r\left(1-\cos\theta\right)\left(3-4\sin^2\phi\right)-r\cos\theta\right]^2+\left(r\sin\phi\sin\theta+z\right)^2-L^2.
\end{aligned}
$$

From Eq. (3.132), we may see that, for a given pose $(\phi,\ \theta,\ z)$, there are at most eight inverse kinematic solutions, corresponding to eight working modes of the mechanism. We are concerned here with the working mode shown in Fig. 3.23b, which can be reached when the "±" sign in Eq. (3.132) is "−" for all three legs.

3.10 Position Analysis of a Simplified 6-RUS Parallel Mechanism

The spatial in-parallel actuated 6-RUS (or 6-RSS) mechanisms have attracted many researchers' attention in the past decade (Chi 1999; Takeda et al. 1997; Takeda 2000; Uchiyama 2001; Anco Engineers 2000; Servos & Simulation 2000; Bonev and Gosselin 2000). The mechanism can be applied to the fields of machine tools (Takeda 2000), 6-axis positioners (Chi 1999), and motion simulators (Servos &Simulation 2000). A simplified 6-RUS parallel mechanism discussed in this chapter is shown in Fig. 3.24a, which consists of two rigid bodies connected by six legs OB_iP_i $(i = 1, 2, \ldots, 6)$. The stationary rigid body is referred to as the base, and the moving rigid body is referred to as the mobile platform. The center of the universal joint connecting the ith leg to the link linking to the base will be denoted as B_i, whereas the center of the spherical joint connecting the same leg to the platform will be denoted as P_i. Each leg of the mechanism connects to the base through a revolute joint, whose center is denoted as O. Vectors $\boldsymbol{p}_i$ $(i = 1, \ldots, 6)$ will be defined as the position vectors of the platform joints, and vectors $\boldsymbol{b}_i$ $(i = 1, \ldots, 6)$ will be defined as the position vectors of the universal joints. The geometric parameters of the mechanism are $O'P_i = R_3$, $P_iB_i = R_2$, $OB_i = R_1$ and the angle ϕ as shown in Fig. 3.24b.

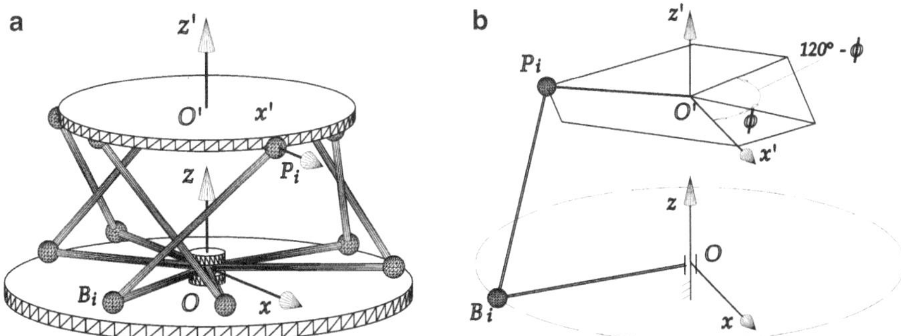

Fig. 3.24 Geometric architecture of the 6-DOF 6-RUS parallel mechanism

Let us consider a fixed coordinate frame $\boldsymbol{R}$: O-xyz attached to the base of the mechanism and a mobile coordinate frame $\boldsymbol{R}'$: O'-$x'y'z'$ attached to the platform, where O' is the point to be positioned by the mechanism and the position of point O' with respect to the fixed frame $\boldsymbol{R}$ is denoted as vector

$$\boldsymbol{r}_R = \left[x,\ y,\ z\right]^{\mathrm{T}} \tag{3.133}$$

where x, y, and z are non-dimensional variables, respectively. Furthermore, let $\boldsymbol{Q}$ be the rotation matrix describing the orientation of $\boldsymbol{R}'$ with respect to $\boldsymbol{R}$. The matrix can be expressed by three angles noted as ϕ_1, ϕ_2, and ϕ_3, respectively:

$$\boldsymbol{Q} = \begin{bmatrix} q_{11} & q_{12} & q_{13} \\ q_{21} & q_{22} & q_{23} \\ q_{31} & q_{32} & q_{33} \end{bmatrix} \tag{3.134}$$

where

$$\begin{aligned}
q_{11} &= -\sin\phi_1 \sin\phi_3 + \cos\phi_1 \cos\phi_3 \cos\phi_2 \\
q_{12} &= -\sin\phi_1 \cos\phi_3 - \cos\phi_1 \sin\phi_3 \cos\phi_2 \\
q_{13} &= \cos\phi_1 \sin\phi_2 \\
q_{21} &= \cos\phi_1 \sin\phi_3 + \sin\phi_1 \cos\phi_3 \cos\phi_2 \\
q_{22} &= \cos\phi_1 \cos\phi_3 - \sin\phi_1 \sin\phi_3 \cos\phi_2 \\
q_{23} &= \sin\phi_1 \sin\phi_2 \\
q_{31} &= -\cos\phi_3 \sin\phi_2 \\
q_{32} &= \sin\phi_3 \sin\phi_2 \\
q_{33} &= \cos\phi_2.
\end{aligned}$$

The vectors $\boldsymbol{p}_i$ in the fixed frame $\boldsymbol{R}$ can be written as

$$\boldsymbol{p}_{iR} = \boldsymbol{r}_R + \boldsymbol{Q}\,\boldsymbol{p}_{iR'} \quad i = 1, \ldots, 6 \tag{3.135}$$

where

$$\boldsymbol{p}_{iR'} = \begin{bmatrix} r_3 \cos \eta_i \\ r_3 \sin \eta_i \\ 0 \end{bmatrix} \quad i = 1, \ldots, 6 \tag{3.136}$$

and

$$\eta = \begin{bmatrix} \eta_1 \\ \eta_2 \\ \eta_3 \\ \eta_4 \\ \eta_5 \\ \eta_6 \end{bmatrix} = \begin{bmatrix} 0 \\ \phi \\ 2\pi/3 \\ 2\pi/3 + \phi \\ -2\pi/3 \\ \phi - 2\pi/3 \end{bmatrix}. \tag{3.137}$$

As shown in Fig. 3.24, the axes of the six revolute joints are aligned with the y-axis. Therefore, the vectors $\boldsymbol{b}_i$ in the fixed frame $\boldsymbol{R}$ can be written as

$$\boldsymbol{b}_{iR} = \begin{bmatrix} r_1 \cos \theta_i \\ r_1 \sin \theta_i \\ 0 \end{bmatrix} \quad i = 1, \ldots, 6 \tag{3.138}$$

where θ_i is the actuated angle of the ith leg. Then the inverse kinematics problem of the mechanisms can be solved by writing the following constraint equation:

$$\| \boldsymbol{p}_{iR} - \boldsymbol{b}_{iR} \| = r_2 \quad i = 1, \ldots, 6 \tag{3.139}$$

that is,

$$\| \boldsymbol{r}_R + \boldsymbol{Q}\, \boldsymbol{p}_{iR'} - \boldsymbol{b}_{iR} \| = r_2 \quad i = 1, \ldots, 6 \tag{3.140}$$

from which we may obtain

$$\theta_i = 2 \tan^{-1}(t_i) \tag{3.141}$$

where

$$t_i = \frac{-v_i \pm \sqrt{v_i^2 - 4u_i w_i}}{2u_i} \tag{3.142}$$

and

$$\begin{aligned} u_i &= r_1^2 + a_i^2 + b_i^2 + c_i^2 - r_2^2 + 2r_1 a_i \\ v_i &= -4r_1 b_i \end{aligned}$$

$$
\begin{aligned}
w_i &= r_1^2 + a_i^2 + b_i^2 + c_i^2 - r_2^2 - 2r_1a_i \\
a_i &= q_{11}r_3 \cos \eta_i + q_{12}r_3 \sin \eta_i + x \\
b_i &= q_{21}r_3 \cos \eta_i + q_{22}r_3 \sin \eta_i + y \\
c_i &= q_{31}r_3 \cos \eta_i + q_{32}r_3 \sin \eta_i + z.
\end{aligned}
$$

Hence, for a given mechanism and for prescribed values of the position and orientation of the platform, the required actuated angles θ_i can be directly computed from Eq. (3.141). From Eq. (3.142), we may see that there are 2^6 branch sets for the mechanism under study.

References

Anco Engineers (2000) http://www.ancoengineers.com/Motion1.html. Accessed 20 Dec 2012

Bonev IA (2002) Geometric analysis of parallel mechanisms. Ph.D. thesis, Laval University, Quebec

Bonev IA, Gosselin CM (2000) A geometric algorithm for the computation of the constant-orientation workspace of 6-RUS parallel manipulators. In: Proceedings of the ASME design engineering technical conferences and computers and information in engineering conference, Baltimore, MD, DETC2000/MECH-14106, ASME press, New York

Chi YL (1999) Systems and methods employing a rotary track for machining and manufacturing. WIPO Patent, No. WO99/38646

Clavel R (1988) Delta: a fast robot with parallel geometry. In: Proceedings of the 18th international symposium on industrial robot, 26–28 April 1988, Lausanne, IFS Publications, Kempston, Bedford, UK, pp 91–100

Denavit J, Hartenberg RS (1955) A kinematic notation for lower pair mechanisms based on matrices. ASME J Appl Mech 22:215–221

Gosselin CM, Sefrioui J, Richard MJ (1994) On the direct kinematics of spherical three-degree-of-freedom parallel manipulators of general architecture. ASME J Mech Des 116(2):594–598

Huang Z, Yao YL (1999) A new closed-form kinematics of the generalized 3-dof spherical parallel manipulator. Robotica 17(5):475–485

Liu X-J, Bonev IA (2008) Orientation capability, error analysis, and dimensional optimization of two articulated tool heads with parallel kinematics. ASME J Manuf Sci Eng 130(1), Article Number: 011015

Liu X-J, Wang J, Pritschow G (2006) On the optimal kinematic design of the PRRRP 2-DOF parallel mechanism. Mech Mach Theory 41:1111–1130

Servos & Simulation (2000) http://www.servos.com/products/motion-platforms.html. Accessed 20 Dec 2012

Söylemez E (2002) Classical transmission-angle problem for slider-crank mechanisms. Mech Mach Theory 37:419–425

Takeda Y (2000) http://www.mech.titech.ac.jp/~mech0000_msd/en/?page=themes. Accessed 1 Feb 2013

Takeda Y, Funabashi H, Ichimaru H (1997) Development of spatial in-parallel actuated manipulators with six degrees of freedom with high motion transmissibility. JSME Int J Series C 40(2):299–308

Uchiyama M (2001) http://www.space.mech.tohoku.ac.jp/research/hexa/hexa.jpg. Accessed 20 Dec 2012

Chapter 4
Velocity and Jacobian Analysis of Parallel Mechanisms

Abstract This chapter extends the study of parallel mechanisms from the position analysis to the velocity analysis. Each velocity equation is firstly obtained by differentiating the position equation with respect to time, and Jacobian matrix respecting the relationship between the input and output is further established. Several corresponding examples are illustrated.

Keywords Velocity analysis • Jacobian matrix • Parallel mechanism

In the field of manipulator and complex mechanism, Jacobian matrix is an important parameter. It is usually used to define the performance index. The Jacobian matrix is defined as the matrix map of the velocity of the end-effector into the vector of actuated joint rates. Generally, position equation can be differentiated with respect to time to obtain the velocity equations. In the following subsections, we will introduce several examples.

4.1 The Planar 5R Parallel Mechanism

For the planar 5R parallel mechanism shown in Fig. 3.1, Eqs. (3.4) and (3.5) can be differentiated with respect to time to obtain the velocity equations, which yields

$$r_1\left[y\cos\theta_1-(x+r_3)\sin\theta_1\right]\dot{\theta}_1=(x+r_3-r_1\cos\theta_1)\,\dot{x}+(y-r_1\sin\theta_1)\,\dot{y} \tag{4.1}$$

$$r_1\left[y\cos\theta_2+(r_3-x)\sin\theta_2\right]\dot{\theta}_2=(x-r_3-r_1\cos\theta_2)\,\dot{x}+(y-r_1\sin\theta_2)\,\dot{y}. \tag{4.2}$$

X.-J. Liu and J. Wang, *Parallel Kinematics: Type, Kinematics, and Optimal Design*, Springer Tracts in Mechanical Engineering, DOI 10.1007/978-3-642-36929-2_4,

Rearranging Eqs. (4.1) and (4.2) leads to an equation of the form

$$\boldsymbol{A}\dot{\boldsymbol{\theta}} = \boldsymbol{B}\,\dot{\boldsymbol{p}} \tag{4.3}$$

where $\dot{\boldsymbol{p}}$ is the vector of output velocities defined as

$$\dot{\boldsymbol{p}} = \begin{pmatrix} \dot{y} & \dot{z} \end{pmatrix}^{\mathrm{T}} \tag{4.4}$$

and $\dot{\boldsymbol{\theta}}$ is the vector of input velocities defined as

$$\dot{\boldsymbol{\theta}} = \begin{pmatrix} \dot{\theta}_1 & \dot{\theta}_2 \end{pmatrix}^{\mathrm{T}}. \tag{4.5}$$

Matrices $\boldsymbol{A}$ and $\boldsymbol{B}$ are, respectively, the 2×2 matrices of the mechanism and can be expressed as

$$\boldsymbol{A} = \begin{bmatrix} y\cos\theta_1 - (x + r_3)\sin\theta_1 & 0 \\ 0 & y\cos\theta_2 + (r_3 - x)\sin\theta_2 \end{bmatrix} r_1 \tag{4.6}$$

$$\boldsymbol{B} = \begin{bmatrix} x + r_3 - r_1\cos\theta_1 & y - r_1\sin\theta_1 \\ x - r_3 - r_1\cos\theta_2 & y - r_1\sin\theta_2 \end{bmatrix}. \tag{4.7}$$

The Jacobian matrix of the mechanism can be written as

$$\boldsymbol{J} = \boldsymbol{A}^{-1}\boldsymbol{B} \tag{4.8}$$

From Eqs. (4.6) and (4.7), one may see that the elements in the Jacobian matrix of the 5R parallel mechanism have the same dimension.

4.2 The Translational RRR&PRRR Parallel Mechanism

Equations (3.28) and (3.31) being differentiated with respect to time lead to

$$\dot{s} = \dot{x} \tag{4.9}$$

$$R_1\left[y\cos\theta - (x - R_3)\sin\theta\right]\dot{\theta} = (x - R_3 - R_1\cos\theta)\,\dot{x} + (y - R_1\sin\theta)\,\dot{y} \tag{4.10}$$

which can be written in an equation of the form

$$\boldsymbol{A}\dot{\boldsymbol{q}} = \boldsymbol{B}\,\dot{\boldsymbol{p}}. \tag{4.11}$$

where $\dot{\boldsymbol{q}} = (\dot{s} \ \ \dot{\theta})^{\mathrm{T}}$ and $\dot{\boldsymbol{p}} = (\dot{x} \ \ \dot{y})^{\mathrm{T}}$ are the joint and Cartesian space velocity vectors, respectively, and $\boldsymbol{A}$ and $\boldsymbol{B}$ are, respectively, the 2×2 matrices and can be expressed as

$$\boldsymbol{A} = \begin{bmatrix} 1 & 0 \\ 0 & R_1 y \cos\theta - R_1 (x - R_3) \sin\theta \end{bmatrix} \text{ and } \boldsymbol{B} = \begin{bmatrix} 1 & 0 \\ x - R_3 - R_1 \cos\theta & y - R_1 \sin\theta \end{bmatrix}. \tag{4.12}$$

If matrix $\boldsymbol{A}$ is nonsingular, the Jacobian matrix of the mechanism can be obtained as

$$\boldsymbol{J} = \boldsymbol{A}^{-1}\boldsymbol{B} = \begin{bmatrix} 1 & 0 \\ \dfrac{x - R_3 - R_1 \cos\theta}{R_1 y \cos\theta - R_1 (x - R_3) \sin\theta} & \dfrac{y - R_1 \sin\theta}{R_1 y \cos\theta - R_1 (x - R_3) \sin\theta} \end{bmatrix} \tag{4.13}$$

from which one may see that there is no any parameter of L_n $(n = 1, 2, 3)$ in this matrix and the dimension of the elements is different.

4.3 A Parallel Cube-Manipulator

Equations (3.59), (3.60), and (3.61) can be differentiated with respect to time to obtain the velocity equations, which leads to

$$(x - \rho_2 - r)\dot{x} + y\dot{y} + z\dot{z} = (x - \rho_2 - r)\dot{\rho}_2$$

$$x\dot{x} + (y - \rho_3 - r)\dot{y} + z\dot{z} = (y - \rho_3 - r)\dot{\rho}_3$$

$$x\dot{x} + y\dot{y} + (z - \rho_1 - r)\dot{z} = (z - \rho_1 - r)\dot{\rho}_1.$$

Rearranging above three equations leads to an equation of the form

$$\boldsymbol{A}\dot{\boldsymbol{\rho}} = \boldsymbol{B}\dot{\boldsymbol{p}} \tag{4.14}$$

where $\dot{\boldsymbol{p}}$ is the vector of output velocities defined as

$$\dot{\boldsymbol{p}} = [\ \dot{x} \ \ \dot{y} \ \ \dot{z}\ ]^{\mathrm{T}} \tag{4.15}$$

and $\dot{\boldsymbol{\rho}}$ is the vector of input velocities defined as

$$\dot{\boldsymbol{\rho}} = [\ \dot{\rho}_2 \ \ \dot{\rho}_3 \ \ \dot{\rho}_1\ ]^{\mathrm{T}}. \tag{4.16}$$

Matrices $\boldsymbol{A}$ and $\boldsymbol{B}$ are, respectively, the 3×3 inverse and forward Jacobian matrices of the manipulator and can be expressed as

$$\boldsymbol{A} = \mathrm{diag}(a_{11},\ a_{22},\ a_{33}) \tag{4.17}$$

$$\boldsymbol{B} = \begin{bmatrix} a_{11} & y & z \\ x & a_{22} & z \\ x & y & a_{33} \end{bmatrix} \tag{4.18}$$

with $a_{11} = x - \rho_2 - r$, $a_{22} = y - \rho_3 - r$, and $a_{33} = z - \rho_1 - r$. And if the matrix $\boldsymbol{A}$ is nonsingular, Eq. (4.14) can be rewritten as

$$\dot{\boldsymbol{\rho}} = \boldsymbol{J}\dot{\boldsymbol{p}} \tag{4.19}$$

where $\boldsymbol{J}$ is the Jacobian matrix of the manipulator,

$$\boldsymbol{J} = \boldsymbol{A}^{-1}\boldsymbol{B} = \begin{bmatrix} 1 & y/a_{11} & z/a_{11} \\ x/a_{22} & 1 & z/a_{22} \\ x/a_{33} & y/a_{33} & 1 \end{bmatrix}. \tag{4.20}$$

Since both inputs and outputs of the manipulator are linear, the elements in the Jacobian matrix have the same dimension and are non-dimensional.

4.4 The 2-PRU&1-PR(Pa)R Parallel Mechanism

Equation (3.79) being differentiated with respect to time yields

$$(z - z_1)\,\dot{z}_1 = (-r + y + R)\,\dot{y} + (z - z_1)\,\dot{z} \tag{4.21}$$

$$(z - z_2)\,\dot{z}_2 = (r + y - R)\,\dot{y} + (z - z_2)\,\dot{z} \tag{4.22}$$

$$(r\sin\phi + z - z_3)\,\dot{z}_3 = y\dot{y} + (r\sin\phi + z - z_3)\,\dot{z} + [Rr\sin\phi + r\,(z - z_3)\cos\phi]\,\dot{\phi}. \tag{4.23}$$

Rearranging Eqs. (4.21), (4.22), and (4.23) leads to an equation of the form

$$\boldsymbol{A}\dot{\boldsymbol{\rho}} = \boldsymbol{B}\dot{\boldsymbol{p}} \tag{4.24}$$

where $\dot{\boldsymbol{p}}$ is the vector of output velocities defined as

$$\dot{\boldsymbol{p}} = \begin{pmatrix} \dot{y} & \dot{z} & \dot{\phi} \end{pmatrix}^{\mathrm{T}} \tag{4.25}$$

and $\dot{\boldsymbol{\rho}}$ is the vector of input velocities defined as

$$\dot{\boldsymbol{\rho}} = \begin{pmatrix} \dot{z}_1 & \dot{z}_2 & \dot{z}_3 \end{pmatrix}^{\mathrm{T}}. \tag{4.26}$$

Matrices $\boldsymbol{A}$ and $\boldsymbol{B}$ are, respectively, the 3×3 inverse and forward Jacobian matrices of the manipulator and can be expressed as

$$\boldsymbol{A} = \begin{bmatrix} z - z_1 & 0 & 0 \\ 0 & z - z_2 & 0 \\ 0 & 0 & r\sin\phi + z - z_3 \end{bmatrix} \tag{4.27}$$

$$\boldsymbol{B} = \begin{bmatrix} -r + y + R & z - z_1 & 0 \\ r + y - R & z - z_2 & 0 \\ y & r\sin\phi + z - z_3 & Rr\sin\phi + r(z - z_3)\cos\phi \end{bmatrix} \tag{4.28}$$

from which one may see that the elements in the Jacobian matrix are different in dimension.

4.5 The 2-RRU&1-RR(Pa)R Parallel Mechanism

Equations (3.92), (3.93), and (3.94) can be differentiated with respect to time to obtain the velocity equations, which leads to

$$\begin{aligned} L_b\left[z\cos\theta_1 - (r + y - R)\sin\theta_1\right]\dot{\theta}_1 &= (r + y - R - L_b\cos\theta_1)\,\dot{y} \\ &\quad + (z - L_b\sin\theta_1)\,\dot{z} \end{aligned} \tag{4.29}$$

$$\begin{aligned} L_b\left[z\cos\theta_2 - (R + y - r)\sin\theta_2\right]\dot{\theta}_2 &= (R + y - r - L_b\cos\theta_2)\,\dot{y} \\ &\quad + (z - L_b\sin\theta_2)\,\dot{z} \end{aligned} \tag{4.30}$$

$$\begin{aligned} L_b\left[(z{+}r\sin\phi)\cos\theta_3{+}\,(r\cos\phi{-}R)\sin\theta_3\right]\dot{\theta}_3 &= y\,\dot{y}{+}\,(r\sin\phi{+}z{-}L_b\sin\theta_3)\,\dot{z} \\ + r\left[(z - L_b\sin\theta_3)\cos\phi + (R - L_b\cos\theta_3)\sin\phi\right]\dot{\phi}. \end{aligned} \tag{4.31}$$

Rearranging Eqs. (4.29), (4.30), and (4.31) leads to an equation of the form

$$\boldsymbol{A}\dot{\boldsymbol{\theta}} = \boldsymbol{B}\,\dot{\boldsymbol{p}} \tag{4.32}$$

where $\dot{\boldsymbol{p}}$ is the vector of output velocities defined as

$$\dot{\boldsymbol{p}} = \begin{pmatrix} \dot{y} & \dot{z} & \dot{\phi} \end{pmatrix}^{\mathrm{T}} \tag{4.33}$$

and $\dot{\boldsymbol{\theta}}$ is the vector of input velocities defined as

$$\dot{\boldsymbol{\theta}} = \left(\dot{\theta}_1 \quad \dot{\theta}_2 \quad \dot{\theta}_3 \right)^{\mathrm{T}}. \tag{4.34}$$

Matrices $\boldsymbol{A}$ and $\boldsymbol{B}$ are, respectively, the 3×3 inverse and forward Jacobian matrices of the manipulator and can be expressed as

$$\boldsymbol{A} = \operatorname{diag}(a_{11}, \quad a_{22}, \quad a_{33}) \tag{4.35}$$

$$\boldsymbol{B} = \begin{bmatrix} b_{11} & b_{12} & 0 \\ b_{21} & b_{22} & 0 \\ y & b_{32} & b_{33} \end{bmatrix} \tag{4.36}$$

in which

$$\begin{aligned}
a_{11} &= L_b \left[z \cos\theta_1 - (r + y - R) \sin\theta_1 \right] \\
a_{22} &= L_b \left[z \cos\theta_2 - (R + y - r) \sin\theta_2 \right] \\
a_{33} &= L_b \left[(z + r \sin\phi) \cos\theta_3 + (r \cos\phi - R) \sin\theta_3 \right] \\
b_{11} &= r + y - R - L_b \cos\theta_1 \\
b_{12} &= z - L_b \sin\theta_1 \\
b_{21} &= R + y - r - L_b \cos\theta_2 \\
b_{22} &= z - L_b \sin\theta_2 \\
b_{32} &= r \sin\phi + z - L_b \sin\theta_3 \\
b_{33} &= r \left[(z - L_b \sin\theta_3) \cos\phi + (R - L_b \cos\theta_3) \sin\phi \right].
\end{aligned}$$

The Jacobian matrix of the parallel manipulator can be written as

$$\boldsymbol{J} = \boldsymbol{A}^{-1} \boldsymbol{B} \tag{4.37}$$

in which the dimension of the element b_{33}/a_{33} is different from those of others.

4.6 The Simplified 6-RUS Parallel Mechanism

Differentiating Eq. (3.140) with respect to time and then rearranging terms lead to

$$d_i \dot{\theta}_i = e_i \dot{x} + f_i \dot{y} + g_i \dot{z} + l_i \dot{\phi}_1 + m_i \dot{\phi}_2 + n_i \dot{\phi}_3, \quad i = 1, \ldots, 6 \tag{4.38}$$

where

$$
\begin{aligned}
d_i &= r_1 b_i \cos\theta_i - r_1 a_i \sin\theta_i \\
e_i &= a_i - r_1 \cos\theta_i \\
f_i &= b_i - r_1 \sin\theta_i \\
g_i &= c_i \\
l_i &= (a_i - r_1\cos\theta_i)\,(r_3\cos\eta_i q_{11a} + r_3\sin\eta_i q_{12a}) \\
&\quad + (b_i - r_1\sin\theta_i)\,(r_3\cos\eta_i q_{21a} + r_3\sin\eta_i q_{22a}) \\
m_i &= (a_i - r_1\cos\theta_i)\,(r_3\cos\eta_i q_{11b} + r_3\sin\eta_i q_{12b}) \\
&\quad + (b_i - r_1\sin\theta_i)(r_3\cos\eta_i q_{21b} + r_3\sin\eta_i q_{22b}) + c_i\,(r_3\cos\eta_i q_{31b} + r_3\sin\eta_i q_{32b}) \\
n_i &= (a_i - r_1\cos\theta_i)\,(r_3\cos\eta_i q_{11c} + r_3\sin\eta_i q_{12c}) \\
&\quad + (b_i - r_1\sin\theta_i)(r_3\cos\eta_i q_{21c} + r_3\sin\eta_i q_{22c}) + c_i\,(r_3\cos\eta_i q_{31c} + r_3\sin\eta_i q_{32c})
\end{aligned}
$$

and

$$
\begin{aligned}
q_{11a} &= q_{22c} = -\cos\phi_1\sin\phi_3 - \sin\phi_1\cos\phi_3\cos\phi_2 \\
q_{11b} &= -\cos\phi_1\cos\phi_3\sin\phi_2 \\
q_{11c} &= q_{22a} = -\sin\phi_1\cos\phi_3 - \cos\phi_1\sin\phi_3\cos\phi_2 \\
q_{12a} &= \sin\phi_1\sin\phi_3\cos\phi_2 - \cos\phi_1\cos\phi_3 \\
q_{12b} &= \cos\phi_1\sin\phi_3\sin\phi_2 \\
q_{12c} &= \sin\phi_1\sin\phi_3 - \cos\phi_1\cos\phi_3\cos\phi_2 \\
q_{21a} &= -q_{12c} \\
q_{21b} &= -\sin\phi_1\cos\phi_3\sin\phi_2 \\
q_{21c} &= -q_{12a} \\
q_{22b} &= \sin\phi_1\sin\phi_3\sin\phi_2 \\
q_{31b} &= -\cos\phi_3\cos\phi_2 \\
q_{31c} &= \sin\phi_3\sin\phi_2 \\
q_{32b} &= \sin\phi_3\cos\phi_2 \\
q_{32c} &= \cos\phi_3\sin\phi_2.
\end{aligned}
$$

The kinematic equation of the manipulators can be written as

$$
\begin{bmatrix} \dot\theta_1 & \dot\theta_2 & \dot\theta_3 & \dot\theta_4 & \dot\theta_5 & \dot\theta_6 \end{bmatrix}^{\mathrm{T}} = \boldsymbol{J} \begin{bmatrix} \dot x & \dot y & \dot z & \dot\phi_1 & \dot\phi_2 & \dot\phi_3 \end{bmatrix}^{\mathrm{T}} \tag{4.39}
$$

where

$$
\boldsymbol{J} = \begin{bmatrix} \frac{e_1}{d_1} & \frac{f_1}{d_1} & \frac{g_1}{d_1} & \frac{l_1}{d_1} & \frac{m_1}{d_1} & \frac{n_1}{d_1} \\ \vdots & \vdots & \vdots & \vdots & \vdots & \vdots \\ \frac{e_6}{d_6} & \frac{f_6}{d_6} & \frac{g_6}{d_6} & \frac{l_6}{d_6} & \frac{m_6}{d_6} & \frac{n_6}{d_6} \end{bmatrix}_{6\times 6} \tag{4.40}
$$

is the Jacobian matrix of the simplified 6-RUS parallel manipulators, and in the matrix, the elements have different dimensions.

Chapter 5
Singularity of Parallel Mechanisms

Abstract This chapter covers the singularity analysis of parallel mechanisms. The singularities of parallel mechanisms are classified into three types. For the first type of singularity, the mechanism loses 1 or more DOFs, while for the second type of singularity, the mechanism gains 1 or more DOFs. The third type of singularity occurs when the first and second types of singularity occur simultaneously. The physical meanings of each type of singularity are illustrated by two parallel mechanisms including a planar 5R type and a 3-DOF type. What is more important, a new approach to singularity analysis of parallel mechanisms taking into account the motion and force transmissibility is presented based on screw theory.

Keywords Singularity analysis • Motion and force transmissibility • Screw theory • Parallel mechanism

There are several issues involved in the field of parallel manipulators. A most important one is singularity analysis. As is well known, a parallel manipulator will gain or lose one or more degrees of freedom (DOFs) at its singular configurations. In other words, a manipulator will be out of control when the singularity occurs. Thus, singular configurations should usually be avoided in the design and practical use of parallel manipulators.

Without entire identification of singularity, a machine will be dangerous, since the machine will be out of control in singular configuration. In the parallel manipulator, singularities occur whenever the matrices $\boldsymbol{A}$, $\boldsymbol{B}$, or both become singular. As singularity leads to a loss of the controllability and degradation of the natural stiffness of manipulators, the analysis of parallel manipulators has drawn considerable attention (Gosselin and Angeles 1990; Daniali et al. 1995; Mayer St-Onge and Gosselin 2000; Park and Kim 1999; Liu et al. 2003). Based on the forward and inverse Jacobian matrices, a classification of the singularities pertaining to parallel manipulators into three main groups was suggested (Daniali et al. 1995).

X.-J. Liu and J. Wang, *Parallel Kinematics: Type, Kinematics, and Optimal Design*, Springer Tracts in Mechanical Engineering, DOI 10.1007/978-3-642-36929-2_5,

The *first kind of singularity* occurs when $\boldsymbol{A}$ becomes singular but $\boldsymbol{B}$ is invertible, i.e.,

$$\det(\boldsymbol{A}) = 0 \text{ and } \det(\boldsymbol{B}) \neq 0. \tag{5.1}$$

This kind of singularity corresponds to the configuration in which the chain reaches either a boundary of its workspace or an internal boundary, limiting the different subregions of the workspace where the number of branches is not the same.

The *second kind of singularity*, occurring only in closed kinematics chains, arises when $\boldsymbol{B}$ becomes singular but $\boldsymbol{A}$ is invertible, i.e., when

$$\det(\boldsymbol{A}) \neq 0 \text{ and } \det(\boldsymbol{B}) = 0. \tag{5.2}$$

In such configuration, the output link is locally movable even when all the actuated joints are locked, and the output link cannot resist one or more forces or moments even when all actuators are locked.

The *third kind of singularity* occurs when both $\boldsymbol{A}$ and $\boldsymbol{B}$ become simultaneously singular. This situation corresponds to configurations in which the chain can undergo finite motions when its actuators are locked or in which a finite motion of the inputs produces no motion of the output.

The first three subsections will introduce the above-mentioned three kinds of singularities of three parallel mechanisms. In the last subsection, singularity analysis will be further discussed by using the concept of motion/force transmissibility based on screw theory. Such an approach is able to not only identify all possible singularities but also explain their physical meanings in terms of motion/force transmission.

5.1 Singularity of the 5R Parallel Mechanism

The singularity of the 5R symmetrical parallel mechanism has been studied by many researchers. Here, we summarize this issue as following.

5.1.1 *The First Kind of Singularity*

This kind of singularity occurs when $\boldsymbol{A}$ of Eq. (4.6) becomes singular but $\boldsymbol{B}$ of Eq. (4.7) remains invertible. Physically, this corresponds to the configuration whenever one of the legs A_1B_1P and A_2B_2P is completely extended or folded. From the analysis of the workspace, this singularity occurs when the output point P reaches its limit or is at the boundary of the workspace. This singularity is also referred to as the *serial singularity*. For example, for the mechanism with the parameters

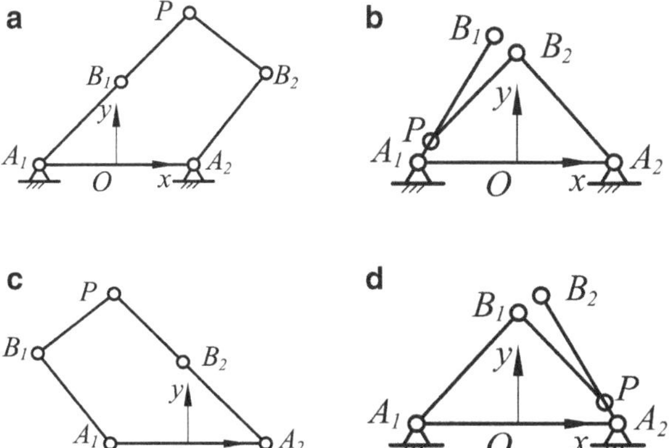

Fig. 5.1 Some configurations of the first kind of singularity: (**a**) the first leg is completely extended; (**b**) the first leg is completely folded; (**c**) the second leg is completely extended; (**d**) the second leg is completely folded

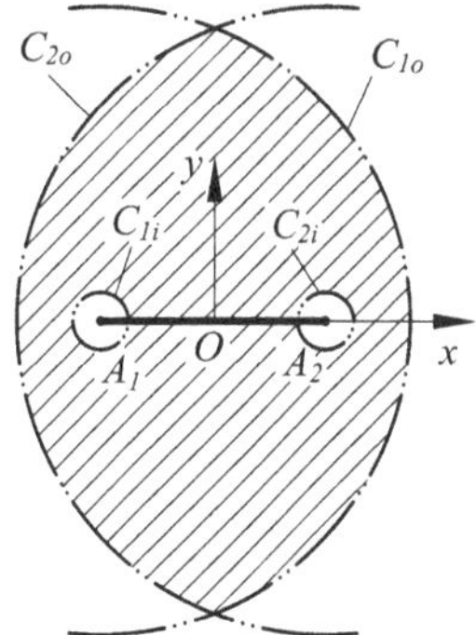

Fig. 5.2 The theoretical workspace (shown as the hatched region) and the first kind of singular loci

$r_1 = 1.2, r_2 = 1.0$, and $r_3 = 0.8$, some configurations of this kind of singularity are shown in Fig. 5.1. The loci of this kind of singularity are actually given by circles C_{1o}, C_{1i}, C_{2o}, and C_{2i} as follows:

$$C_{1o}: \ (x + r_3)^2 + y^2 = (r_1 + r_2)^2 \tag{5.3}$$

$$C_{1i}: \ (x + r_3)^2 + y^2 = (r_1 - r_2)^2 \tag{5.4}$$

$$C_{2o}: \ (x - r_3)^2 + y^2 = (r_1 + r_2)^2 \tag{5.5}$$

$$C_{2i}: \ (x - r_3)^2 + y^2 = (r_1 - r_2)^2 \tag{5.6}$$

which are actually the boundaries of the theoretical workspace as shown in Fig. 5.2.

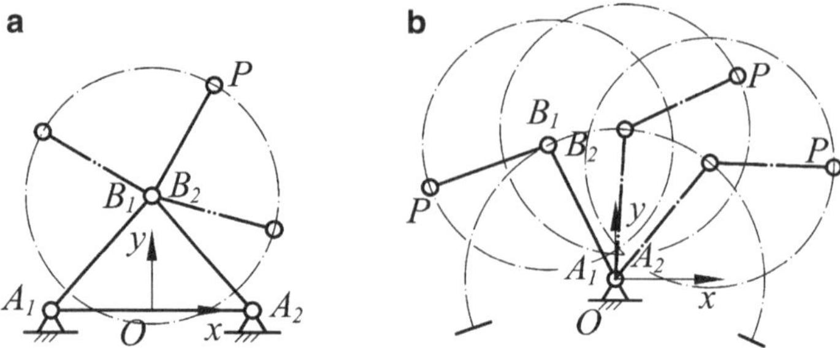

Fig. 5.3 Configurations when B_1 and B_2 are coincident

Note that $r_1 = 0$ leads to $\det(\boldsymbol{A}) = 0$ as well. But this also results the mechanism in an unmovable status.

5.1.2 *The Second Kind of Singularity*

Usually, there are two cases for this singularity of the mechanism. The first case is that B_1PB_2 is completely folded, i.e., the points B_1 and B_2 are coincident. Such a configuration is shown in Fig. 5.3a, from which one may see that when this condition occurs the locus of the point P is a circle centered at point $\left(0,\ \sqrt{r_1^2 - r_3^2}\right)$ or $\left(0, -\sqrt{r_1^2 - r_3^2}\right)$ with a radius of r_2, namely,

$$x^2 + \left(y - \sqrt{r_1^2 - r_3^2}\right)^2 = r_2^2 \text{ or } x^2 + \left(y + \sqrt{r_1^2 - r_3^2}\right)^2 = r_2^2. \tag{5.7}$$

Please note that, if $r_3 = 0$, this type of singularity can occur easily when links A_1B_1 and A_2B_2 are coincident as shown in Fig. 5.3b. For this case, the loci of point P are an annulus, which is bounded by two circles

$$x^2 + y^2 = (r_1 + r_2)^2 \text{ and } x^2 + y^2 = (r_1 - r_2)^2. \tag{5.8}$$

The second case is that when B_1PB_2 is completely extended. Such a singular configuration is shown in Fig. 5.4. The locus of point P for this type of singularity can be described by

$$\begin{cases} x = r_1\left(\cos\theta_2 + \cos\theta_1\right)/2 \\ y = r_1\left(\sin\theta_2 + \sin\theta_1\right)/2 \end{cases} \tag{5.9}$$

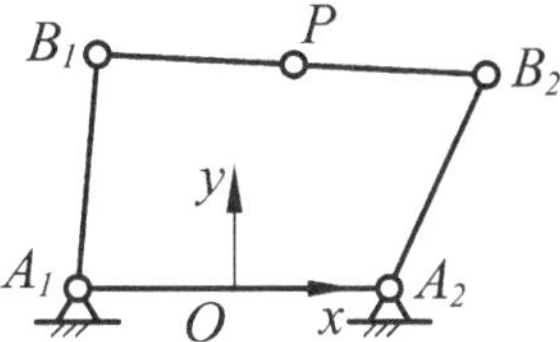

Fig. 5.4 The configuration when B_1PB_2 is completely extended

where

$$\theta_1 \in [0,\ \ 2\pi]$$
$$\theta_2 = 2\tan^{-1}(z)$$
$$z = \left(-b \pm \sqrt{b^2 - 4ac}\right)/(2a)$$
$$a = 4r_2^2 + 4r_1r_3\cos\theta_1 + 4r_1r_3 - 4r_3^2 - 2r_1^2 - 2r_1^2\cos\theta_1$$
$$b = 4r_1^2\sin\theta_1$$
$$a = 4r_2^2 + 4r_1r_3\cos\theta_1 - 4r_1r_3 - 4r_3^2 - 2r_1^2 + 2r_1^2\cos\theta_1.$$

Based on the analysis in (Bonev and Gosselin 2001), Eq. (5.9) can be also rewritten as the following form:

$$\begin{aligned}&(x + 2r_3)^2\left(x^2 + 4r_3x + 3r_3^2 + r_2^2 - r_1^2 + y^2\right)^2\\&\quad + y^2\left(x^2 + 4r_3x + 5r_3^2 + r_2^2 - r_1^2 + y^2\right)^2 - 4r_3^2r_2^2y^2 = 0.\end{aligned} \tag{5.10}$$

For the second kind of singularity, if $r_1 < r_3$, the singularity that B_1 and B_2 are coincident will not occur. If $r_2 > r_1 + r_3$, there is no the singularity that B_1PB_2 is completely extended.

5.1.3 The Third Kind of Singularity

From the analysis on the first and second kinds of singularities, one may see that the third kind of singularity occurs when the five points A_1, B_1, P, B_2, and A_2 are collinear. There are six cases for this kind of singularity, as shown in Fig. 5.5. The parameter conditions to those singularities are as follows:

(a) $r_1 = r_3$; one singularity is shown in Fig. 5.5a.
(b) $r_2 = r_3$; a corresponding singular configuration is shown in Fig. 5.5b.
(c) $r_3 = r_1 + r_2$; see Fig. 5.5c for the singular configuration.
(d) $r_2 = r_1 + r_3$; Fig. 5.5d shows such a singularity.
(e) $r_1 = r_3 + r_2$; see Fig. 5.5e.
(f) $r_3 = 0$; one singular configuration is shown in Fig. 5.5f.

The mechanism satisfies any one of the five parameter conditions (a)–(e) is referred to as *a change point mechanism*.

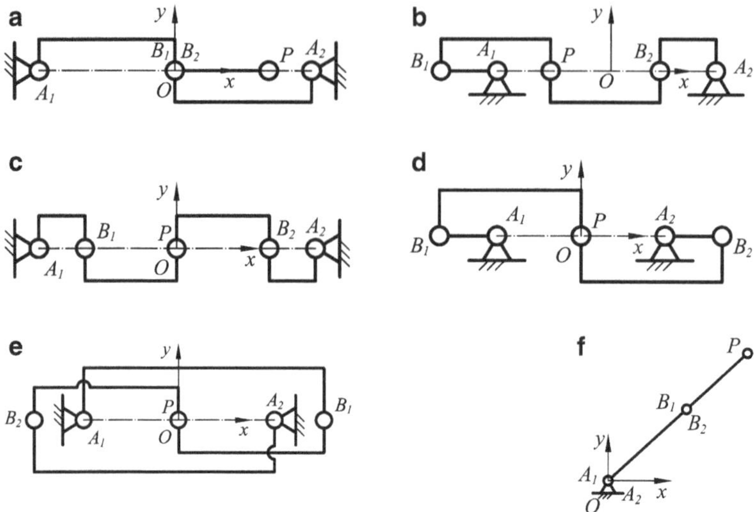

Fig. 5.5 Five configurations of the third kind of singularity

The analysis on the kinematics of the mechanism shows that there are four solutions for the inverse kinematics and two solutions for the forward kinematics. Any one of the singularities will result in the change of solution number of the kinematics. For example, the first kind of singularity leads to the loss of solution number of the inverse kinematics. While in the uncertainty singular configuration, the solution number of the forward kinematics can be less or more than two. In the third kind of singularity, both the inverse and forward kinematics will be different. Then the first kind of singularity can be called the inverse kinematic singularity, the second kind of singularity the forward kinematic singularity, and the third kind of singularity the inverse and forward kinematic singularity. Notably, the loci of the second kind of singularity must be within the workspace.

5.2 Singularity of the 2-RRU&1-RR(Pa)R Parallel Mechanism

5.2.1 The First Kind of Singularity

This singularity is encountered here when one of the diagonal entries of $\boldsymbol{A}$ of Eq. (4.34) vanishes, that is,

$$z\cos\theta_1 - (r + y - R)\sin\theta_1 = 0 \tag{5.11}$$

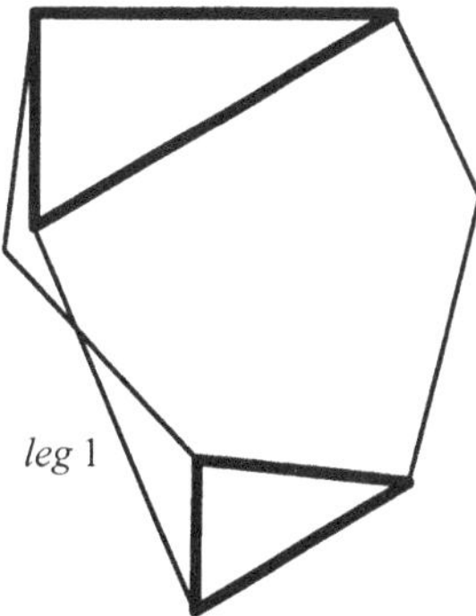

Fig. 5.6 One singular configuration of the first kind of singularity: leg 1 is completely extended

or

$$z \cos \theta_2 - (R + y - r) \sin \theta_2 = 0 \tag{5.12}$$

or

$$(z + r \sin \phi) \cos \theta_3 + (r \cos \phi - R) \sin \theta_3 = 0 \tag{5.13}$$

which means that this type of configuration for our mechanism is reached whenever the lower and upper links for each one of the three legs $P_1E_1B_1$, $P_2E_2B_2$, and $P_3E_3B_3$ are completely extended or folded. One of such singularity is shown in Fig. 5.6.

5.2.2 *The Second Kind of Singularity*

From Eq. (4.35), one obtains

$$\det(\boldsymbol{B}) = b_{33}\left(b_{11}b_{22} - b_{12}b_{21}\right) \tag{5.14}$$

from which one may see that $\det(\boldsymbol{B}) = 0$ can be satisfied if (a) $b_{33} = 0$, (b) $b_{21} = b_{11} = 0$, (c) $b_{22} = b_{12} = 0$, and (d) both $b_{22} = b_{12}$ and $b_{11} = b_{21}$. The mechanism will be in the configuration that lower link E_3P_3 of the third leg $P_3E_3B_3$ is in the plane defined by P_1, P_2 (in this chapter, it is called the mobile platform plane) and P_3 if $b_{33} = 0$, the configuration that $E_1P_1 \perp P_1P_2$ if $b_{11} = 0$, $E_2P_2 \perp P_1P_2$ if $b_{21} = 0$. When $b_{12} = 0$ or $b_{22} = 0$, $E_1P_1P_2$ or $P_1P_2E_2$ will be completely extended or folded. From Eq. (4.35), one can also find that if $b_{12} = b_{11} = 0$ or $b_{22} = b_{21} = 0$, the mechanism can be in a singular configuration as well. But $b_{11} = 0$ and $b_{12} = 0$ correspond to different configurations of the first leg. Because a leg cannot be in two different configurations at the same time, the cases $b_{12} = b_{11} = 0$ and $b_{22} = b_{21} = 0$ will not occur actually.

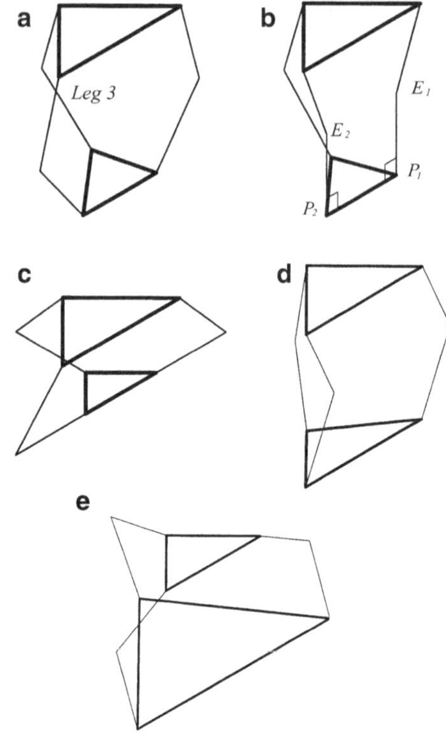

Fig. 5.7 Some singular configurations of the second kind of singularity: (**a**) lower links of the first and second legs are collinear; (**b**) $E_1P_1 \perp P_1P_2$ and $E_2P_2 \perp P_1P_2$; (**c**) the lower link of the leg 3 is in the mobile platform plane; (**d**) $R = r$ and $\theta_2 = \theta_1$; (**e**) $R + L_b \cos\theta_1 = r$ and $\theta_2 = \pi - \theta_1$

If $b_{22} = b_{12}$, i.e., $z - L_b \sin\theta_1 = z - L_b \sin\theta_2$, there is

$$\theta_2 = \theta_1 \quad \text{or} \quad \theta_2 = \pi - \theta_1, \quad (\theta_1, \quad \theta_2 \in [0, \quad 2\pi])\,. \tag{5.15}$$

Substitution into $b_{11} = b_{21}$ leads to

$$R = r \quad \text{or} \quad R + L_b \cos\theta_1 = r. \tag{5.16}$$

From above analysis, we may see that the second kind of singularity for the mechanism will arise if any one of the following cases occurs:

- The lower link E_3P_3 of the third leg $P_3E_3B_3$ is in the mobile platform plane. In this case, the mechanism will be in the second kind of singularity, as shown in Fig. 5.7a.
- $E_1P_1 \perp P_1P_2$ and $E_2P_2 \perp P_1P_2$, as shown in Fig. 5.7b.
- $E_1P_1P_2$ and $P_1P_2E_2$ are completely extended or folded, i.e., four points E_1, P_1, E_2, P_2 are collinear, as shown in Fig. 5.7c.
- $R = r$ and $\theta_2 = \theta_1$, as shown in Fig. 5.7d.
- $R + L_b \cos\theta_1 = r$ and $\theta_2 = \pi - \theta_1$, as shown in Fig. 5.7e.
- $r = 0$, which leads to $b_{33} = 0$.

from which one may see that the singularities as shown in Fig. 5.7b, d are different from others, because they correspond to a different assembling mode, i.e., the assembling mode as shown in Fig. 3.12 cannot reach such singular configurations.

5.2.3 The Third Kind of Singularity

From the analysis in Sect. 4.2.2, we may see if $R = r$ and $\theta_2 = \theta_1$, the determinant of matrix $\boldsymbol{B}$ will be zero. In the geometric condition $R = r$, from Eq. (4.34), $\det(\boldsymbol{A}) = 0$ can also equal to zero at point $(y = 0,\ \ z = 0)$. Then $R = r$ and $\theta_2 = \theta_1$ with $(y = 0,\ \ z = 0)$ are the architecture and configuration conditions, respectively, for the third kind of singularity.

Additionally, due to the architecture of the mechanism, as shown in Fig. 3.12, when the relationship between the parameters is

$$R = r + L_a + L_b \tag{5.17}$$

or

$$r = R + L_a + L_b, \tag{5.18}$$

the mechanism can be assembled but cannot move anymore; this is the so-called architecture singularity.

Generally, all possible singular configurations of a parallel mechanism with three DOFs can be listed easily, which is one of the reasons that such mechanisms can be applied to industrial applications as quickly as possible. As shown in this section, the singularity of the parallel mechanism with revolute actuators can be roughly classified into eleven cases, including one case of architecture singularity that $R = r + L_a + L_b$. From the view of singularity, a 2-PRU&1-PR(Pa)R parallel mechanism is better than a 2-RRU&1-RR(Pa)R mechanism. In the mechanism with prismatic actuators, the second kind of singularity occurs only in the cases of $R = r$ and the third leg being in its mobile platform plane, which will be discussed in the subsequent section.

5.3 Singularity of the 2-PRU&1-PR(Pa)R Parallel Mechanism

5.3.1 The First Kind of Singularity

This condition is encountered here when one of the diagonal entries of matrix $\boldsymbol{A}$ given in Eq. (5.27) vanishes, that is,

$$(z - z_1)\,(z - z_2)\,(r \sin\phi + z - z_3) = 0 \tag{5.19}$$

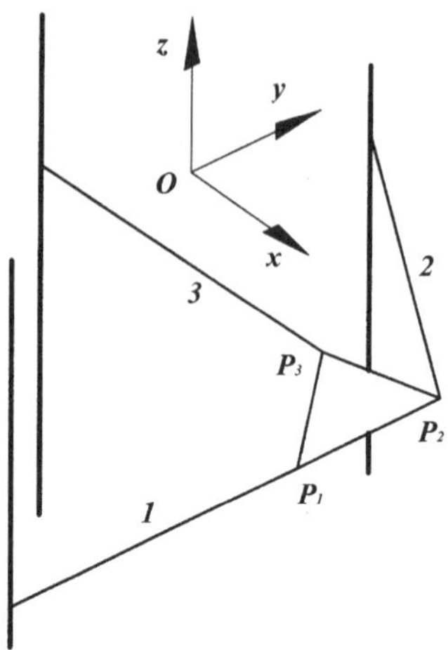

Fig. 5.8 One singular configuration of the first kind of singularity: link 1 is paralleling to y-axis

which means that this type of configuration for the mechanism is reached whenever one of the legs P_1B_1, P_2B_2, and P_3B_3 is in the plane paralleling to the O-xy plane. Figure 5.8 is one singular configuration of the kind of singularity.

5.3.2 The Second Kind of Singularity

From Eq. (4.28), $\det(\boldsymbol{B}) = 0$ results in

$$Rr\sin\phi + r(z - z_3)\cos\phi = 0 \tag{5.20}$$

and

$$(R - r)(2z - z_1 - z_2) + y(z_1 - z_2) = 0. \tag{5.21}$$

If Eq. (5.20) is verified and $\cos\phi \neq 0$, one obtains

$$\tan\phi = \frac{(z_3 - z)}{R}. \tag{5.22}$$

From Fig. 3.11, we may see that Eq. (5.22) corresponds the configuration that the third leg P_3B_3 is in the plane defined by P_1, P_2, and P_3. Figure 5.9 shows one of this kind singularity.

Considering only the configuration shown in Fig. 3.11, from Eqs. (3.80) and (3.81), we may obtain

$$z_1 = \sqrt{L^2 - (-r + y + R)^2} + z \tag{5.23}$$

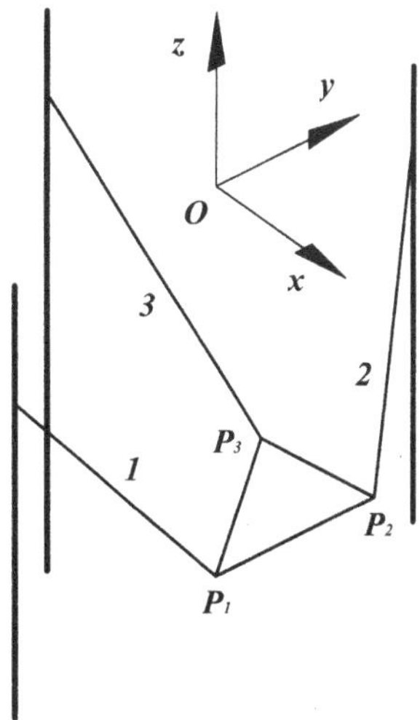

Fig. 5.9 One singular configuration of the second kind of singularity: link 3 is in the plane defined by points P_1, P_2, and P_3

$$z_2 = \sqrt{L^2 - (r + y - R)^2} + z. \tag{5.24}$$

Substituting Eqs. (5.23) and (5.24) into Eq. (5.21) leads to

$$\begin{aligned} &y\left(\sqrt{L^2 - (-r + y + R)^2} - \sqrt{L^2 - (r + y - R)^2}\right) \\ &\quad - (R - r)\left(\sqrt{L^2 - (-r + y + R)^2} + \sqrt{L^2 - (r + y - R)^2}\right) = 0 \end{aligned} \tag{5.25}$$

from which one may see that if Eq. (5.21) is verified, there must be

$$R = r. \tag{5.26}$$

5.3.3 The Third Kind of Singularity

From Eqs. (5.19) and (5.21), we may see that if

$$z = z_1 = z_2, \tag{5.27}$$

matrices $\boldsymbol{A}$ and $\boldsymbol{B}$ will become singular simultaneously. And the conditions on the linkage parameters are

$$R = r + L \quad \text{or} \quad R = r. \tag{5.28}$$

Additionally, when $L = 0$, the mechanism can be assembled only if $R = r$, which will lead to the third kind of singularity.

Finally, let's consider the singularity of the planar four-bar parallelogram used in the mechanism. As well known, the singularity of planar four-bar parallelogram occurs only when the four bars are parallel to each other or are aligned. For the mechanism proposed in this chapter, the condition to arise this singularity is $R = r$, which is the parameter condition to the third kind of singularity of the mechanism. Therefore, the singularity of the planar four-bar parallelogram need not be considered in the singularity analysis of the mechanism.

From the three examples, one may see that there are more singularities for the parallel mechanisms with revolute actuators than for mechanisms with prismatic actuators.

5.4 Singularity Analysis of Parallel Manipulators Taking into Account the Motion and Force Transmissibility

Singularity is always of importance in the process of kinematic analysis of a mechanism and the design of a machine. It is undoubted that there is no uniform method to the singularity analysis. In this section, this issue will be further discussed taking into account the motion and force transmissibility of parallel mechanisms.

5.4.1 An Approach to Singularity Analysis of Parallel Manipulators

In this subsection, an approach that takes into account motion/force transmissibility will be introduced for singularity analysis of parallel manipulators. First, we shall declare that the parallel manipulators considered in this study are non-redundant, and gravity and friction will not be taken into account.

The flowchart in Fig. 5.10 shows the process of singularity analysis of an n-DOF parallel manipulator. The details may be presented as follows:

Step 1: *Determine the wrenches in each kinematic leg of the manipulator.*
For a parallel manipulator, its moving platform (or end-effector) usually connects to the base by two or more kinematic legs. There must be some internal wrenches in the kinematic legs to balance the external wrenches exerted on the moving platform. Thus, it is necessary to determine the wrenches in kinematic legs. In this step, a method based on screw theory is introduced to determine the wrenches in kinematic legs. The mathematical foundation of screw theory has already been presented in Chap. 2; therefore, it will be no longer presented here.

Suppose that a kinematic leg contains m kinematic pairs, and the ith ($i = 1, 2, \ldots, m$) pair has κ_i-DOF. Then, we may get a twist screw system of the leg as

$$\boldsymbol{U} = \begin{bmatrix} \$_1 & \$_2 & \cdots & \$_t \end{bmatrix} \tag{5.29a}$$

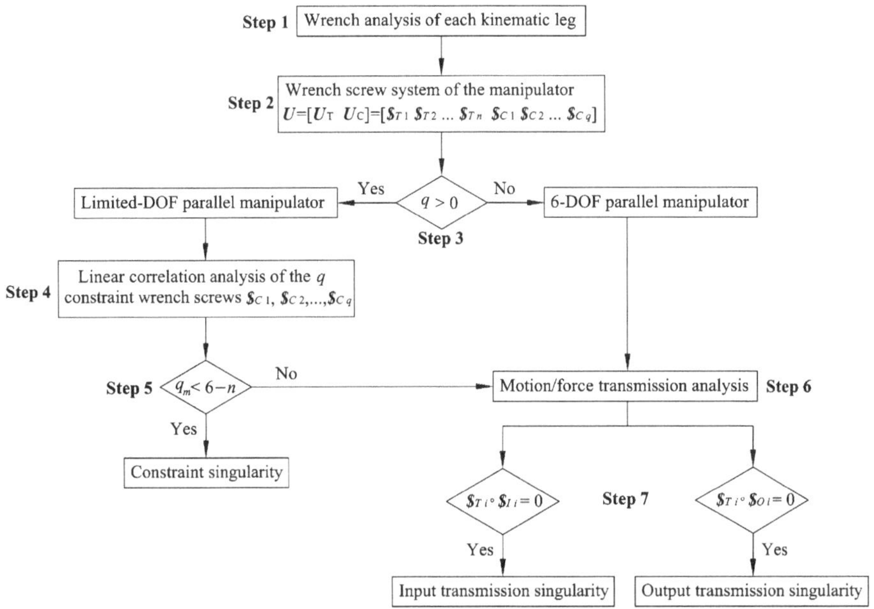

Fig. 5.10 Process of singularity analysis of an n-DOF parallel manipulator

where

$$t = \sum_{i=1}^{m} \kappa_i. \tag{5.29b}$$

Such a screw system can be considered as a $t \times 6$ matrix. If the order of the matrix is equal to n ($n \leq t$), the kinematic leg can be regarded as an n-DOF leg. By picking $n \leq 6$ linearly independent screws out from $\boldsymbol{U}$, the twist screw system of the leg can be rewritten as

$$\boldsymbol{U}_n = \boldsymbol{\$}_1 \ \boldsymbol{\$}_2 \cdots \boldsymbol{\$}_n \tag{5.30}$$

When $n < 6$, we may obtain $6-n$ linearly independent screws $\boldsymbol{\$}_j^r$ ($j = 1, 2, \ldots, 6-n$) that are reciprocal to all the screws in $\boldsymbol{U}_n$, i.e.,

$$\boldsymbol{\$}_i \circ \boldsymbol{\$}_j^r = 0 \; (i = 1, 2, \ldots, n; \; j = 1, 2, \ldots, 6-n) \tag{5.31}$$

Since the screw $\boldsymbol{\$}_j^r$ can represent a constraint wrench that the kinematic leg provides for the moving platform, it is referred to as the *constraint wrench screw* (CWS). Here, $\boldsymbol{\$}_C$ is used to denote CWS. All the constraint wrench screws form a $(6-n)$-order constraint wrench screw system which is represented by

$$\boldsymbol{U}_C = \left[\boldsymbol{\$}_1^r \ \boldsymbol{\$}_2^r \cdots \boldsymbol{\$}_{6-n}^r\right] = \left[\boldsymbol{\$}_{C1} \ \boldsymbol{\$}_{C2} \cdots \boldsymbol{\$}_{C(6-n)}\right] \tag{5.32}$$

A kinematic leg may contain one or more actuators. It is assumed that the twist of the actuator corresponds to a twist screw ($\boldsymbol{\$}_A$) in $\boldsymbol{U}_n$. When the actuator is locked, $\boldsymbol{\$}_A$ will not belong to the twist screw system ($\boldsymbol{U}_n$) of the leg, and $\boldsymbol{U}_n$ will become an $(n-1)$-order system $\boldsymbol{U}_{n-1}$. Then we may get a screw $\boldsymbol{\$}_T$ reciprocal to all the screws in $\boldsymbol{U}_{n-1}$, i.e.,

$$\boldsymbol{\$}_T \circ \boldsymbol{\$}_i = 0 \ (i = 1, 2, \ldots, n, \text{ and } i \neq A) \tag{5.33}$$

$\boldsymbol{\$}_T$ should not belong to the constraint wrench screw system $\boldsymbol{U}_C$. In other words, $\boldsymbol{\$}_T$ should be linearly independent with all the constraint wrench screws in $\boldsymbol{U}_C$. Such a screw $\boldsymbol{\$}_T$ represents a wrench, by which motion/force from the actuator is transmitted to the moving platform; thus, it is called the *transmission wrench screw* (TWS). Notably, the number of TWS is the same as that of actuator in the leg.

When $n = 6$, there exists no screw reciprocal to all the screws in $\boldsymbol{U}_6$. This means that there is no CWS in a 6-DOF leg. In other words, a 6-DOF leg provides no constraint wrench to the moving platform.

A 6-DOF leg without an actuator is usually used for pose feedback of the moving platform. However, such a leg has no effect on the motion/force transmission of parallel manipulators. Therefore, there is always at least one actuator in a 6-DOF leg. Assume that the actuator in the leg corresponds to the twist screw $\boldsymbol{\$}_A$ in $\boldsymbol{U}_6$. When the actuator is locked, $\boldsymbol{\$}_A$ will be eliminated from $\boldsymbol{U}_6$, and $\boldsymbol{U}_6$ will become a 5-order system $\boldsymbol{U}_5$. Then, the screw reciprocal to all the screws in $\boldsymbol{U}_5$ is the TWS of the leg. It should be noted that there is a one-to-one correspondence between transmission wrench screws and actuators in a leg; in other words, the number of transmission wrench screws in a 6-DOF leg is the same as that of actuators.

Step 2: All the transmission and constraint wrench screws together form the wrench screw system of the manipulator.

It is assumed that an n-DOF non-redundant parallel manipulator contains p legs, and the ith ($i = 1, 2, \ldots, p$) leg provides σ_i constraint wrenches to the moving platform. Then, the numbers of TWS and CWS are equal to n and $q = \sum_{i=1}^{p} \sigma_i$, respectively. Thus, the wrench screw system of the manipulator can be expressed by

$$\left[\boldsymbol{U} = \boldsymbol{U}_T \quad \boldsymbol{U}_C\right] = \left[\boldsymbol{\$}_{T1} \ \boldsymbol{\$}_{T2} \cdots \boldsymbol{\$}_{Tn} \ \boldsymbol{\$}_{C1} \ \boldsymbol{\$}_{C2} \cdots \boldsymbol{\$}_{Cq}\right] \tag{5.34}$$

where $\boldsymbol{U}_T$ and $\boldsymbol{U}_C$ represent the transmission and constraint wrench screw systems, respectively.

Step 3: Classify the parallel manipulators into two types by the condition $q > 0$, where q represents the number of constraint wrench screw.

If $q > 0$, it means that the legs in the manipulator provide at least one constraint wrench to the moving platform. In this case, the moving platform loses at least one

DOF, and the manipulator is referred to as a limited-DOF one. When two or more constraint wrenches exist in the manipulator, there might be a constraint singularity. Once the constraint singularity occurs, the manipulator will be out of control even if all the actuators are locked. Thus, the constraint singularity should first be analyzed in the singularity analysis of limited-DOF parallel manipulators.

Obviously, q is a nonnegative integer. Thus, the other case is that $q = 0$, which means that all the legs provide no constraint wrench to the moving platform. In this case, the manipulator has six DOFs, and the constraint singularity will never occur in such a 6-DOF manipulator. The constraint singularity analysis in **Steps 4 and 5** can then be skipped, and we may go directly to **Step 6**.

Steps 4 and 5: For a limited-DOF parallel manipulator, all the q constraint wrenches together should restrict exactly $6-n$ DOFs of the moving platform. Therefore, $\boldsymbol{U}_C$ should be a $(6-n)$-order system. In other words, the maximal number (q_m) of linearly independent constraint wrench screws should be equal to $6-n$. If $q_m < 6-n$, the manipulator will gain one or more DOFs, which means that the motions corresponding to the increased DOF cannot be achieved by driving the n actuators. In this case, the manipulator is said to have the *constraint singularity*. If $q_m > 6-n$, *constraint singularity* will not occur, because the manipulator actually is redundant.

As mentioned in **Step 3**, when the constraint singularity occurs, the manipulator will be out of control even if all the actuators are locked. Thus, the designer should change the assembly mode of some kinematic pair or modify the type of the manipulator to make sure that there is no constraint singularity in the desired workspace. Then, the singularity analysis of the manipulator may proceed to **Step 6**.

Steps 6 and 7: The function of a parallel manipulator is to transmit motion/force from the actuators to the moving platform by means of transmission wrenches. Thus, the relationship between TWS and *input twist screw* (ITS) and that between TWS and *output twist screw* (OTS) should be both investigated in the motion/force transmission analysis of parallel manipulators.

As to an n-DOF parallel manipulator without constraint singularity, its constraint wrench screw system $\boldsymbol{U}_C$ should be a $(6-n)$-order one. Then, its wrench screw system can be represented by

$$\begin{bmatrix}\boldsymbol{U} = \boldsymbol{U}_T & \boldsymbol{U}_C\end{bmatrix} = \begin{bmatrix}\$_{T1} & \$_{T2} \cdots \$_{Tn} & \$_{C1} & \$_{C2} \cdots \$_{C(6-n)}\end{bmatrix} \tag{5.35}$$

in which there are n transmission wrench screws and $6-n$ constraint wrench screws. Such a screw system can be considered as a 6×6 matrix. An easy way for singularity identification is to calculate the determinant of the matrix. If $\det(\boldsymbol{U}) = 0$, some singularity will occur in the manipulator. Another approach to identify the singularity is presented below.

In an n-DOF parallel manipulator, each actuator corresponds to an ITS and a TWS. Here, we use $\$_{Ti}$ and $\$_{Ii}$ to represent the corresponding TWS and ITS of the ith actuator, respectively. If the reciprocal product of $\$_{Ti}$ and $\$_{Ii}$ is equal to zero, i.e., $\$_{Ti} \circ \$_{Ii} = 0$, the power between the transmission wrench and the input twist will vanish. It means that the motion from the ith actuator cannot be transmitted by its corresponding transmission wrench. In this case, the manipulator has a transmission singularity. Such a kind of transmission singularity is referred to as the *input transmission singularity*. It should be noted that, if any one of the n reciprocal products, i.e., $\$_{Ti} \circ \$_{Ii}$ $(i = 1, 2, \ldots, n)$, equals zero, the input transmission singularity will occur.

By locking all the actuators except the ith one, the manipulator becomes a single-DOF one, and the instantaneous motion of the moving platform can be represented by an OTS $\$_{Oi}$ $(i = 1, 2, \ldots, n)$. In this case, only the ith transmission wrench represented by $\$_{Ti}$ might do work on the moving platform, while all the other $n-1$ transmission wrenches become constraint wrenches. Then we get

$$\$_{Tj} \circ \$_{Oi} = 0 \; (j = 1, 2, \ldots, n \;\&\; j \neq i) \tag{5.36}$$

However, if $\$_{Ti} \circ \$_{Ii} = 0$, the reciprocal products between $\$_{Oi}$ and all the transmission wrench screws ($\$_{Ti}$, $j = 1, 2, \ldots, n$) will be equal to zero. It means that no transmission wrench can transmit power from any actuator to enable the moving platform achieve the motion represented by $\$_{Oi}$; or it can be said that an external wrench along the axis of $\$_{Oi}$ cannot be counterbalanced by any transmission wrench. In this case, the manipulator is considered having a transmission singularity. This kind of singularity is called *output transmission singularity*. Notably, if any one of the n reciprocal products, i.e., $\$_{Ti} \circ \$_{Oi}$ $(i = 1, 2, \ldots, n)$, equals zero, the output transmission singularity will occur. Thus, the input and output transmission singularities can be obtained by means of $\$_{Ti} \circ \$_{Ii} = 0$ and $\$_{Ti} \circ \$_{Oi} = 0$, respectively.

5.4.2 An Example

Singular analysis of a 3-$\underline{\text{R}}$UU parallel manipulator shown in Fig. 5.11a is taken as an example to present the application of the introduced approach. The moving platform is connected to the base by three identical $\underline{\text{R}}$UU legs. The three R pairs attached to the base are actuated, and they are equally distributed at a nominal angle of 120° on a circle whose radius is r_1. Parameter r_2 represents the distance from the center (O') of the moving platform to $U_{i,2}$ $(i = 1,2,3)$. The parameter r_3 and r_4 represent the lengths of the input links $R_iU_{i,1}$ and the coupler links $U_{i,1}U_{i,2}$ $(i = 1,2,3)$, respectively. The origin of the global coordinate system O-xyz lies at the center of the base, and all the axes of the three R pairs lie in the plane Oxy. The axis of the R_1 pair is parallel to the x-axis and perpendicular to the y-axis.

The first step of singularity analysis is determining the constraint and transmission wrenches in the $\underline{\text{R}}$UU leg. As shown in Fig. 5.11b, the origin of the local coordinate system O''-$x''y''z''$ lies at the center of the $U_{i,1}$ pair; the x''-axis, y''-axis, and z''-axis are parallel to the x-axis, y-axis, and z-axis, respectively. With respect to O''-$x''y''z''$, the twists in the ith leg can be expressed by screws as follows:

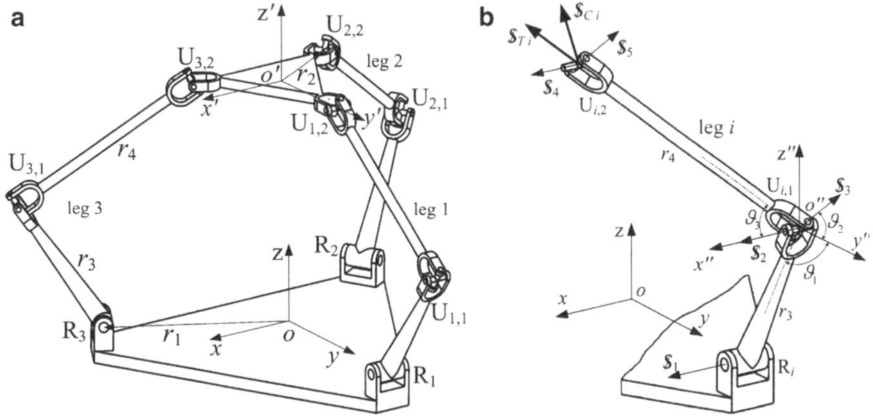

Fig. 5.11 A 3-RUU parallel manipulator: (**a**) kinematic structure; (**b**) an RUU leg

$$\$_{Ii} = \$_1 = (1,\ 0,\ 0;\ 0,\ -r_3 \sin\vartheta_1,\ -r_3 \cos\vartheta_1) \tag{5.37a}$$

$$\$_2 = (1,\ 0,\ 0;\ 0,\ 0,\ 0) \tag{5.37b}$$

$$\$_3 = (0,\ \cos\vartheta_2,\ \sin\vartheta_2;\ 0,\ 0,\ 0) \tag{5.37c}$$

$$\$_4 = (1,\ 0,\ 0;\ 0,\ r_4 \sin\vartheta_3 \cos\vartheta_2,\ r_4 \sin\vartheta_3 \sin\vartheta_2) \tag{5.37d}$$

$$\$_5 = (0,\ \cos\vartheta_2,\ \sin\vartheta_2;\ -r_4 \sin\vartheta_3,\ -r_4 \cos\vartheta_3 \sin\vartheta_2,\ r_4 \cos\vartheta_3 \cos\vartheta_2) \tag{5.37e}$$

where ϑ_1 represents the angle between the y''-axis and the input link $R_iU_{i,1}$, ϑ_2 stands for the angle between the y''-axis and the axis of $\$_3$, and ϑ_3 denotes the angle between the x''-axis and coupler link $U_{i,1}U_{i,2}$.

Based on Eq. (5.33), the constraint wrench screw of the ith RUU leg can be obtained as

$$\$_{Ci} = (0;\ \boldsymbol{\tau}_i) = (0,\ 0,\ 0;\ 0,\ -\sin\vartheta_2,\ \cos\vartheta_2) \tag{5.38}$$

$\$_{Ci}$ represents a constraint moment, the axis of which is perpendicular to the joint plane of the U pair in the leg.

Since the input twist screw is equivalent to $\$_1$, we may get the TWS as

$$\$_{Ti} = (\boldsymbol{f}_i;\ 0) = (\cos\vartheta_3,\ -\sin\vartheta_3 \sin\vartheta_2,\ \sin\vartheta_3 \cos\vartheta_2;\ 0,\ 0,\ 0) \tag{5.39}$$

which denotes a pure force along the coupler link $U_{i,1}U_{i,2}$. With respect to the global coordinate system *O-xyz*, the TWS in the ith leg can be expressed by

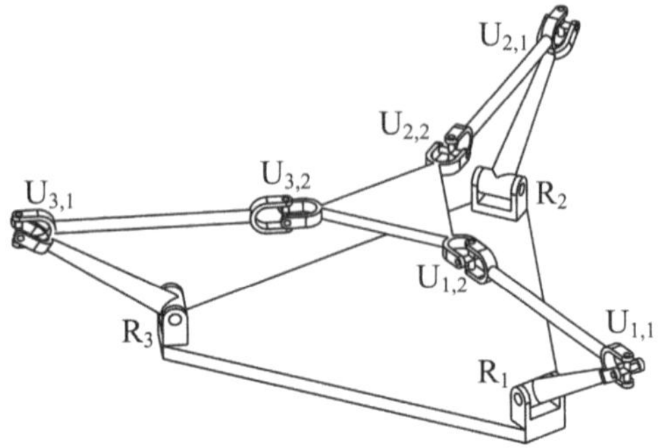

Fig. 5.12 A singular configuration of the 3-RUU manipulator

$$\$_{Ti} = (\boldsymbol{f}_i;\ \boldsymbol{r}_{\mathrm{U}_i} \times \boldsymbol{f}_i) = (\cos\vartheta_3,\ -\sin\vartheta_3\sin\vartheta_2,\ \sin\vartheta_3\cos\vartheta_2;\ \boldsymbol{r}_{\mathrm{U}_i} \times \boldsymbol{f}_i) \tag{5.40}$$

where $\boldsymbol{r}_{\mathrm{U}_i}$ denotes the vector from origin to the joint center of $\mathrm{U}_{i,1}$.

According to the above wrench analysis, each RUU leg provides one constraint moment and one transmission force to the moving platform. Then, there are three constraint moments and three transmission forces in the manipulator. The three constraint moments together restrict three rotational DOFs of the moving platform; thus, the 3-RUU manipulator is a three-DOF translational one. The motions or forces from the actuators are transmitted to achieve the three-DOF translations of the moving platform by the three transmission forces.

(a) Constraint singularity. When the maximal number (q_m) of linearly independent constraint moments is smaller than 3, i.e., $q_m < 3$, the constraint singularity occurs.

For the 3-RUU manipulator, when the axes of the three constraint moments become coplanar, q_m will be smaller than 3; then, the constraint singularity occurs, and the manipulator gains one more rotational DOF. According to Eq. (5.38), vector $\boldsymbol{\tau}_i$ represents the unit vector along the axis of constraint moment $\$_{Ci}$. The scalar triple product of $\boldsymbol{\tau}_i$ $(i = 1, 2, 3)$ can be used to judge whether the axes of the three constraint moments are coplanar. If $\boldsymbol{\tau}_1 \cdot (\boldsymbol{\tau}_2 \times \boldsymbol{\tau}_3) = 0$, it means that vectors $\boldsymbol{\tau}_1$, $\boldsymbol{\tau}_2$, and $\boldsymbol{\tau}_3$ lie in the same plane. Then, an external moment, the axis of which is perpendicular to the plane, cannot be counterbalanced by the constraint moments; the manipulator is said to have the constraint singularity.

Figure 5.12 shows a constraint singular configuration of the 3-RUU manipulator, in which all the coupler links $\mathrm{U}_{i,1}\mathrm{U}_{i,2}$ $(i = 1,2,3)$ are coplanar with the moving platform. In this case, all the axes of constraint moments lie in the same plane. Then, we can obtain $\boldsymbol{\tau}_1 \cdot (\boldsymbol{\tau}_2 \times \boldsymbol{\tau}_3) = 0$, and the constraint singularity occurs at the configuration.

(b) Input transmission singularity. If any one of $\$_{Ti} \circ \$_{Ii}$ $(i = 1, 2, \ldots, n)$ is equal to zero, the input transmission singularity occurs in the manipulator.

Based on Eqs. (5.37a) and (5.39), the reciprocal product of $\$_{Ti}$ and $\$_{Ii}$ can be obtained as

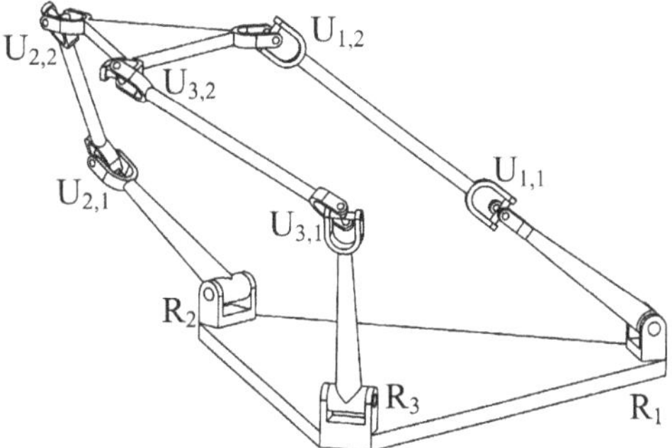

Fig. 5.13 An input transmission singular configuration of the 3-RUU manipulator

$$\begin{aligned}\$_{Ti} \circ \$_{Ii} &= r_3 \sin\vartheta_1 \sin\vartheta_3 \sin\vartheta_2 - r_3 \cos\vartheta_1 \sin\vartheta_3 \cos\vartheta_2 \\ &= -r_3 \sin\vartheta_3 \cos(\vartheta_1 + \vartheta_2)\end{aligned} \tag{5.41}$$

Then, the condition of input transmission singularity of the ith leg becomes $\sin\vartheta_3 = 0$ or $\cos(\vartheta_1 + \vartheta_2) = 0$. Case 1: $\sin\vartheta_3 = 0$ means that the angle between the coupler link $U_{i,1}U_{i,2}$ and the x''-axis is equal to 0 or 180°, i.e., $\vartheta_3 = 0°$ or 180°. In this case, the x''-axis and the axis of coupler link $U_{i,1}U_{i,2}$ are collinear. Case 2: it can be derived from $\cos(\vartheta_1 + \vartheta_2) = 0$ that $\vartheta_1 + \vartheta_2 = 90°$ or 270°, which means that the axes of $\$_3$ and input link $R_iU_{i,1}$ (see Eq. (5.37c) and Fig. 5.12) are perpendicular to each other. By summing up Cases 1 and 2, it can be concluded that when the coupler link $U_{i,1}U_{i,2}$ lies in the plane that is identified by the input link $R_iU_{i,1}$ and the joint axis of R_i, motion or force from the actuator R_i cannot be transmitted by the transmission force, and the input transmission singularity occurs. Figure 5.13 shows an input transmission singular configuration of the 3-RUU parallel manipulator.

(c) Output transmission singularity. When any one of $\$_{Ti} \circ \$_{Oi} = 0$ $(i = 1, 2, \ldots, n)$ equals to zero, the output transmission singularity occurs in the manipulator.

The 3-RUU manipulator has three translational DOFs; thus, when the actuators R_2 and R_3 are locked, the manipulator becomes a single-DOF one. In this case, the transmission forces represented by $\$_{T2}$ and $\$_{T3}$ can be considered as two constraint forces, and the unit motion of the moving platform can be expressed by

$$\$_{O1} = (0;\ \boldsymbol{f}_2 \times \boldsymbol{f}_3 / |\boldsymbol{f}_2 \times \boldsymbol{f}_3|) \tag{5.42}$$

Then, based on Eqs. (5.40) and (5.42), the reciprocal product of $\$_{T1}$ and $\$_{O1}$ can be obtained as

$$\$_{T1} \circ \$_{O1} = \boldsymbol{f}_1 \cdot (\boldsymbol{f}_2 \times \boldsymbol{f}_3) / |\boldsymbol{f}_2 \times \boldsymbol{f}_3| \tag{5.43a}$$

Similarly, the reciprocal product of $\$_{Ti}$ and $\$_{Oi}$ $(i = 2,3)$ can be obtained as

$$\$_{T2} \circ \$_{O2} = \boldsymbol{f}_2 \cdot (\boldsymbol{f}_1 \times \boldsymbol{f}_3) / |\boldsymbol{f}_1 \times \boldsymbol{f}_3| \tag{5.43b}$$

$$\$_{T3} \circ \$_{O3} = \boldsymbol{f}_3 \cdot (\boldsymbol{f}_1 \times \boldsymbol{f}_2) / |\boldsymbol{f}_1 \times \boldsymbol{f}_2| \tag{5.43c}$$

According to Eqs. (5.43a), (5.43b), and (5.43c), the scalar triple product of $\boldsymbol{f}_i$ $(i = 1, 2, 3)$, i.e., $\boldsymbol{f}_1 \cdot (\boldsymbol{f}_2 \times \boldsymbol{f}_3)$, $\boldsymbol{f}_2 \cdot (\boldsymbol{f}_1 \times \boldsymbol{f}_3)$, or $\boldsymbol{f}_3 \cdot (\boldsymbol{f}_1 \times \boldsymbol{f}_2)$, can be used as the judgment condition of output transmission singularity. If $\boldsymbol{f}_1 \cdot (\boldsymbol{f}_2 \times \boldsymbol{f}_3) = \boldsymbol{f}_2 \cdot (\boldsymbol{f}_1 \times \boldsymbol{f}_3) = \boldsymbol{f}_3 \cdot (\boldsymbol{f}_1 \times \boldsymbol{f}_2)$, we may get $\$_{Ti} \circ \$_{Oi}$ $(i = 1,2,3)$, and the output transmission singularity occurs in the manipulator. In this case, all the three transmission forces are coplanar, and no transmission force can transmit power to enable the moving platform move along the vector $\boldsymbol{f}_1 \times \boldsymbol{f}_2$ (or $\boldsymbol{f}_2 \times \boldsymbol{f}_3, \boldsymbol{f}_1 \times \boldsymbol{f}_3$); or it can be said that an external force along the vector $\boldsymbol{f}_1 \times \boldsymbol{f}_2$ cannot be counterbalanced by any transmission force. Take the configuration in Fig. 5.13 as an example. Since all the coupler links $U_{i,1}U_{i,2}$ $(i = 1,2,3)$ are coplanar with the moving platform, all the axes of transmission forces lie in the same plane. Then, we can obtain $\boldsymbol{f}_1 \cdot (\boldsymbol{f}_2 \times \boldsymbol{f}_3) = 0$, and the output transmission singularity occurs. Therefore, both the constraint singularity and the output transmission singularity occur at the configuration in Fig. 5.12.

From the above analysis in this section, it can be seen that the approach introduced in this chapter is able to not only identify all possible singularities of the 3-RUU parallel manipulator but also explain their physical meanings in terms of motion/force transmission.

References

Bonev IA, Gosselin CM (2001) Singularity loci of planar parallel manipulators with revolute joints. In: Proceedings of the 2nd workshop on computational kinematics, Seoul, South Korea, pp 291–299

Daniali HRM, Zsombor-Murray PJ, Angeles J (1995) Singularity analysis of a general class of planar parallel manipulators. In: Proceedings of IEEE ICRA, pp 1547–1552

Gosselin CM, Angeles J (1990) Singularity analysis of closed loop kinematic chains. IEEE Trans Robot Autom 6(3):281–290

Liu GF, Lou YJ, Li ZX (2003) Singularities of parallel manipulators: a geometric treatment. IEEE Trans Robot Autom 19(4):579–594

Mayer St-Onge B, Gosselin CM (2000) Singularity analysis and representation of the General Gough-Stewart Platform. Int J Robot Res 19(3):271–288

Park FC, Kim JW (1999) Singularity analysis of closed kinematic chains. J Mech Des 121:32–38

Chapter 6
Workspace of Parallel Mechanisms

Abstract This chapter deals with a geometric determination of the theoretical and usable workspace for a parallel mechanism. Several parallel mechanisms including a planar 5R parallel mechanism, a DELTA mechanism, several 3-DOF spatial parallel mechanisms, and 6-RUS parallel mechanism are illustrated. In particular, the orientational workspace of 3-[PP]S type of parallel mechanism with a high rotational capability is described in detail. The workspaces presented in this chapter are actually a foundation on determining other two types of workspaces, i.e., the good-condition workspace and good-transmission workspace discussed in the following chapter.

Keywords Workspace • Orientational workspace • High rotational capability • Parallel mechanism • 3-[PP]S type

Workspace is an important index to evaluate the working performance of a mechanism. How to determine the workspace is one of the most important issues in the design of a mechanism, especially for a parallel mechanism which is well known to have relatively small useful workspace. What is more, the geometric determination of the workspace for a parallel mechanism attracts more and more researcher's attention (Gosselin 1990; Bonev and Ryu 2001a; Merlet 1999a).

There are several types of workspaces, such as theoretical workspace, researchable workspace, orientational workspace, constant-orientation workspace, dexterous workspace, and usable workspace. All of these workspaces can be thought of as some kind of maximal workspace. However, for any one of these workspaces, there has more or less singularity within or on the boundary of the workspace. Therefore, these workspaces usually cannot be applied into the design of mechanisms that need a singularity-free or even good-condition working space. In this section, we consider only the theoretical workspace and the usable workspace. In Chap. 7, two kinds of workspaces, i.e., good-condition workspace (GCW) and good-transmission

X.-J. Liu and J. Wang, *Parallel Kinematics: Type, Kinematics, and Optimal Design*, Springer Tracts in Mechanical Engineering, DOI 10.1007/978-3-642-36929-2_6,

workspace (GTW) that can be directly used in the design of a parallel mechanism, will be introduced. The workspaces presented in this chapter are the analysis foundation of the GCW and GTW.

6.1 The Planar 5R Parallel Mechanism

6.1.1 Theoretical Workspace

Theoretical workspace is defined as the region that the output point can reach if θ_i changes from 0 to 2π without the consideration of interference between links and the singularity.

From Eq. (4.1), one may see that if θ_1 is specified, the workspace of the first leg is a circle centered at the point $\left(r_1 \cos\theta_1 - r_3,\ r_1 \sin\theta_1\right)$ with a radius of r_2. The circle is denoted as C_{11}. If θ_i changes from 0 to 2π, the center point is located at a circle centered at point $A_1 \left(-r_3\ 0\right)$ with a radius of r_1. The circle is denoted as C_{12}. Then, the workspace of the leg is the enveloping region of the circle C_{11} when its center rolls at the circle C_{12}. Actually, the enveloping region is an annulus bounded by two circles given by Eqs. (5.3) and (5.4). For the second leg, the workspace is an annulus bounded by circles given by Eqs. (5.5) and (5.6). When $r_1 = r_2$ the workspaces are two circles defined by Eqs. (5.3) and (5.5), respectively. The theoretical workspace of the mechanism is the intersection of the two annuluses. For example, the workspace of a mechanism with $r_1 = 1.2$, $r_2 = 1.0$, and $r_3 = 0.8$ is shown as the hatched region in Fig. 6.1, where the boundaries are also illustrated.

From the above analysis, one may see that if $r_3 > r_1 + r_2$, there is no intersection, i.e., no workspace. Especially, if $r_3 = r_1 + r_2$, the workspace is only one point, which is the center of the line segment A_1A_2 of Fig. 6.1.

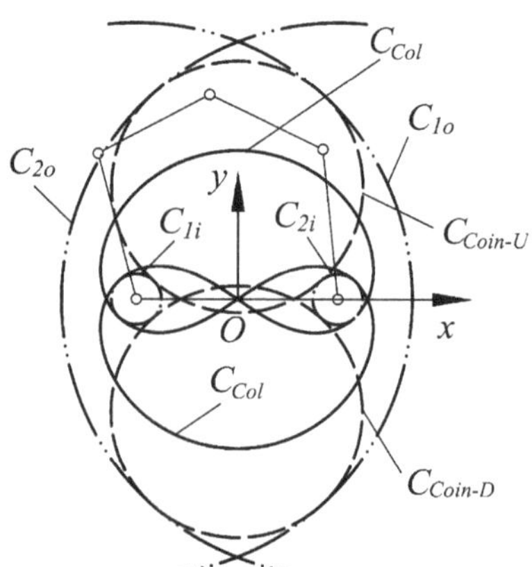

Fig. 6.1 The singular loci of a mechanism

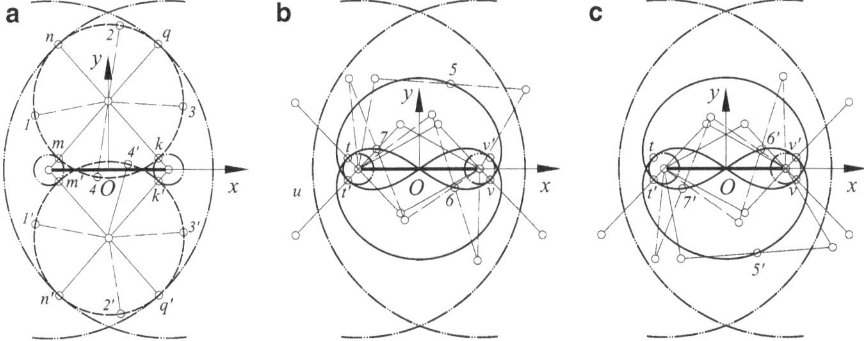

Fig. 6.2 The second kind of loci of a mechanism with different working modes

6.1.2 Usable Workspace

6.1.2.1 Definition

As there exist singular loci inside the theoretical workspace, a mechanism wants to move from one point to another it maybe should pass a singular configuration. That means it should change from one working mode to another. In practice, changing working mode during the working process is definitely impossible. Therefore, we should find out a working space without singularity.

The *usable workspace* is defined as the maximum continuous workspace that contains no singular loci inside but bounded by singular loci outside. According to this definition, not every point within the usable workspace can be available for a practical mechanism. The mechanism will be out of control at the points on the boundaries and their neighborhoods. But within this region, the mechanism with a specified working mode can move freely.

In Sect. 5.1, the first and second kinds of singularities of the mechanism have been presented, and singular loci have been obtained. There are two cases for this kind of singular locus. One is that when the points B_1 and B_2 are coincident. The loci are actually two circles given by Eq. (5.7), which are denoted as $C_{\mathrm{Coin-U}}$ and $C_{\mathrm{Coin-D}}$. The second case occurs when B_1PB_2 is completely extended. The locus is presented by Eq. (5.9) or Eq. (5.10), which is denoted as C_{Col}. The first kind of singularity is actually the boundary of a theoretical workspace. Then, a mechanism with every working mode can have such singular loci. However, as the second kind of singularity occurs inside the workspace, not every working mode has all such singularities. Normally, for most 5R parallel mechanisms, there are four tangent points between the first and second kinds of singular loci. At these points, the mechanism is in the change point's mechanism. The points can be used to identify which singular loci a specified working mode can have. For example, all singular loci of the mechanism $r_1 = 1.2$, $r_2 = 1.0$ and $r_3 = 0.8$ are shown in Fig. 7.1. And Fig. 6.2 shows some singular configurations and singular loci of the mechanism. As

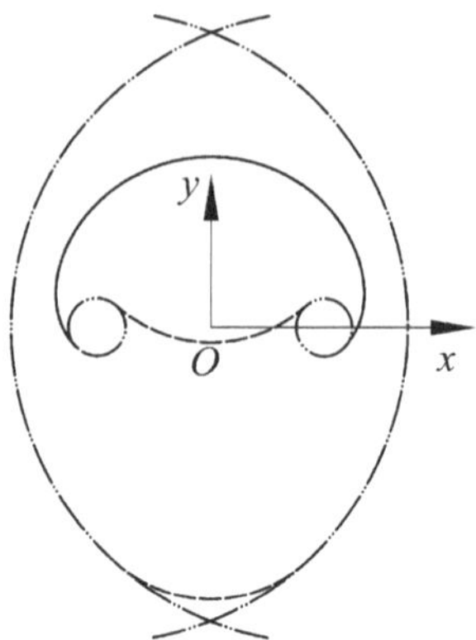

Fig. 6.3 The singular loci of a 5R mechanism with the working mode "+−"

shown in Fig. 6.2a, there are four tangent points m, n, q and k between the singular curve $C_{\text{Coin-U}}$ and the first kind of singular loci. At these four points, both of the first and second kinds of singularities occur. The four points divide the singular curve $C_{\text{Coin-U}}$ into four parts. At the arc $m1n$, the mechanism is in singular only when it is with the "− −" mode. At the arcs $n2q$, $q3k$, and $k4m$, the working modes "− +", "++", and "+ −" are in singular, respectively. Similarly, at $C_{\text{Coin-D}}$, arcs $m'1'n'$, $n'2'q'$, $q'3'k'$, and $k'4'm'$ arc the singular curves for the working modes "++", "+ −", "− −", and "− +", respectively. For the singular locus C_{Col}, as shown in Fig. 6.2b, c, curves $t'5v$, $t76v$, $t5'v'$, and $t'7'6'v'$ are the singular loci for the working modes "+−", "++", "−+", and "− −", respectively. For the mechanisms with the parameter condition like $r_1 > r_2 + r_3$ and $r_2 > r_3$, the condition does not allow the case that the singular locus C_{Col} and the first kind of singular locus are tangent to occur. These mechanisms with the working modes "++" and "− −" do not have the singularity that B_1PB_2 is completely extended.

In this chapter, we are just concerned about the mechanism with the working mode "+−". Figure 6.3 shows the singular loci of the example with the working mode.

The singular loci shown in Fig. 6.3 can be used to determine the usable workspace of a 5R parallel mechanism with both the working mode "+ −" and the up-configuration. The workspace of a mechanism should be continuous. As shown in Fig. 6.3, the singular loci divide the theoretical workspace into several regions. The regions for the mechanisms are usually separated. A practical mechanism cannot access the singular loci from one region to another. Even though, the two actuated links of a mechanism may be both crank links. Normally, only the region above the x-axis will be used in practice. For some mechanisms, there exist several regions with good performance. However, they are usually separated by the region with bad performance. In order to present a workspace closer to the practical workspace of the mechanism, the usable workspace defined here will not be the whole workspace. What is more, in the workspace under the x-axis, the two actuated links of a mechanism with the working mode "+ −" usually intersect. Practically, we prefer the configuration as shown in Fig. 4.1. For such reasons, we define the

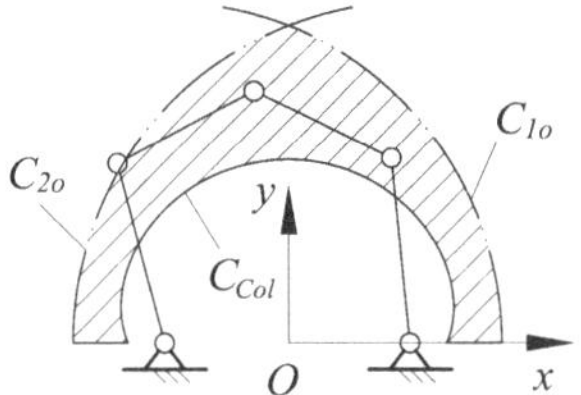

Fig. 6.4 The usable workspace of a mechanism

region above the x-axis as the usable workspace of a 5R parallel mechanism. For example, the usable workspace of the mechanism with $r_1 = 1.2$, $r_2 = 1.0$, and $r_3 = 0.8$ is shown as the hatched region in Fig. 6.4. Within this workspace, the mechanism has the configuration with both the inverse working mode "+ −" and the forward up-configuration.

6.2 DELTA

We may see that there is no determination for orientation but only for position of the mobile platform of a DELTA. The workspace of a DELTA is embedded in a three-dimensional space. Based on the inverse kinematics of the robot, the geometric determination of the workspace for a DELTA can be obtained, from Eq. (3.53), as follows

$$(x - x_i)^2 + (y - y_i)^2 + (z - z_i)^2 = R_2^2 \quad i = 1, \ldots 3 \tag{6.1}$$

where

$$x_i = (R_B - R_P) \cos \alpha_i + R_1 \cos \alpha_i \cos \theta_i \tag{6.2}$$

$$y_i = (R_B - R_P) \sin \alpha_i + R_1 \sin \alpha_i \cos \theta_i \tag{6.3}$$

$$z_i = R_1 \sin \theta_i . \tag{6.4}$$

In fact, Eq. (6.1) represents three spherical faces centered at the points $O_i\,(x_i,\ y_i,\ z_i)$, and each of the spherical radius is R_2. Equations (6.2), (6.3), and (6.4) can be rewritten as

$$R_1 \cos \alpha_i \cos \theta_i = x_i - (R_B - R_P) \cos \alpha_i \tag{6.5}$$

$$R_1 \sin \alpha_i \cos \theta_i = y_i - (R_B - R_P) \sin \alpha_i \tag{6.6}$$

$$R_1 \sin \theta_i = z_i . \tag{6.7}$$

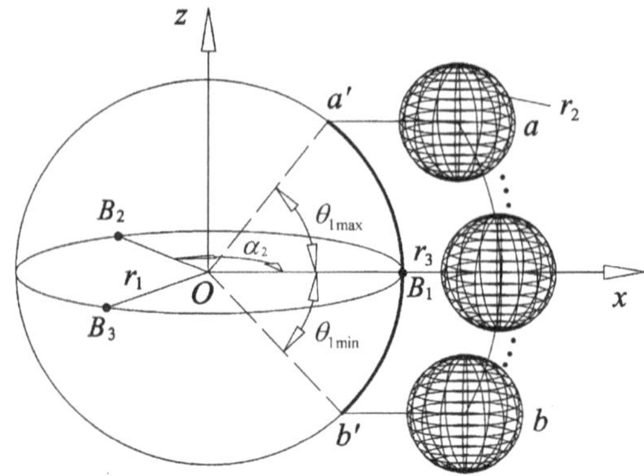

Fig. 6.5 The workspace of the first leg is the enveloping solid of a sphere rolling on the arc of ab

Equations (6.5), (6.6), and (6.7) represent an arc on a sphere centered at the point O. For the first leg of a DELTA (i.e., $i = 1$), it is the arc ab as shown in Fig. 6.5. The length of the arc is related to the input angle of θ_1, i.e., the values $\theta_{1\,max}$ and $\theta_{1\,min}$. The center point O_1 $(x_1, \ y_1, \ z_1)$ is then located at the arc ab, where $aa' = bb' = r_3$. The workspace of the first leg is the enveloping solid of a sphere rolling on the arc ab. The radius of the sphere is R_2. Especially, if the distance between point a' and b', which is also depended on $\theta_{1\,max}$ and $\theta_{1\,min}$, is less than $2r_2$, there is cavity in this solid. The enveloping solid is a torus solid, which is closed when $\theta_1 = 2\pi$. The solid volume has no relation to whether R_3 is negative or positive. The workspace of a DELTA is the intersection of such three identical torus solids, in which the two other torus solids for the second and third legs are distributed with an angle $2\pi/3$ about the one for the first leg.

One may use the software AutoCAD (other software could also do such things) to obtain the workspace of a DELTA based on above-mentioned analysis. Theoretically, the inputs $\theta_i = 2\pi$, which means that the workspace of each leg for a DELTA is a closed torus. In this case, generally, one can use the following sequence to obtain the workspace of a DELTA based on AutoCAD R14:

- From origin O $(0, \ 0, \ 0)$ draw three rays oc, od, and oe along x-, y-, and z-axes, respectively, using the Cartesian coordinate O-xyz.
- Draw a circle centered at the point $(r_1, \ 0, \ 0)$ with the radius r_2.
- If $r_2 > r_1$, a section of the circle lies in counterpart of ray oc, mirror this part with respect to od the side of oc, using the "MIRROR" command, and edit the two arcs as one entity by using the command "PEDIT"; if $r_2 \le r_1$, the circle only lies in the side oc.
- Using the "REVOLVE" command, a torus can be obtained by revolving the entity (a crescent or a circle generated in last step) with respect to the ray od.
- Move the torus the distance r_3 along oc by using the "MOVE" command. The moved torus is the workspace of the first leg of a DELTA.

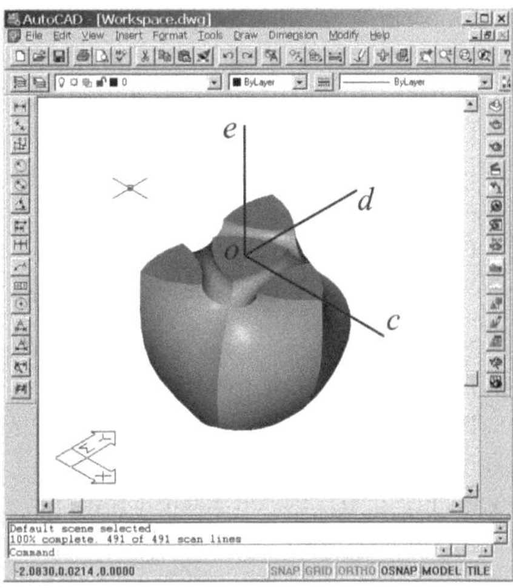

Fig. 6.6 Generation of the workspace for a DELTA based on AutoCAD R14

- Copy two other tori with respect to the point $O\,(0,\ 0,\ 0)$ using the "ARRAY" command. The workspace of two other legs is obtained.
- The intersection of the three tori can be obtained with the command "INTERSECT". The result is the workspace of a DELTA. The lower hemi-part of the workspace can be obtained by the "SLICE" command. This is the workspace with the assembly as shown in Fig. 6.6. The volume can also be obtained from the pull-down menu "TOOLS-INQUIRY-MASS PROPERTIES".

In this process, only the kinematic constraints are taken into account.

As an example, a DELTA with the non-dimensional parameters $r_1 = 1.3$, $r_2 = 1.2$, and $r_3 = 0.5$ if $\theta_i \in [0,\ 2\pi]$ is considered here. The detailed operating sequence could be:

1. As shown in Fig. 6.6, in the Southeast Isometric System of AutoCAD R14, draw three rays oc, od, and oe by using the "LINE" command, where $O\,(0,\ 0,\ 0)$, $c\,(3,\ 0,\ 0)$, $d\,(0,\ 3,\ 0)$, and $e\,(0,\ 0,\ 3)$.
2. Using the "CIRCLE" command, draw a circle centered at point $(1.3,\ 0,\ 0)$ with the radius 1.2.
3. Using the "REVOLVE" command, revolve the circle with respect to ray od with the angle 360° and a torus entity is obtained.
4. Move the entity from point $(0,\ 0,\ 0)$ to $(0.5,\ 0,\ 0)$ by using command "MOVE". This is then the workspace of the first leg.
5. Using the "ARRAY" command with the number 3, copy two other tori with respect to $O\,(0,\ 0,\ 0)$. They are the workspaces of the second and third legs.
6. Using the "INTERSECT" command, one may obtain the intersection of the three tori, which is the workspace of the robot with $r_1 = 1.3$, $r_2 = 1.2$, and $r_3 = 0.5$.

One can find that the part above the plane o-cd is identical with that under the plane. Using the "SLICE" command, divide the intersection to two identical parts with respect to plane o-cd and erase the part above the plane. The left solid entity shown in Fig. 6.6 is actually the workspace of the robot with the assembly as shown in Fig. 3.10. All points that the shaded solid contains can be accessible if each of $\theta_i \quad (i = 1, 2, 3)$ varies from 0 to 2π.
7. One can also know its volume, 8.6382, by using the pull-down menu "TOOLS-INQUIRY-MASS PROPERTIES".

What is more, this process can be programmed using AutoLisp language (Rawls and Hagen 1998) of AutoCAD.

Practically, θ_i cannot be from 0 to 2π because of joint limits and leg interference constraints. Then, the method to obtain its workspace will be a little different. For example, if $\theta_i \quad (i = 1, 2, 3)$ are specified as $\left[-\frac{\pi}{2}(-90°),\ \frac{\pi}{4}(45°)\right]$, in spite of other steps, steps (2) and (3) in above sequence should be replaced by:

2. As shown in Fig. 6.7a, define two reference points $f(1.3\cos(-\pi/2),\ 0,\ 1.3\sin(-\pi/2))$ and $g(1.3\cos(\pi/4),\ 0,\ 1.3\sin(\pi/4))$. In the plane defined by points o, d, and g, draw a circle centered at point g with the radius 1.2.
3. Revolve the circle with respect to ray od with the angle 135°, i.e., 45° – (–90°), a solid entity obtained. Draw two solid spheres with the radii 1.2 centered at points f and g, respectively, using the command "SPHERE". With the "UNION" command, unify the solid entity and the two spheres into one entity, which is shown in Fig. 6.7b.

Then, the workspace of the DELTA with the non-dimensional parameters $r_1 = 1.3$, $r_2 = 1.2$, and $r_3 = 0.5$ if $\theta_i \in [-\pi/2,\ \pi/4]$ can be obtained as shown in Fig. 6.7c following steps (4), (5) and (6). Its workspace volume is 2.3686.

6.3 A Parallel Cube-Mechanism

From Eqs. (3.59), (3.60), and (3.61), one may see that if the inputs of the cube-mechanism are given, workspaces of the three legs are three spheres, which are centered at points $(0,\ 0,\ \rho_1 + r)$, $(\rho_2 + r,\ 0,\ 0)$, and $(0,\ \rho_3 + r,\ 0)$, respectively. The radius of each of the three spheres is L. If the inputs ρ_i are specified as

$$\rho_i \in [\rho_{\min}, \rho_{\max}], \tag{6.8}$$

the workspace of each of the three legs of the cube-mechanism is then the enveloping solid of a sphere whose center is moving along a line between two points C_i and D_i, where $C_1(0,\ 0,\ \rho_{\min} + r)$ and $D_1(0,\ 0,\ \rho_{\max} + r)$ are for

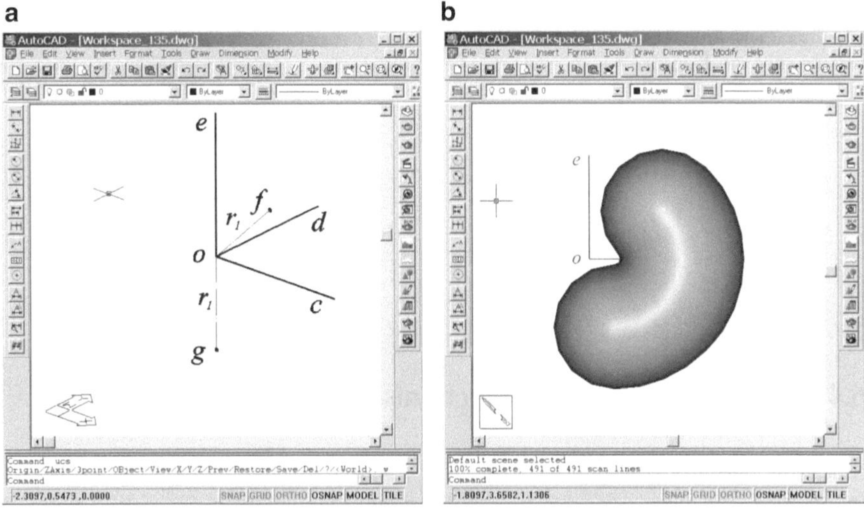

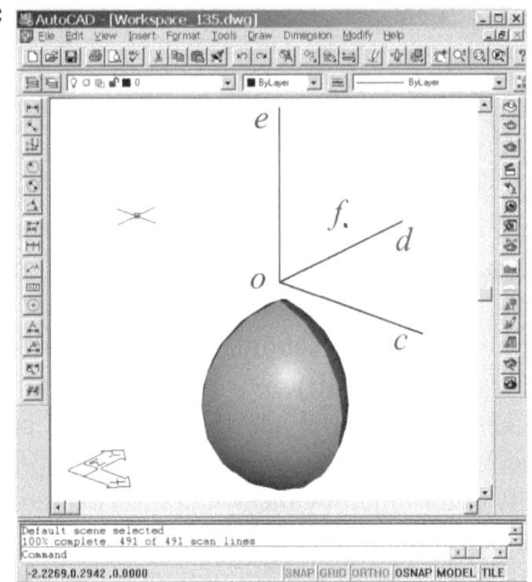

Fig. 6.7 Generation of the workspace for a DELTA with specified inputs: (a) the definition of points f and g; (b) side view of the workspace for the first leg; (c) the workspace of the robot

the first leg, $C_2\,(\rho_{\min}+r,\ \ 0,\ \ 0)$ and $D_2\,(\rho_{\max}+r,\ \ 0,\ \ 0)$ for the second leg, and $C_3\,(0,\ \ \rho_{\min}+r,\ \ 0)$ and $D_3\,(0,\ \ \rho_{\max}+r,\ \ 0)$ for the third leg. And the workspace of the cube-mechanism is, then, the intersection of the three enveloping solids.

For example, the workspace of a cube-mechanism with the geometric parameters $r = 260$, $L = 1,000$, $\rho_{\max} = -774$ and $\rho_{\min} = -1,746$ ($|\rho_{\max} - \rho_{\min}| = 972$) is shown in Fig. 6.8, the volume of which is 940022941.13, which can be obtained easily on the AutoCAD R14 platform.

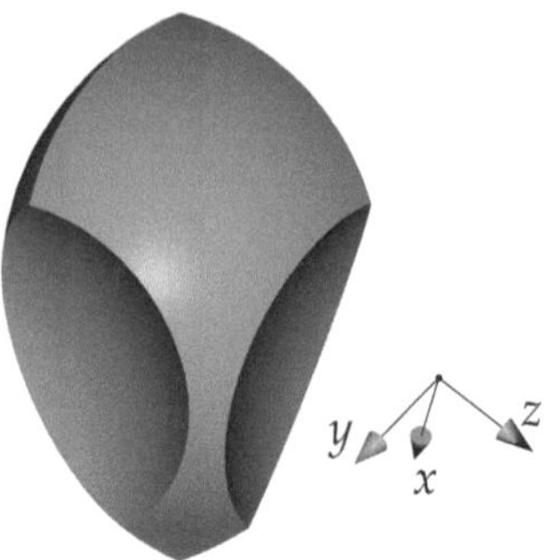

Fig. 6.8 The workspace of a three translational DOFs parallel cube-mechanism

6.4 Two 3-[PP]S Parallel Mechanisms

Since the tilting capability of a 3-[PP]S parallel mechanism is very important in practice, we here consider only the orientational workspace of two parallel mechanisms shown in Fig. 3.23.

6.4.1 *Tilt Angle of a Spherical Joint*

In this chapter, orientational workspace is defined as the maximum tilt angle θ for a given azimuth angle ϕ. In each of the two [PP]S parallel mechanisms shown in Fig. 3.23, there are three spherical joints. It is well known that a spherical joint has a limited tilt angle, but unlimited self-rotation. For this reason, we are only concerned with the tilt angle of each spherical joint. This angle restricts the orientational workspace of the whole mechanism. When the orientation parameters ϕ and θ of the mechanism are specified, the tilt angle of each spherical joint can be obtained. The tilt angle is actually determined by the orientation of lines oP_i and P_iB_i. Most often, instead of using ball joints, spherical joints having the wrist structure shown in Fig. 6.9a are chosen, where the w-axis is defined as the self-rotation axis and the tilt angle is relative to the rotation angles about the v and u axes, which are referred to as the *pitch* and *yaw* angles respectively, denoted here by α_v and α_u.

For the third leg of the 3-$\underline{\text{P}}_\text{V}\text{P}_\text{H}$S mechanism, the two angles are illustrated in Fig. 6.9b, where points o_o and P_{3o} are the projections of points o and P_3 in the O-XY plane. Since the orientation of the w-axis remains constant, the pitch and yaw angles are

$$\alpha_v = \tan^{-1}\left(\frac{z_{3\Re} - z}{|o_o P_{3o}|}\right) = \tan^{-1}\left[\frac{\sin\phi\sin\theta}{\sqrt{1-\sin^2\phi\sin^2\theta}}\right] \tag{6.9}$$

$$\alpha_u = \tan^{-1}\left(\frac{y}{x - x_{3\Re}}\right) = \tan^{-1}\left[\frac{\sin\phi\cos\phi\,(1-\cos\theta)}{\cos^2\phi + \sin^2\phi\cos\theta}\right]. \tag{6.10}$$

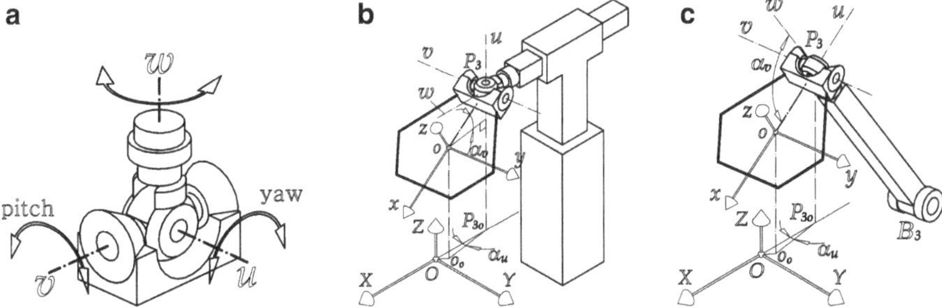

Fig. 6.9 Tilt angle of a spherical joint: (**a**) a type of spherical joint; (**b**) this joint in a 3-$\underline{P}_V P_H S$ mechanism; (**c**) this joint in a 3-$\underline{P}_V RS$ mechanism

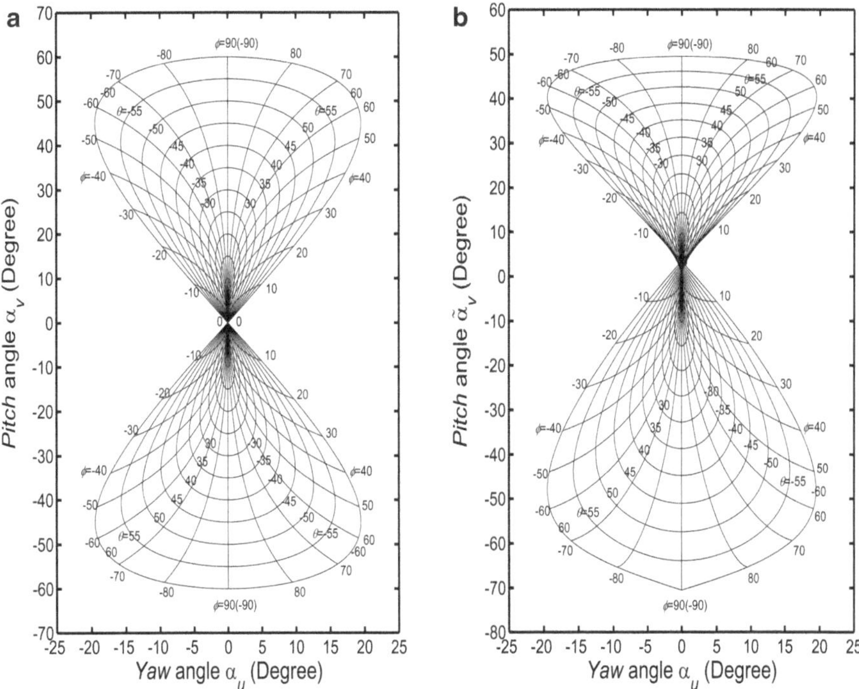

Fig. 6.10 Relationship between the tilt angle of a spherical joint and the orientation of the mobile platform: (**a**) for a 3-$\underline{P}_V P_H S$ parallel mechanism; (**b**) for a 3-$\underline{P}_V RS$ parallel mechanism

The above equations show that the pitch and yaw angles are independent of the parameters r and z. Figure 6.10a illustrates the corresponding angles of the spherical joint at P_3 as function of the orientation of the platform of a 3-$\underline{P}_V P_H S$ mechanism. It shows that, at any orientation, the pitch angle is never less than the yaw angle.

If the tilt angle θ belongs to $\left[-60°, 60°\right]$, the maximum yaw angle is smaller than 20° and the maximum pitch angle is equal to the maximum tilt angle of the mobile platform.

As shown in Fig. 6.9c, since the orientation of the w-axis in the spherical joint of the 3-$\underline{P}_V$RS mechanism varies, the pitch angle α_v from the w-axis can no longer be calculated by Eq. (6.9), while the yaw angle α_u can be still obtained using Eq. (6.10). For the spherical joint in the third leg of the 3-$\underline{P}_V$RS mechanism, angle α_v can be expressed as

$$\alpha_v = \pi - \cos^{-1}\left(\frac{|oP_3|^2 + |P_3B_3|^2 - |oB_3|^2}{2\,|oP_3|\,|P_3B_3|}\right) \tag{6.11}$$

where α_v is independent of z. Since the spherical joint is usually not pitched from zero, the angle given in Eq. (6.11) is not the absolute pitch angle of the spherical joint. In this chapter, the absolute pitch angle is defined as the relative angle between the current orientation and the reference orientation ($\theta = 0$). At $\theta = 0$, the pitch angle is denoted by α_{vo}. Then, the absolute pitch angle of the spherical joint will be $\tilde{\alpha}_v = \alpha_v - \alpha_{vo}$.

For example, if $R = 2.5$, $r = 1.0$, $L = 4.5$, $\alpha_{vo} = 70°$, and $\theta \in \left[-60°, 60°\right]$, then the angles $\tilde{\alpha}_v$ and α_u for a constant orientation (ϕ, θ) are as illustrated in Fig. 6.10b. The figure shows that, in a 3-$\underline{P}_V$RS mechanism, just as is the case in a 3-$\underline{P}_V P_H$S mechanism, the pitch angle is larger than the yaw angle and reaches its maximum when $\phi = 90°$. From Fig. 6.10a, b, we see that, unlike the case of a 3-$\underline{P}_V P_H$S mechanism, the pitch and yaw angles of a 3-$\underline{P}_V$RS mechanism are not symmetrical with respect to $\theta = 0$. For this reason, in assembling a 3-$\underline{P}_V$RS mechanism, the spherical joint should be mounted at an inclination to avoid exceeding its pitch limits.

6.4.2 Orientational Workspace of the 3-$\underline{P}_V P_H$S Parallel Mechanism

As shown in Fig. 6.10a, for the third leg of this parallel mechanism, the pitch angle reaches its maximum when $\phi = 90°$. Using the structure shown in Fig. 6.9a, the pitch capability of such a spherical joint can be as high as ±90° and the yaw capability nearly ±45°. Therefore, the orientational workspace of the mobile platform of a 3-$\underline{P}_V P_H$S mechanism is actually determined by the pitch angle of the spherical joint, and it can be as high as ±90°. However, the mechanism is in a singularity at $\alpha_v = \pm 90°$. Besides, a tilting capability in the ±60° range is high enough for machining applications.

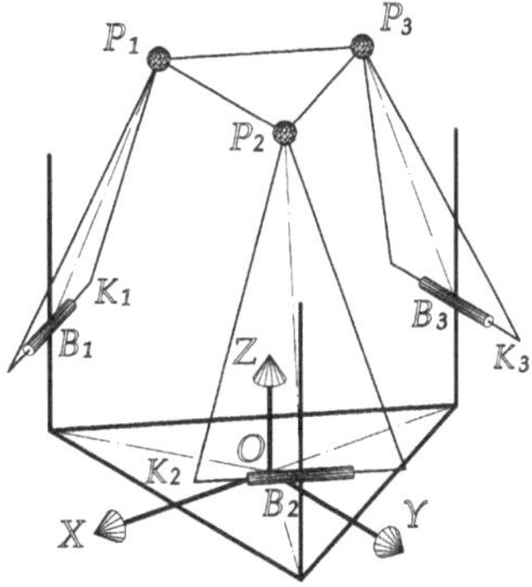

Fig. 6.11 The mobile platform plane and leg planes of a 3-PRS mechanism

6.4.3 Orientational Workspace of the 3-$\underline{P}_V$RS Parallel Mechanism

It is worth noting that, although Fig. 6.10b shows that the mobile platform can reach an orientational workspace of $[-60°, 60°]$, not all 3-$\underline{P}_V$RS mechanisms have such a capability. Indeed, the orientational workspace also depends on the mechanism singularities.

While the 3-$\underline{P}_V P_H$S mechanism has no singularities for tilt angles of less than 90°, the 3-$\underline{P}_V$RS mechanism has two types of singularity. As shown in Fig. 6.11, the first kind of singularity occurs if any one of the three leg planes is parallel to the *O-XY* plane; the second one arises when the mobile platform plane, defined by three points P_i, is in any one of the three leg planes, specified by a triple of points P_i, B_i, and K_i where K_i are any points in the axes of the revolute joints.

In the application of a 3-$\underline{P}_V$RS parallel mechanism, we usually expect its tilt angle limits to be symmetrical with respect to its reference orientation ($\theta = 0$), for a given a-axis (specified by ϕ). For example, we usually say that the parallel mechanism can tilt ±40°, but not +30° and −50°, with respect to any a-axis. Here, for a given ϕ, the tilt angle is denoted by θ_P when the mobile platform is in a first type of singularity, and it is referred to as θ_N when the mobile platform is in a second type of singularity. Therefore, in this chapter, the orientational workspace for a specified ϕ of a 3-$\underline{P}_V$RS mechanism is defined as the absolute value of the smaller of θ_P and θ_N. Then, if the orientational workspace for a specified ϕ is 45°, we say that the mobile platform can tilt ±45° about the a-axis.

For example, the orientational workspace of a 3-$\underline{P}_V$RS mechanism with $R = 2.5$, $r = 1.0$, and $L = 4.5$ is shown in Fig. 6.12, from which we may see that the capability reaches its maximum when $\phi = 30°, -30°$, and ±90° and reaches its minimum when $\phi = 0, -60°$, and 60°. It is worth noting that the orientational workspace of the 3-$\underline{P}_V$RS mechanism is independent of the z coordinate.

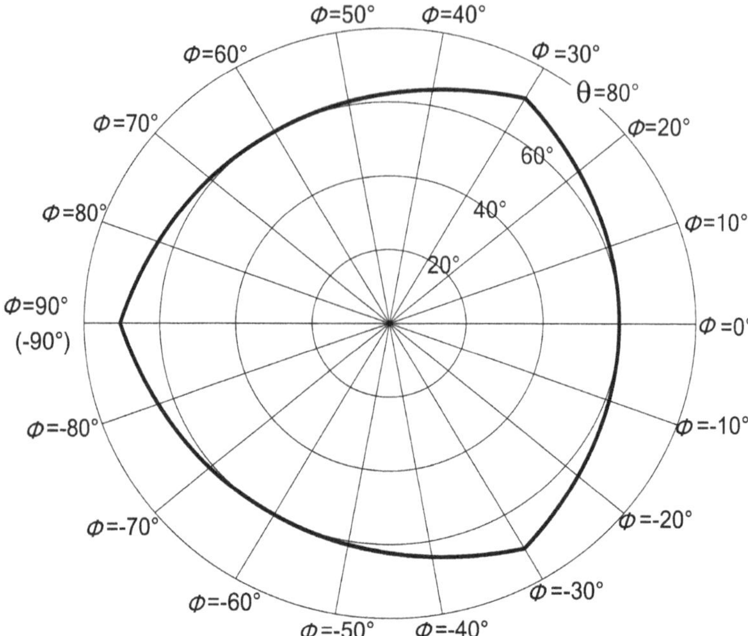

Fig. 6.12 Orientational workspace of a 3-$\underline{P}_V$RS parallel mechanism as limited by singularities only

Thus, from Fig. 6.12, we can pinpoint the largest tilt that can be accomplished in any direction (any ϕ) without reaching a singularity, which for this example is about 60°. Of course, the tilt should be further limited, because the mechanism performance deteriorates significantly when a singularity is close. How close to a singularity can the mechanism safely operates is a question that requires further analyses that are not the object of this chapter. Let us assume that in this example, we may operate safely within a 50° tilt.

Having defined the maximal singularity-free orientational workspace, we should find the corresponding required characteristics for the spherical joints. Figure 6.10b shows that we should be only concerned about the pitch angle $\tilde{\alpha}_v$, which reaches its maximum value of $\tilde{\alpha}_v = 42.57°$ when $\theta = 50°(-50°)$ and $\phi = 90°(-90°)$ and its minimum value of $\tilde{\alpha}_v = -57.43°$ when $\theta = 50°(-50°)$ and $\phi = -90°(90°)$.

6.5 The Spatial 2-RRU&1-RR(Pa)R Parallel Mechanism

The mobile platform of a spatial 2-RRU&1-RR(Pa)R parallel mechanism has two translational DOFs and one rotational DOF. So, in this section, both the constant-orientation workspace and the orientational workspace will be discussed. For a parallel mechanism with a rotational DOF, the rotational capability is usually concerned. So, an index to evaluate the rotational capability of the mobile platform of the mechanism at the point in the reachable workspace will be defined.

From the inverse kinematics Eqs. (3.92), (3.93), and (3.94), we may obtain

$$(y - y_1)^2 + (z - z_1)^2 = L_a^2 \tag{6.12}$$

$$(y - y_2)^2 + (z - z_2)^2 = L_a^2 \tag{6.13}$$

$$y^2 + (z - z_3)^2 = r_3^2 \tag{6.14}$$

where

$$y_1 = R + L_b \cos\theta_1 - r, \quad z_1 = L_b \sin\theta_1 \tag{6.15}$$

$$y_2 = r + L_b \cos\theta_2 - R, \quad z_2 = L_b \sin\theta_2 \tag{6.16}$$

$$z_3 = L_b \sin\theta_3 - r\sin\phi, \quad r_3^2 = L_a^2 - (r\cos\phi - R + L_b\cos\theta_3)^2 \tag{6.17}$$

from which we may see that if inputs θ_i and the orientation ϕ of the mobile platform are specified, Eqs. (6.12), (6.13), and (6.14) represent three circles in the frame O-yz. The centers are $(y_1,\ z_1)$, $(y_2,\ z_2)$, and $(0,\ z_3)$, and the radii are L_a, L_a, and r_3, respectively. Each of these circles denotes the set of points that the end point O' of each of the three legs can reach.

From Eqs. (6.15) and (6.16), one may see that there is no relationship between the orientation ϕ of the mobile platform and workspaces of the first and second legs. But the workspaces are related to the inputs θ_1 and θ_2. Let us consider the Eqs. (6.15) and (6.16) furthermore. One obtains

$$[y_1 - (R - r)]^2 + z_1^2 = L_b^2 \tag{6.18}$$

$$[y_2 - (r - R)]^2 + z_2^2 = L_b^2 \tag{6.19}$$

which represent two circles, actually two arcs when the input ranges of θ_1 and θ_2 are specified, as well. Then, centers of all circles given by Eqs. (6.12) and (6.13) are restricted on arcs given by Eqs. (6.18) and (6.19), respectively. Workspaces of the first and second legs, denoted as Ω_1 and Ω_2, are two enveloping faces generated by two circles; the radii are L_a, whose centers are rolling on another two circles centered at points $(R - r,\ \ 0)$ and $(r - R,\ \ 0)$, respectively. Radii of the latter two circles are L_b. And they can be defined as

$$\Omega_1 = \Big[(y,\ z)\,|(y - y_1)^2 + (z - z_1)^2 = L_a^2,\ \ y_1 = R + L_b \cos\theta_1 - r,$$

$$z_1 = L_b \sin\theta_1,\ \ \theta_1 \in [\theta_{1\min},\ \ \theta_{1\max}]\Big] \tag{6.20}$$

$$\Omega_2 = \Big[(y,\ z)\,|(y - y_2)^2 + (z - z_2)^2 = L_a^2,\ \ y_2 = r + L_b \cos\theta_2 - R,$$

$$z_2 = L_b \sin\theta_2,\ \ \theta_2 \in [\theta_{2\min},\ \ \theta_{2\max}]\Big]. \tag{6.21}$$

The intersection, denoted as Ω_{12}, of the two enveloping faces is the intersection of workspaces for the first and second legs, which can be expressed as

$$\Omega_{12} = \Omega_1 \cap \Omega_2. \tag{6.22}$$

6.5.1 Constant-Orientation Workspace

The constant-orientation workspace is defined as the region that can be reached by the reference point on the mobile platform when the orientation of mobile platform is kept constant. Most research of the issue is focused on 6-DOF Stewart parallel mechanisms (Gosselin 1990; Bonev and Ryu 2001b). The definition and more detailed research for the constant-orientation workspace were presented by Merlet (1999a). For the mechanism considered in this chapter, the workspace is the region that the reference point O' can reach when the orientation ϕ is specified. Equation (6.14) gives the workspace, actually the reachable workspace denoted as Ω_{3R}, of the third leg, which is the set of a tuft of circles centered at point $(0,\ \ z_3)$, and the radius r_3. From Eq. (6.17) one may see that the radius and center of the circle are variable and dependent on the input θ_3. If the input range $\theta_3 \in [\theta_{3\min},\ \ \theta_{3\max}]$ is specified, the constant-orientation workspace of the mechanism is largely dependent on the workspace of the third leg. That is, if the point $(y,\ \ z)$ in the set Ω_{12} belongs to the constant-orientation workspace of the mechanism, the coordinate value of the point must satisfy Eq. (6.14). If the orientation ϕ is specified, the workspace of the third leg is defined the constant workspace of the third leg, which is denoted as Ω_{3O}. For specified value of ϕ, Ω_{3O} can be expressed as

$$\Omega_{30} = \Big[(y,\ z)\,|y^2 + (z - z_3)^2 = r_3^2,\quad z_3 = L_b \sin\theta_3 - r \sin\phi,$$

$$r_3^2 = L_a^2 - (r \cos\phi - R + L_b \cos\theta_3)^2,\ \theta_3 \in [\theta_{3\min}, \theta_{3\max}]\Big]. \tag{6.23}$$

Then the constant-orientation workspace of the mechanism, which is defined as Ω_O, can be written as

$$\Omega_O = \Omega_{12} \cap \Omega_{3O}. \tag{6.24}$$

And the reachable workspace Ω_{3R} of the third leg is the set of Ω_{3O} with all possible orientation ϕ.

6.5.2 Reachable Workspace

The reachable workspace is known as the region that can be reached by the reference point with at least one orientation. For the parallel mechanism studied in this chapter, if the reference point O' $(y,\ z)$ within Ω_{12} belongs to the reachable workspace of the parallel mechanism, $(y,\ z)$ must satisfy Eq. (6.14) for at least one value ϕ, and $\phi \in [0,\ 2\pi]$. The reachable workspace Ω_R of the mechanism can be expressed as

$$\Omega_R = \Omega_{12} \cap \Omega_{3R}. \tag{6.25}$$

6.5.3 Rotational Capability of the Mobile Platform

Rotational capability is one index to evaluate whether a device is competent for the task in hand or not. One of the disadvantages in the parallel mechanism is its lower rotational capability (Tonshoff and Grendel 1999), which limits the further application in the industry. The analysis of rotational capability of the output link (mobile platform) in the workspace is one of the most important issues in the design and application of a parallel kinematics (Merlet 1999b; Chrisp and Gindy 1999). According to the kinematics of the spatial 3-DOF parallel mechanism, the rotational DOF of the mechanism is the rotation of the mobile platform with respect to the y'-axis as shown in Fig. 3.12. The position vector of point P_3 in frame $\Re$ can be written as

$$(\boldsymbol{p}_3)_{\Re} = \begin{pmatrix} -r\cos\phi \\ y \\ r\sin\phi + z \end{pmatrix}. \tag{6.26}$$

Let $x_p = -r\cos\phi$, $y_p = y$, and $z_p = r\sin\phi + z$; Eq. (3.94) can be rewritten as

$$[x_P - (L_b\cos\theta_3 - R)]^2 + y_P^2 + (z_P - L_b\sin\theta_3)^2 = L_a^2 \tag{6.27}$$

which stands for a spherical surface centered at $(L_b\cos\theta_3 - R,\ 0,\ L_b\sin\theta_3)$, and the radius is L_a. If the input θ_3 of the third actuator is specified and the workspace point $(y,\ z)$ is given, Eq. (6.27) represents a circle centered at point $(L_b\cos\theta_3 - R,\ y,\ L_b\sin\theta_3)$, and the radius is $\sqrt{L_a^2 - y^2}$. The equation can be rewritten as

$$[x_P - (L_b\cos\theta_3 - R)]^2 + (z_P - L_b\sin\theta_3)^2 = L_a^2 - y_P^2 \tag{6.28}$$

which is located on the plane paralleling to O-xz plane. If θ_3 is specified as $\theta_3 \in [\theta_{3\,\min},\ \theta_{3\,\max}]$, Eq. (6.28) represents a tuft of circles of radius $\sqrt{L_a^2 - y^2}$. The set, denoted as Ω_{3M}, of points on the circles is the locus of point P_3 in the plane. As we have known, for a given position $(y,\ z)$ of the mobile platform, the locus of the point P_3 is the subset of a circle of radius r. The equation can be written as

$$x_p^2 + \left(z_p - z\right)^2 = r^2 \tag{6.29}$$

which is located on the plane paralleling to O-xz plane, defined by y. From Eqs. (6.28) and (6.29), we may see that all the circles defined by Eqs. (6.28) and (6.29) are located on a same plane. When $(y,\ z)$ and θ_3 are specified, the point of intersection $\left(x_p,\ y_p,\ z_p\right)$ of the two circles can be obtained.

Equation (6.28) minus Eq. (6.29) produces

$$z_p = ax_p + b \tag{6.30}$$

where

$$a = (L_b \cos\theta_3 - R)\,/(z - L_b \sin\theta_3)$$
$$b = \left[L_b^2 - y_p^2 - r^2 + z^2 - (L_b \cos\theta_3 - R)^2 - (L_b \sin\theta_3)^2\right] / \left[2\,(z - L_b \sin\theta_3)\right].$$

Substituting Eq. (6.30) into Eq. (6.27) leads to

$$A'x_p^2 + B'x_p + C' = 0 \tag{6.31}$$

where $A' = 1 + a^2$, $B' = 2a\,(b - z)$, and $C' = (b - z)^2 - r^2$. Then there is

$$x_p = \frac{-B' \pm \sqrt{B'^2 - 4A'C'}}{2A'}. \tag{6.32}$$

z_p can be also obtained by substituting Eq. (6.32) into Eq. (6.30). From Eqs. (6.30), (6.31), and (6.32), we may see that there are two points of intersection. The central angle of the arc defined by the two points is denoted as ζ, which can indicate rotational capability of the mobile platform at the given position $(y,\ z)$.

The central angle ζ, actually, the orientation ϕ of the mobile platform, represents configurations of the mobile platform at a given point of the workspace. What we should notice is that these configurations include that of singularities, which will separate the arc into two mechanism-inaccessible ones. In actually, the rotational capability of a real mechanism device can only be valued by the corresponding angle of one of the two arcs, which is denoted as ς, and there is $0 \le \varsigma \le 360°$. For the mechanism as shown in Fig. 3.12, the singularity is that of the third leg B_3P_3, when the lower link of the leg is in the mobile platform plane. And the index of rotational capability, which can evaluate the rotational capability of the mobile platform at a point within the workspace, is defined as

$$\mu = \varsigma\,/360°. \tag{6.33}$$

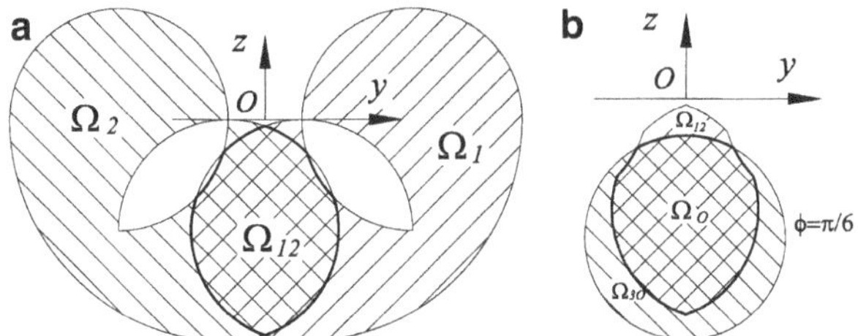

Fig. 6.13 The workspace of the mechanism for an example: (**a**) the reachable workspace Ω_{12}; (**b**) the constant-orientation workspace Ω_O, when $\phi = \pi/6$

And there is

$$0 \leq \mu \leq 1 \tag{6.34}$$

which means that the larger is the index μ, the better will the rotational capability be. When $\mu = 1$, the point is that of the dexterous workspace. The detailed presentation of the rotational capability for the parallel mechanism will be discussed by an example in the following section.

6.5.4 Example

As an example of application of the workspace analysis for the parallel mechanism proposed in this chapter, we here consider a spatial 3-DOF parallel mechanism with parameters $R = 1.0$, $r = 0.6$, and $L_a = L_b = 1.2$. The actuated motion is specified as

$$\theta_1 \in [-\pi/2,\ 0],\quad \theta_2 \in [-\pi,\ -\pi/2],\quad \theta_3 \in [-\pi,\ -\pi/2]. \tag{6.35}$$

According to the analysis of the workspace, the intersection Ω_{12} of the workspaces of the first and second legs is shown as in Fig. 6.13a. If the orientation of the mechanism is given as $\phi = \pi/6$, from Eq. (6.17), for the third leg there should be

$$L_a^2 - (r\cos\phi - R + L_b\cos\theta_3)^2 \geq 0 \tag{6.36}$$

from which the available input range can be obtained as

$$\theta_3 \in \left[-\cos^{-1}\left[(R - L_a - r\cos\phi)/L_b\right],\ -\pi/2\right] \tag{6.37}$$

and there are $z_{3\max} = -1.26$, $r_{3\min} = 0.0$, $z_{3\min} = -1.5$, and $r_{3\min} = 1.1$. The constant-orientation workspace, i.e., Ω_O, of the mechanism is illustrated in Fig. 6.13b. We can also obtain the constant-orientation workspace of the mechanism

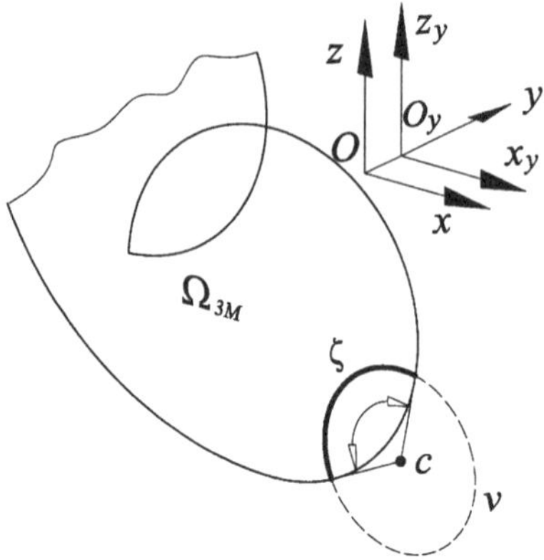

Fig. 6.14 The central angle ζ at point c

when $\phi = 0$, which is just the intersection of workspaces of the first and second legs, Ω_{12}. Therefore, we can reach the result that, for the parallel mechanism considered in this chapter, the reachable workspace is just the intersection of workspaces of the first and second legs.

As an example to present the rotational capability index μ, point O' on the mobile platform is specified as point c ($y_c = 0.37$, $z_c = -2.0$) in the reachable workspace. The circle described by Eq. (6.29) is shown as ν centered at point c in the plane O_y-$x_y y_y$, which is parallel to plane O-xy and $OO_y = y_c$, as illustrated in Fig. 6.14. The region Ω_{3M}, which is the locus of point P_3 in the plane O_y-$x_y y_y$ given by Eq. (6.28) when $\theta_3 \in [-\pi, \ -\pi/2]$, is shown as Fig. 6.14. The central angle of the arc of circle ν embodied in the region is $\zeta = 126.04°$ at the point c.

As mentioned in the last section, the central angle contains singular configurations of the mechanism. For the parallel mechanism studied here, to investigate the rotational capability of the mobile platform, only the singularity when the lower link of the third leg $P_3E_3B_3$ is in the mobile platform plane should be considered. And the singular configurations separate the arc of the corresponding central angle into two mechanism-inaccessible ones. In order to find the singular configurations, we can mirror the third leg $P_3E_3B_3$ and the mobile platform or $O'P_3$ into the plane O-xz. As illustrated in Fig. 6.15, the mirror of configuration P_3E_3 of the third leg into plane O-xz is P'_mE_3, and point O_m is the corresponding point of O' in plane O-xz. Then, $O_mP'_m$ is the mirror of $O'P_3$. Conditions for the mechanism being in singular configurations are

$$\alpha = 180° \quad \text{and} \quad \alpha = 0 \tag{6.38}$$

where α can be obtained through the inverse kinematics of the third leg, e.g., for the assemble mode as shown in Fig. 3.12, there are

$$\alpha = \arccos\left(\frac{L_a^2 - y_c^2 + r^2 - (L_b\cos\theta_3 - R)^2 - (L_b\sin\theta_3 - z_c)^2}{2r\sqrt{L_a^2 - y_c^2}}\right) \tag{6.39}$$

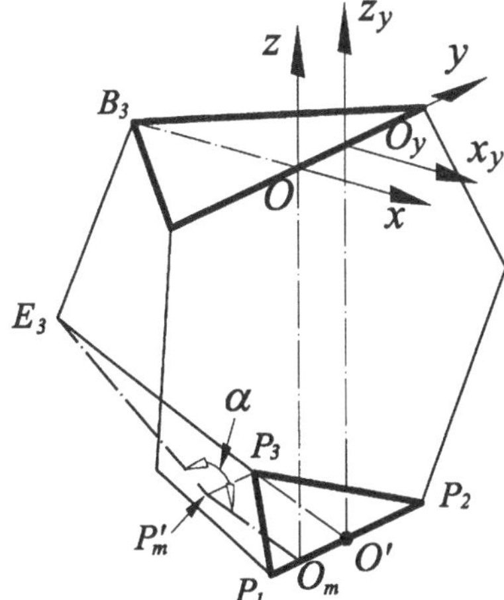

Fig. 6.15 The mirror of the rotational configuration to O-xz plane

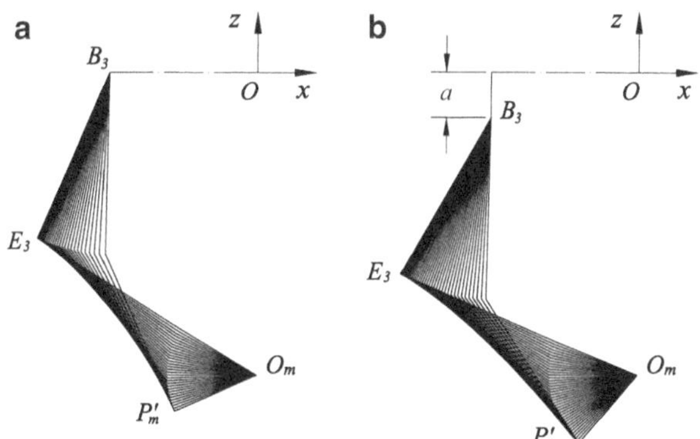

Fig. 6.16 Simulation of configurations of the third leg and the mobile platform

in which θ_3 can be obtained by Eq. (3.97) when $i = 3$ and sign "±" is "−" in Eq. (3.98). Therefore, the angle ς of a parallel mechanism at a specified point can be expressed as

$$\varsigma \in \zeta \tag{6.40}$$

in which for any value $\phi \in \varsigma$, there must be

$$\theta_3 \in [-\pi, \ -\pi/2], \quad \text{and} \quad 0 < \alpha < 180^\circ. \tag{6.41}$$

In this case, Fig. 6.16a shows the simulation of configurations of the third leg and the mobile platform when $\phi \in \varsigma$ in plane O-xz. At the same time, the angle ς of

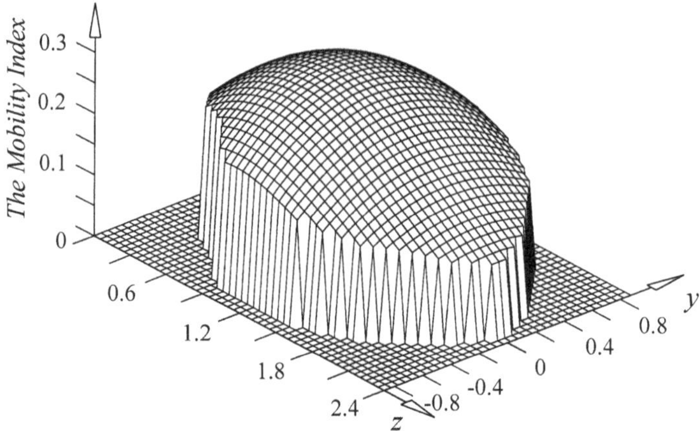

Fig. 6.17 Distributing of the rotational capability index on the reachable workspace

the mechanism at the point c can be obtained as $\varsigma = 55.60^\circ$, and the rotational capability index is just $\mu = 0.154$.

From Fig. 6.16a, we may see that it is easier to reach singular configuration for the mobile platform when $\phi > 0$ than that when $\phi < 0$, for which we can specify the original position of the base joint B_3 for the third leg in such way that its z coordinate is different from that of the first and second legs, i.e., point B_3 is not in the plane O-xy, which will make the rotational capability index better. For example, if we specify the offset a as $a = -0.3$, as shown in Fig. 6.16b, the angle ς of point $(y_c = 0.37, \quad z_c = -2.0)$ will be $\varsigma = 72.7^\circ$ and the rotational capability index can be increased as $\mu = 0.202$. What we should notice is that not every rotational capability index value in the workspace can be increased by this means, but we can just increase the rotational capability of the point within a usable workspace. The simulation of configurations is shown in Fig. 6.16b. The result is very useful for the actual design to increase the rotational capability of such device.

Figure 6.17 represents the distributing of the rotational capability index on the reachable workspace, which shows that:

- The rotational capability index μ is symmetric about $y = 0.0$.
- The maximum value of ς can reach 106.50°, $\mu_{\max} = 0.296$, and the minimum value is 43.00°, $\mu_{\min} = 0.119$.
- In the region $\Omega \in [(y, \ z) \mid -0.45 \le y \le 0.45, -1.45 \le z \le -0.50]$, the index ς is very high $\mu \ge 0.229$, i.e., $\varsigma \ge 82.30^\circ$.

Therefore, the rotational capability of the mechanism can be very high. If point B_3 is in a right position under the plane O-xy, the index values for some points in the workspace will be better, which will be depended on designer's demand.

6.6 The Spatial 2-PRU&1-PR(Pa)R Parallel Mechanism

6.6.1 Constant-Orientation Workspace

From Eqs. (3.80) and (3.81), we may obtain

$$[y-(r-R)]^2+(z-z_1)^2=L^2 \tag{6.42}$$

$$[y-(R-r)]^2+(z-z_2)^2=L^2 \tag{6.43}$$

$$y^2+[z-(z_3-r\sin\phi)]^2=L^2-(R-r\cos\phi)^2 \tag{6.44}$$

from which we may see that if z_1, z_2, z_3, and ϕ are specified, Eqs. (6.42), (6.43), and (6.44) represent three circles in the frame O-yz. For the first leg, the circle is centered at $(r-R,\ \ z_1)$ and the radius is L. For the second leg, the circle is centered at $(R-r,\ \ z_2)$ and the radius is L. And the third one, $(0,\ \ z_3-r\sin\phi)$ and $\sqrt{L^2-(R-r\cos\phi)^2}$ are the center and radius of the circle, respectively. Indeed, if mechanical interferences are neglected, the boundary of the workspace for each leg is attained whenever at least one of the actuators reaches one of its limits. If we assume that ϕ is specified and the range of motion of the actuators is given by

$$z_1,\ z_2,\ z_3\in[z_{\min},\ \ z_{\max}], \tag{6.45}$$

the workspace for each leg is the enveloping face of innumerable circles. The constant-orientation workspace of the parallel mechanism is the intersection of the three enveloping faces.

6.6.2 Reachable Workspace

For the parallel mechanism studied in this chapter, if the reference point $O'(y,\ \ z)$ within the intersection of the above-mentioned enveloping faces belongs to the reachable workspace of the parallel mechanism, $(y,\ \ z)$ must satisfy Eq. (6.44) for at least one value of $\phi\in[0,\ \ 2\pi]$.

The rotational DOF of the mechanism is the rotation of the mobile platform with respect to the y'-axis. The DOF has no effect on the workspaces of the first and second legs. From the above analysis, we may see that the reachable workspaces for the first and second legs are also the enveloping faces of innumerable circles centered at the line segments $y=r-R\,(z\in[z_{\min},\ \ z_{\max}])$ and $y=R-r\ (z\in[z_{\min},\ \ z_{\max}])$, respectively. And the radius of each circle is L.

If $z_1 = z_{\min}$ and $z_2 = z_{\min}$, Eqs. (6.42) and (6.43) can be rewritten as

$$[y - (r - R)]^2 + (z - z_{\min})^2 = L^2 \tag{6.46}$$

$$[y - (R - r)]^2 + (z - z_{\min})^2 = L^2 \tag{6.47}$$

Points of intersection of the above two circles can be written in the frame $O\text{-}yz$ as

$$\left(0,\ \sqrt{L^2 - (R - r)^2} + z_{\min}\right),\quad \left(0, -\sqrt{L^2 - (R - r)^2} + z_{\min}\right) \tag{6.48}$$

which are the nearest points, among the intersection of the reachable workspaces for the first and second legs, far from the y-axis.

If $z_1 = z_{\max}$ and $z_2 = z_{\max}$, the points of intersection of the two circles can also be obtained from Eqs. (6.42) and (6.43) as

$$\left(0,\ \sqrt{L^2 - (R - r)^2} + z_{\max}\right),\quad \left(0, -\sqrt{L^2 - (R - r)^2} + z_{\max}\right) \tag{6.49}$$

which are the farthest points, among the intersection of the reachable workspaces for the first and second legs, away from the y-axis.

Let us consider the third leg of the mechanism. When $\phi = 0$, Eq. (6.44) can be rewritten as

$$y^2 + [z - z_3]^2 = L^2 - (R - r)^2. \tag{6.50}$$

Let $z_3 \in [z_{\min},\ z_{\max}]$; the constant-orientation workspace of the leg is the enveloping face of innumerable circles centered at the line segment $y = 0 (z \in [z_{\min},\ z_{\max}])$. The radius of each circle is $\sqrt{L^2 - (R - r)^2}$. If $y = 0$ and $z_3 = z_{\min}$, from Eq. (6.50), one obtains

$$z = \pm\sqrt{L^2 - (R - r)^2} + z_{\min}. \tag{6.51}$$

If $y = 0$ and $z_3 = z_{\max}$, there is

$$z = \pm\sqrt{L^2 - (R - r)^2} + z_{\max}. \tag{6.52}$$

From Eqs. (6.48), (6.49), (6.51), and (6.52), we may see that the constant-orientation ($\phi = 0$) workspace of the third leg embodies the intersection of reachable workspaces of the first and second legs. Therefore, we can conclude that the intersection of reachable workspaces of the first and second legs is then the reachable workspace of the parallel mechanism.

6.6.3 Rotational Capability of the Mobile Platform

According to the kinematics of the spatial 3-DOF parallel mechanism, the rotational DOF of the mechanism is the rotation of the mobile platform with respect to the y'-axis as shown in Fig. 3.11. The position vector of point P_3 in frame $\Re$ can be written as

$$(\boldsymbol{p}_3)_{\Re} = \begin{pmatrix} -r\cos\phi \\ y \\ r\sin\phi + z \end{pmatrix}. \tag{6.53}$$

Let $x_p = -r\cos\phi$, $y_p = y$, and $z_p = r\sin\phi + z$; Eq. (6.44) can be rewritten as

$$\left(x_p + R\right)^2 + y_p^2 + \left(z_p - z_3\right)^2 = L^2 \tag{6.54}$$

which stands for a spherical surface centered at $(-R,\ 0,\ z_3)$, and the radius is L. If the range of motion z_3 of the third actuator is specified and the workspace point $(y,\ z)$ is given, Eq. (6.54) represents a circle centered at point $(-R,\ y,\ z_3)$, and the radius is $\sqrt{L^2 - y_p^2}$. The equation can be rewritten as

$$\left(x_p + R\right)^2 + \left(z_p - z_3\right)^2 = L^2 - y_p^2 \tag{6.55}$$

which is located on the plane paralleling to O-xz plane. If z_3 is specified as $z_3 \in [z_{\min},\ z_{\max}]$, Eq. (6.55) represent circles of radius $\sqrt{L^2 - y^2}$. As we have known, for a given position $(y,\ z)$ of the mobile platform, the locus of the point P_3 is the subset of a circle of radius r. The equation can be written as

$$x_p^2 + \left(z_p - z\right)^2 = r^2 \tag{6.56}$$

which is located on the plane paralleling to O-xz plane, defined as y. From Eqs. (6.55) and (6.56), we may see that all the circles defined by Eqs. (6.55) and (6.56) are located on a same plane. When $(y,\ z)$ and z_3 are specified, the point of intersection $\left(x_p,\ y_p,\ z_p\right)$ of the two circles can be obtained.

Equation (6.55) minus Eq. (6.56) produces

$$z_p = ax_p + b \tag{6.57}$$

where $a = -R/(z - z_3)$ and $b = \left(L^2 - y_p^2 - r^2 + z^2 - z_3^2 - R^2\right)/2\,(z - z_3)$. Substituting Eq. (6.57) into Eq. (6.54) leads to

$$A'x_p^2 + B'x_p + C' = 0 \tag{6.58}$$

where $A' = 1 + a^2$, $B' = 2a\,(b - z)$, and $C' = (b - z)^2 - r^2$. Then there is

$$x_p = \frac{-B' \pm \sqrt{B'^2 - 4A'C'}}{2A'}. \tag{6.59}$$

z_p can be also obtained by substituting Eq. (6.59) into Eq. (6.57). From Eqs. (6.57), (6.58), and (6.59), we may see that there are two points of intersection. The central angle of the arc defined by the two points is defined as the global rotational capability index ζ to evaluate the rotational capability of the mobile platform at the given position $(y,\ z)$ globally.

The central angle, actually, the orientation ϕ of the mobile platform, represents configurations of the mobile platform at a given point of the workspace. What we should notice is that these configurations include that of singularities, which will separate the arc into two mechanism-inaccessible ones. In actually, the rotational capability of a real mechanism device can only be valued by one of the two arcs, which will be defined as local rotational capability index ς. For the mechanism as shown in Fig. 3.11, only the second kind of singularity will occur in a usable workspace of a mechanism. According to the analysis of Sect. 6.3, what should be considered here is the singularity of the third leg B_3P_3. This singularity occurs when the leg and the mobile platform are completely extended or folded. The detailed presentation of the rotational capability will be discussed by an example in the following section.

6.6.4 Example

In this section, we investigate the workspace of a spatial 3-DOF parallel mechanism with parameters $r = 1.0$, $R = 2.5$, and $L = 4.0$. The actuated motion is specified as

$$z_1, z_2, z_3 \in [-2.0,\ \ 0]. \tag{6.60}$$

There is no condition to arise the first and third kinds of singularities, and only the second kind of singularity will occur in the workspace. According to the analysis of the workspace, the reachable workspace of the example can be obtained as the shade region of Fig. 6.18, which is the intersection of workspaces of the first and second legs. The workspace of the first leg is the region between the boundaries shown as the dash arcs in Fig. 6.18, and the workspace for the second leg is the region between long-dash double-dot arc boundaries. In order to prove the results of Sect. 6.6.2, the orientation workspace of the third leg when $\phi = 0$ is also illustrated in Fig. 6.18, which is the region between solid arc boundaries, from which we may see that the intersection of workspaces for the first and second legs is embodied in the region.

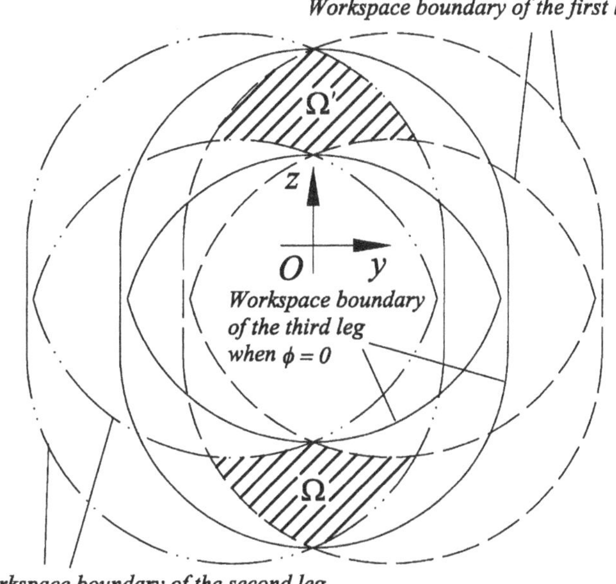

Fig. 6.18 Workspace of the mechanism with $r = 1.0, R = 2.5$, and $L = 4.0$

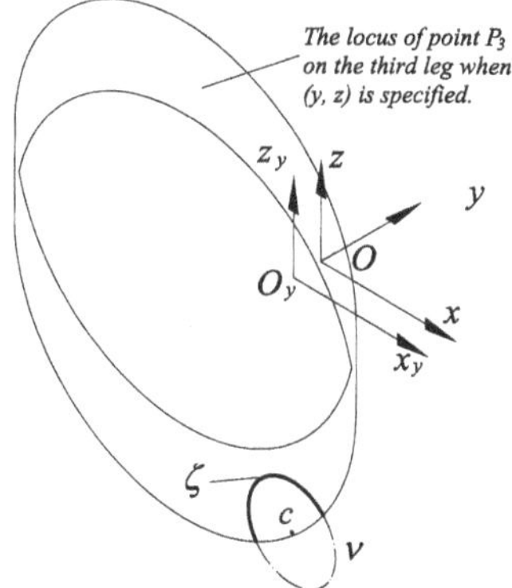

Fig. 6.19 Global rotational capability index ζ

As an example to present the global and local rotational capability indices, point O' on the mobile platform is specified as point c ($y_c = -0.63,\ \ z_c = -5.0$) in the reachable workspace. The circle described by Eq. (6.56) is shown as ν centered at point c in the plane O_y-$x_y z_y$, which is parallel to plane O-xz and $OO_y = y_c$, as illustrated in Fig. 6.19. The region, in the plane O_y-$x_y z_y$, which is the locus of point P_3 given by Eq. (6.55) when $z_3 \in [-2.0,\ \ 0]$ is shown as Fig. 6.19. The central angle of the arc of circle ν embodied in the region is the global rotational capability index, and $\zeta = 170.5°$ at the point c.

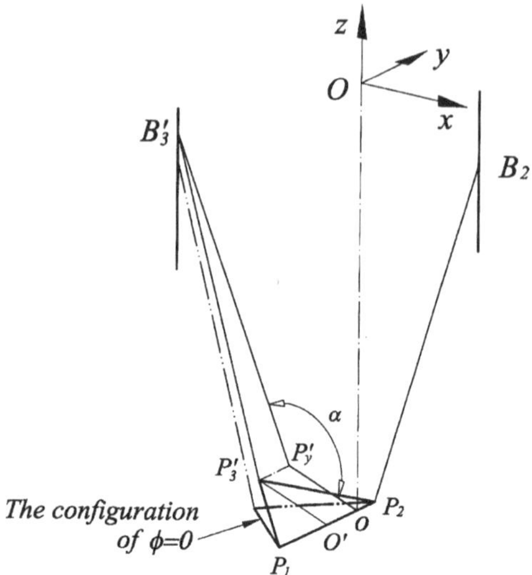

Fig. 6.20 Mirror of the rotational configuration to O-xz plane

As mentioned in the last section, the central angle contains singular configurations of the mechanism. And the singular configurations separate the arc of the corresponding central angle into two mechanism-inaccessible ones. In order to find the singular configurations, we can mirror the third leg P_3B_3 and the mobile platform or $O'P_3$ into the plane O-xz. As illustrated in Fig. 6.20, the mirror of configuration $P_3'B_3'$ of the third leg into plane O-xz is $P_y'B_3'$, and point o is the corresponding point of O' in plane O-xz. Then, oP_y' is the mirror of $O'P_3'$. Conditions for the mechanism being in singular configurations are

$$\alpha = 180^\circ \quad \text{and} \quad \alpha = 0 \tag{6.61}$$

where α can be obtained through the inverse kinematics of the third leg, e.g., for the assemble mode as shown in Fig. 3.11, there are

$$\alpha = \cos^{-1}\left(\frac{L^2 - y_c^2 + r^2 - R^2 - (z_3 - z_c)^2}{2r\sqrt{L^2 - y_c^2}}\right) \tag{6.62}$$

in which $z_3 = \sqrt{L^2 - (R - r\cos\phi)^2 - y^2} + r\sin\phi + z$. Therefore, the local rotational capability ς of a mechanism at a specified point can be expressed as

$$\varsigma \in \zeta \tag{6.63}$$

in which for any value $\phi \in \varsigma$, there must be

$$z_3 = \sqrt{L^2 - (R - r\cos\phi)^2 - y^2} + r\sin\phi + z \in [-2.0,\ 0],$$
$$\text{and} \quad 0 < \alpha < 180^\circ. \tag{6.64}$$

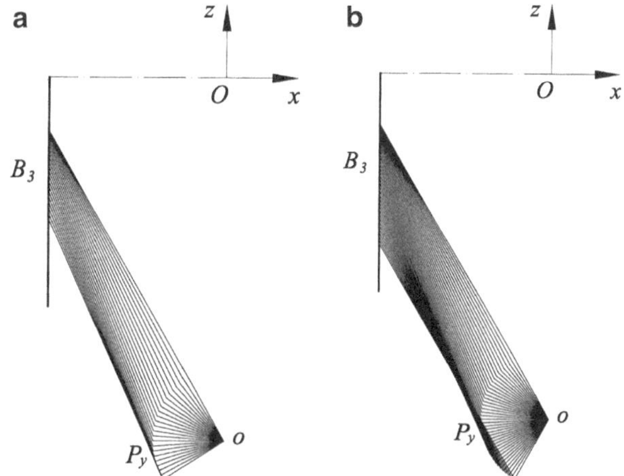

Fig. 6.21 Simulation of configurations of the third leg and the mobile platform

In this case, Fig. 6.21a shows the simulation of configurations of the third leg and the mobile platform when $\phi \in \varsigma$ in plane O-xz. At the same time, the local rotational capability index of the mechanism at the point c can be obtained as $\varsigma = 94.70°$.

From Fig. 6.21a, we may see that it is easier to reach singular configuration for the mobile platform when $\phi > 0$ than that when $\phi < 0$, for which we can specify the actuated input z_3 of the third leg in such way that is different from that of the first and second legs, which will make the local rotational capability index better. For example, if we specify z_3 as $z_3 \in [-2.5, -0.5]$, the local rotational capability index of point $(y_c = -0.63, \ z_c = -5.0)$ will be $\varsigma = 121.95°$, and the simulation of configurations is shown in Fig. 6.21b. The result is very useful for the actual design to increase the rotational capability of such device.

Figure 6.22 represents the distributing of the local rotational capability index on the reachable workspace, which shows that:

- The local rotational capability index is symmetric about $y = 0.0$.
- In the region $z \in [-4.66, -3.76]$, the index ς is very high $\varsigma \geq 125.45°$. The maximum value of ς can reach 135.62°, and the minimum value is 62.78°.

Therefore, the rotational capability of the mechanism can be very high. If the actuated input z_3 of the third leg is different from that of the first and second legs z_1 and z_2, the index will be better.

Although both the 2-RRU&1-RR(Pa)R and the 2-PRU&1-PR(Pa)R parallel mechanisms have high rotational capability, it is easier for the mechanism with prismatic actuators to reach higher rotational capability since the mechanism with revolute actuators has more singularities.

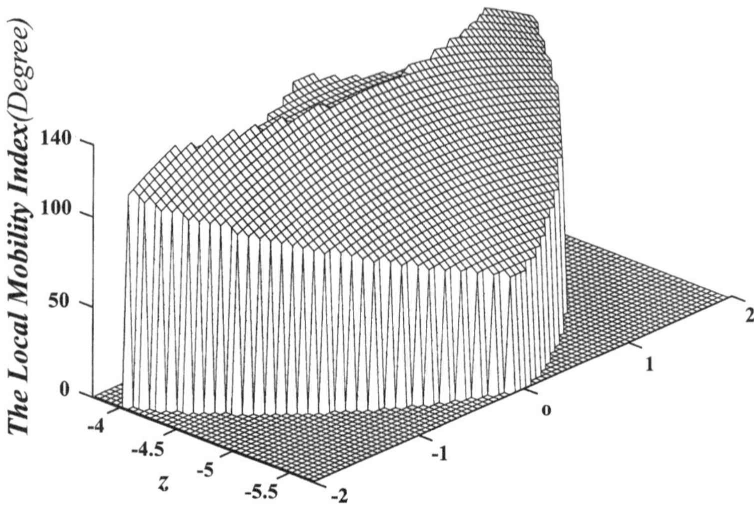

Fig. 6.22 Distributing of the local rotational capability index on the reachable workspace

6.7 The Spatial 6-RUS Parallel Mechanism

Since the workspace of a 6-DOF parallel mechanism is embedded in a six-dimensional space, it is very difficult to represent. The constant-orientation workspace is the region of the three-dimensional Cartesian space that can be attained by a point from the platform of the mechanism for a given orientation of the platform. This is in fact a three-dimensional section of the six-dimensional complete workspace. The constant-orientation workspace of a 6-RUS parallel mechanism can be easily obtained geometrically (Bonev and Gosselin 2000) based on the inverse kinematic problem of the mechanism.

From Eq. (3.140), one may obtain

$$(x - x_i)^2 + (y - y_i)^2 + (z - z_i)^2 = r_2^2 \quad i = 1, \ldots, 6 \tag{6.65}$$

where

$$x_i = r_1 \cos\theta_i - q_{11} r_3 \cos\eta_i - q_{12} r_3 \sin\eta_i \tag{6.66}$$

$$y_i = r_1 \sin\theta_i - q_{21} r_3 \cos\eta_i - q_{22} r_3 \sin\eta_i \tag{6.67}$$

$$z_i = -q_{31} r_3 \cos\eta_i - q_{32} r_3 \sin\eta_i. \tag{6.68}$$

From which we may see that, Eq. (6.65) represents six spheres centered at the points $O_i\left(x_i, y_i, z_i\right)$ and the radius r_2. Let us consider Eqs. (6.66) and (6.67), which can be rewritten as

$$r_1 \cos\theta_i = x_i + x_{oi} \tag{6.69}$$

$$r_1 \sin\theta_i = y_i + y_{oi} \tag{6.70}$$

that is,

$$(x_i + x_{oi})^2 + (y_i + y_{oi})^2 = r_1^2 \tag{6.71}$$

where

$$x_{oi} = q_{11} r_3 \cos\eta_i + q_{12} r_3 \sin\eta_i \tag{6.72}$$

$$y_{oi} = q_{21} r_3 \cos\eta_i + q_{22} r_3 \sin\eta_i . \tag{6.73}$$

x_{oi} and y_{oi} are related to the value of η_i. When η_i is specified, the circle of Eq. (6.71) can be determined. The circle is centered at point $O_{Oi}\,(-x_{oi}, -y_{oi})$ and is with radius r_1. Therefore, the center of the sphere centered at point $O_i\left(x_i, y_i, z_i\right)$ is on the circle centered at $O_{Oi}\,(-x_{oi}, -y_{oi})$.

From the above analysis, we may see that the workspace of each of the six legs of a simplified 6-RUS mechanism is the enveloping solid of a sphere whose center is moved along a circle centered at point of $O_{Oi}\,(-x_{oi}, -y_{oi})$. The enveloping solid is a torus solid whose center is $O_{ti}\,(-x_{oi}, -y_{oi}, \; z_i)$. The constant-orientation workspace of a 6-RUS parallel mechanism is the intersection of six identical tori. From Eqs. (6.65), (6.66), (6.67), (6.72), and (6.73), one obtains that the position vector of the center point $O_{ti}\,(-x_{oi}, -y_{oi}, \; z_i)$ is related to matrix $\boldsymbol{Q}$ and the position vectors $\boldsymbol{p}_{iR'}$ and $\boldsymbol{b}_{iR}$. For a given mechanism, the position vectors $\boldsymbol{p}_{iR'}$ and $\boldsymbol{b}_{iR}$ are constant. Hence, points $O_{ti}\,(-x_{oi}, -y_{oi}, \; z_i)$ are only related to the rotation matrix $\boldsymbol{Q}$. Therefore, if the orientation of the mechanism is specified, the constant-orientation workspace and its volume can be obtained geometrically by using Eqs. (6.65) and (6.71) using the commercial CAD system AutoCAD, for example.

As an example of application of the method described above, the workspace of a 6-RUS parallel mechanism, whose geometric parameters are $r_1 = 0.8$, $r_2 = 1.2$, $r_3 = 1.0$, and $\phi = 0.0$, will now be studied. The constant-orientation workspace of the simplified 6-RUS parallel mechanism will be obtained, when $\boldsymbol{Q} = 1$, where 1 is the identity matrix. As mentioned previously, for the first leg, $i = 1$, the workspace is a torus generated by the arc of *abcd* revolving with respect to the axis Z'', as shown in Fig. 6.23a. One-half of the torus is shown in Fig. 6.23b. The constant-orientation workspace of the mechanism is the intersection of six such tori, as shown in Fig. 6.24. The volume of the workspace is $V = 4.65$, as obtained by AutoCAD.

In this chapter, in the determination of the constant-orientation workspace and reachable workspace, only the kinematic constraints are taken into account. Neither joint limits nor leg interference constraints are verified.

The constant-orientation workspace of a simplified 6-RUS parallel mechanism is determined geometrically. Although the reachable workspace is the union of all constant-orientation workspaces, its determination cannot be solved geometrically.

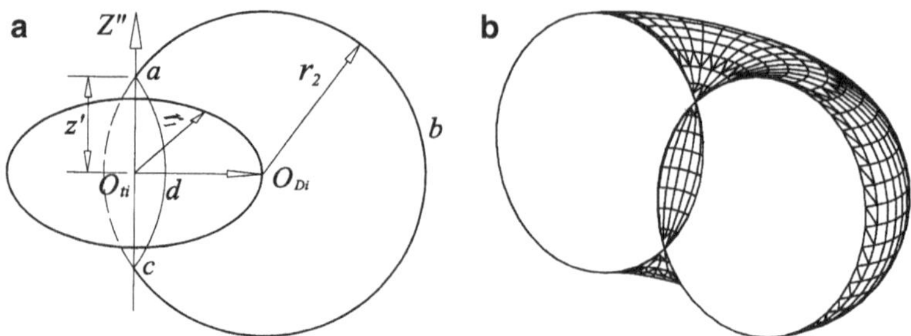

Fig. 6.23 Workspace of each of legs is the torus generated by the closed arc *abcd* with respect to axis of Z''

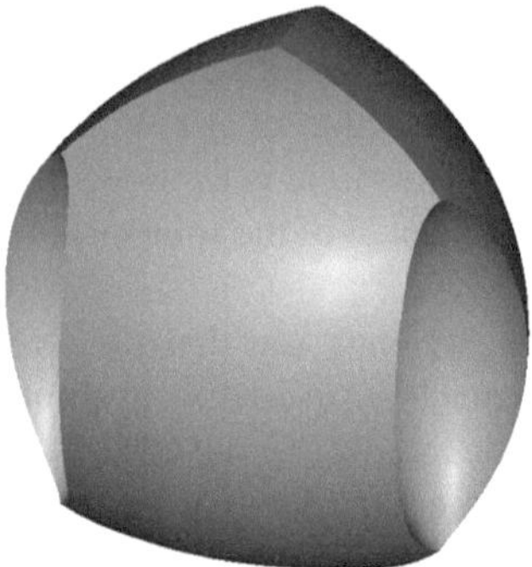

Fig. 6.24 Constant-orientation workspace of a 6-RUS parallel mechanism

References

Bonev IA, Gosselin CM (2000) A geometric algorithm for the computation of the constant-orientation workspace of 6-RUS parallel manipulators. In: Proceedings of the ASME design engineering technical conferences and computers and information in engineering conference, Baltimore, MD, DETC2000/MECH-14106, ASME press, New York

Bonev IA, Ryu J (2001a) A geometrical method for computing the constant-orientation workspace of 6-PRRS parallel manipulators. Mech Mach Theory 36(1):1–13

Bonev IA, Ryu J (2001b) A new approach to orientation workspace analysis of 6-DOF parallel manipulators. Mech Mach Theory 36(1):15–28

Chrisp AG, Gindy NNZ (1999) Parallel link machine tools: simulation, workspace analysis and component positioning. In: Boer CR, Molinari-Tosatti L, Smith KS (eds) Parallel kinematic machines. Springer, London, pp 245–256

Gosselin C (1990) Determination of the workspace of 6-dof parallel manipulators. Trans ASME J Mech Des 112:331–336

Merlet J-P (1999a) Determination of 6D workspace of Gough-type parallel manipulator and comparison between different geometries. Int J Robot Res 18(9):902–916

Merlet J-P (1999b) The importance of optimal design for parallel structures. In: Boer CR, Molinari-Tosatti L, Smith KS (eds) Parallel kinematic machines. Springer, London, pp 99–110

Rawls RR, Hagen MA (1998) AutoLISP programming principles and techniques. Goodheart-Willcox, Tinley Park

Tonshoff HK, Grendel H (1999) A systematic comparison of parallel kinematics. In: Boer CR, Molinari-Tosatti L, Smith KS (eds) Parallel kinematic machines. Springer, London, pp 295–311

Part III
Optimal Kinematic Design

Chapter 7
Performance Evaluation of Parallel Mechanisms

Abstract This chapter concerns several new performance evaluation indices available for parallel mechanisms as well as introducing the use of conventional but the most frequently used indices. The performance evaluation of parallel mechanisms is the foundation of optimal design and also a challenging task. Some Jacobian-based indices such as condition number and manipulability are no longer suitable for those parallel mechanisms with rotation-translation coupling. In this case, it is necessary to address several better performance indices. Therefore, in addition to commonly used indices for evaluating the kinematic performances such as the workspace, stiffness, and accuracy of a parallel mechanism, a series of novel frame-free indices in terms of motion/force transmissibility, including input transmission index (ITI), output transmission index (OTI), constraint transmission index (CTI), and local singularity index (LSI), are presented and illustrated by several parallel mechanisms.

Keywords Kinematic performance • Performance evaluation index • Frame-free index • Parallel mechanism • Condition number

The performance evaluation of parallel mechanisms is the foundation of optimal design; it is a constantly challenging task. The kind of performance that should be considered is of primary importance.

The Jacobian matrix, manipulability, and condition number are concepts that have been proposed and used in serial robots. In particular, the condition number of the Jacobian matrix is used to evaluate the dexterity of a robot. Study results were, in general, directly extended to parallel mechanisms from the very beginning. In the field of parallel mechanisms, the local conditioning index (LCI, the reciprocal of the condition number) was proposed to evaluate the dexterity, control accuracy, isotropy (Gosselin and Angeles 1989), and closeness of a pose to singularity (Voglewede and Ebert-Uphoff 2004). Most researchers typically use the LCI as the foremost criterion in the optimum design of a parallel mechanism. For some applications,

X.-J. Liu and J. Wang, *Parallel Kinematics: Type, Kinematics, and Optimal Design*, Springer Tracts in Mechanical Engineering, DOI 10.1007/978-3-642-36929-2_7,

other indices such as accuracy (Ryu and Cha 2001) and stiffness (Kim and Tsai 2003) may be considered. To evaluate the performance of a mechanism within a workspace, Gosselin and Angeles (1991) defined a global conditioning index (GCI). Thus, the LCI and GCI are two of the most frequently used performance indices in the design of parallel mechanisms. We introduce the use of the indices in this chapter.

Nevertheless, the LCI suffers from problems such as (a) being relative to the coordinate system and (b) the absence of physical significance when the Jacobian matrix has a unit inconsistency problem. The counterparts of serial robots, i.e., parallel mechanisms, considerably perform well in terms of motion/force transmission, and not only at dexterous manipulation. Hence, the approaches to the performance evaluation of such mechanisms should not be the same. To avoid these problems, we introduce the motion/force transmissibility of parallel mechanisms and define their evaluation index in this chapter.

7.1 Jacobian-Matrix-Based Performance Evaluation

7.1.1 Local and Global Conditioning Indices

The Jacobian matrix is heavily dependent on the geometric parameters of a mechanism. Therefore, a subtle relationship exists between performance and link lengths. Consequently, almost all of the proposed indices are based on the Jacobian matrix. The LCI (Gosselin and Angeles 1989) is one such index.

From the standpoint of mathematics, the condition number of a matrix is used in numerical analyses to estimate the error generated in the solution of a linear system of equations by the error on the data (Strang 1976). Suppose that the velocity equation of a parallel mechanism is $\dot{\boldsymbol{q}} = \boldsymbol{J}\dot{\boldsymbol{X}}$, where $\dot{\boldsymbol{q}}$ and $\dot{\boldsymbol{X}}$ are the input and output velocity vectors, respectively. Then, one obtains (Strang 1976)

$$\frac{\|\delta\dot{\boldsymbol{q}}\|}{\|\dot{\boldsymbol{q}}\|} \leq \kappa \frac{\left\|\dot{\delta}\dot{\boldsymbol{X}}\right\|}{\left\|\dot{\boldsymbol{X}}\right\|} \tag{7.1}$$

where κ is the condition number of the Jacobian matrix and $\kappa = \left\|\boldsymbol{J}^{-1}\right\| \|\boldsymbol{J}\|$, in which $\|\cdot\|$ denotes the Euclidean norm of a matrix. When applied to the Jacobian matrix, the condition number determines the accuracy of mechanism control. The LCI is defined as the reciprocal of the condition number of the Jacobian matrix. That is,

$$\text{LCI} = 1/\kappa\,, \quad 0 \leq \text{LCI} \leq 1. \tag{7.2}$$

The LCI may be used to evaluate the dexterity, isotropy, and static stiffness of a parallel mechanism. This number is to be kept as large as possible.

The Jacobian matrix and LCI are dependent on both the coordinate system and pose of a mechanism, making the LCI a local index. To evaluate the global behavior over a workspace of a mechanism, we define a GCI η_J as the LCI integrated over the workspace and divide it by the volume of the workspace (Gosselin and Angeles 1991). Therefore, η_J can be written as

$$\eta_J = \frac{\int_W (1/\kappa)\mathrm{d}W}{\int_W \mathrm{d}W}, \tag{7.3}$$

which is used to measure the global behavior of a mechanism condition number; W is the workspace of the mechanism.

7.1.2 Good-Condition Workspace

In Sect. 6.1.2, the *usable workspace* is defined as the maximum continuous workspace that contains no internal singular loci but is bounded by external singular loci. According to this definition, control over the mechanism will lose at the points on the boundaries and their neighborhoods. In practical applications, these points are not used and should be excluded in the design process. The remaining workspace that is applied can be referred to as the *good-condition workspace* (GCW), which can be bounded by a locus in which the LCI is equal to a specified LCI value, i.e., $1/\kappa$. Therefore, the GCW is the set of poses in which the LCI is no less than a specified LCI value.

7.1.2.1 GCW of the Planar 5R Parallel Mechanism

As an example, we consider the planar 5R parallel mechanism shown in Fig. 3.1. In Fig. 7.1, the usable workspace of a mechanism with parameters $r_1 = 0.85$, $r_2 = 1.6$, and $r_3 = 0.55$ is bounded by singular loci. In this workspace, the LCI value differs with varied points (x, y). The distribution of the LCI is illustrated in the figure. When the LCI is specified as 0.5, the GCW of the mechanism is the region surrounded by the LCI locus with a value of 0.5 (Fig. 7.1).

7.1.2.2 GCW of the Linear Tsai's Mechanism

The parallel mechanism with three translational DOFs, an adaptation of Tsai's parallel mechanism (Tsai and Stamper 1996), is shown in Fig. 7.2; the mobile platform is connected to the base by three identical serial chains. Each chain is connected to the base by a prismatic joint and contains some parallelogram. The three parallelograms are connected to the mobile platform, and the sliders are

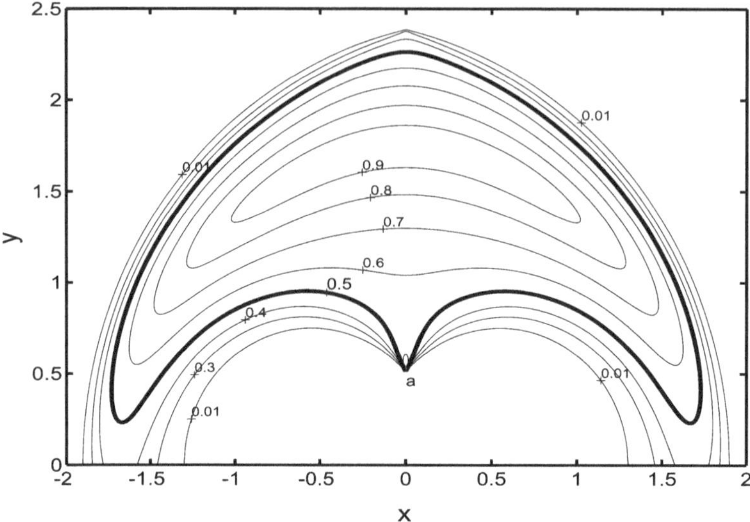

Fig. 7.1 Distribution of the LCI within the usable workspace of a 5R parallel mechanism

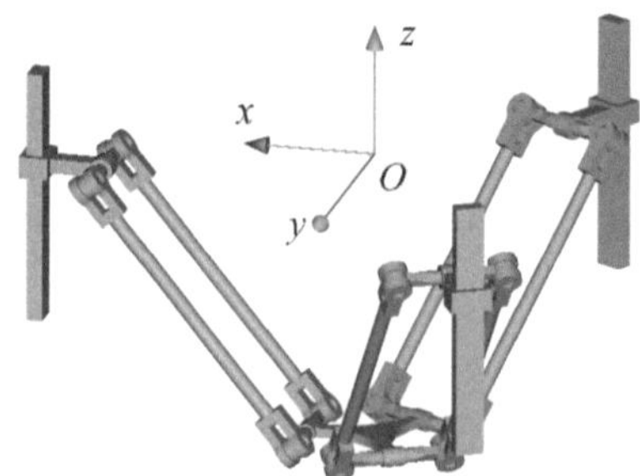

Fig. 7.2 Linear Tsai's parallel mechanism

connected by revolute joints. The mobile platform of the mechanism has three translational DOFs with respect to the base. The output can be achieved through the combination of the actuations to the three prismatic joints.

Kinematically, the parallel mechanism is identical to that of the DELTA parallel robot with linear actuators (Liu et al. 2004). A kinematic model of the mechanism is developed, as shown in Fig. 7.3. The center of the revolute joint that connects the parallelogram to the slider in each of the three chains is denoted as B_i $(i = 1, 2, 3)$, and the center of the revolute joint connected to the mobile platform in each chain is denoted as P_i $(i = 1, 2, 3)$. Figure 7.3 shows a fixed global reference system $\Re$:O-xyz located at the center of regular triangle $b_1b_2b_3$, with the z-axis normal to the base and the x-axis directed along Ob_1. Another reference system, $\Re'$:$\Re$:O'-$x'y'z'$, is located at the center of regular triangle $P_1P_2P_3$. The z$'$-axis is perpendicular to the output platform and the x$'$-axis is directed along $O'P_1$ (Fig. 7.4). Related geometric parameters are $Ob_i = R$, $O'P_i = r$, and $B_iP_i = R_1$, where $i = 1, 2, 3$. The objective of inverse kinematics is to identify the input of the mechanism with the given position of reference point O'.

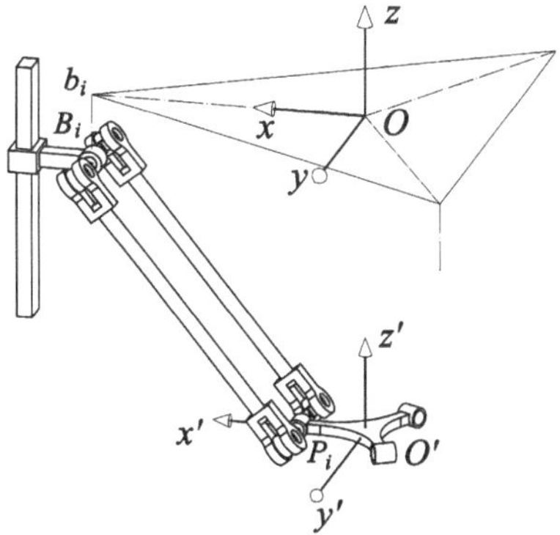

Fig. 7.3 Kinematic model of the mechanism with one leg

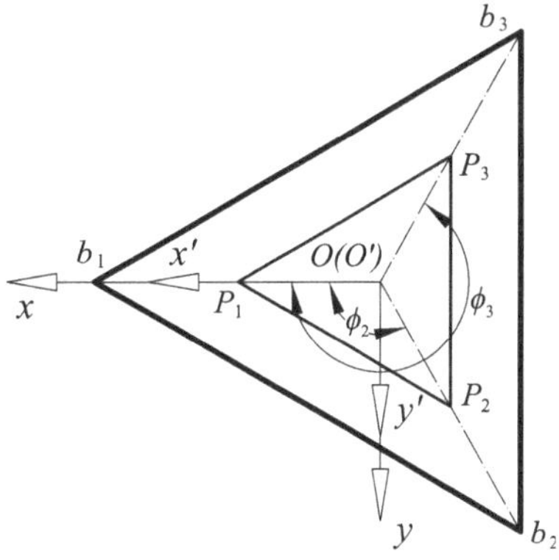

Fig. 7.4 Top view of the base frame and mobile platform

Position vector $[\boldsymbol{c}]_{\Re}$ of point O' in frame $\Re$ can be written as

$$[\boldsymbol{c}]_{\Re} = \begin{pmatrix} x & y & z \end{pmatrix}^{\mathrm{T}}. \tag{7.4}$$

As shown in Fig. 7.4, the coordinate of point P_i in frame $\Re'$ can be described by vector $[\boldsymbol{P}_i]_{\Re'}$ $(i = 1,\ 2,\ 3)$, which can be expressed as

$$[\boldsymbol{P}_i]_{\Re'} = \begin{pmatrix} r\cos\varphi_i & r\sin\varphi_i & 0 \end{pmatrix}^{\mathrm{T}}, \quad i = 1, 2, 3, \tag{7.5}$$

where

$$\phi_i = \frac{2\,(i-1)}{3}\pi, \quad (i = 1, 2, 3)\,, \tag{7.6}$$

which is the angle from the x′-axis to line $O'P_i$ (Fig. 7.4). Thus, vectors $[\boldsymbol{P}_i]_{\Re}$ $(i = 1,\ 2,\ 3)$ in frame O-xyz can be written as

$$[\boldsymbol{P}_i]_{\Re} = \begin{pmatrix} r\cos\varphi_i + x & r\sin\varphi_i + y & z \end{pmatrix}^{\mathrm{T}}. \tag{7.7}$$

As shown in Fig. 7.4, vectors $[\boldsymbol{B}_i]_{\Re}$ $(i = 1,\ 2,\ 3)$ are defined as the position vectors of points B_i in frame $\Re$, and

$$[\boldsymbol{B}_i]_{\Re} = \left(R\cos\varphi_i \ \ R\sin\varphi_i \ \ z_i \right)^{\mathrm{T}}, \quad i = 1,\ 2,\ 3. \tag{7.8}$$

Thus, the inverse kinematics of the translational mechanism can be solved by writing the following constraint equation:

$$\|[\boldsymbol{P}_i - \boldsymbol{B}_i]_{\Re}\| = R_1, \quad i = 1, 2, 3, \tag{7.9}$$

that is,

$$(x - x_i)^2 + (y - y_i)^2 + (z - z_i)^2 = R_1^2 \tag{7.10}$$

in which $x_i = R_2 \cos\varphi_i$, $y_i = R_2 \sin\varphi_i$, $R_2 = R - r$, and z_i represent the input of the mechanism and can be rearranged from Eq. (7.10) as

$$z_i = \pm\sqrt{R_1^2 - (x - x_i)^2 - (y - y_i)^2} + z. \tag{7.11}$$

Equation (7.11) shows eight inverse kinematic solutions for a given position of the parallel mechanism. The eight solutions correspond to the eight kinds of assembly modes of the mechanism. Hence, for a given mechanism and prescribed values of the position of the mobile platform, the required actuated input can be directly computed using Eq. (7.11). The assembly configuration shown in Fig. 7.2 corresponds to the kinematic mode in which the "$\pm$" in Eq. (7.11) is "+".

The Jacobian matrix is defined as the matrix that maps the relationship between the velocity of the mobile platform and the vector of actuated joint rates. The velocity equation of the mechanism is first considered to derive the Jacobian matrix. Equation (7.10) can be differentiated with respect to time to obtain the velocity equations, which leads to

$$(z - z_i)\,\dot{z}_i = (x - x_i)\,\dot{x} + (y - y_i)\,\dot{y} + (z - z_i)\,\dot{z}, \quad i = 1,\ 2,\ 3. \tag{7.12}$$

Rearranging Eq. (7.12) yields an equation of the form

$$\boldsymbol{A}\,\dot{\boldsymbol{\rho}} = \boldsymbol{B}\,\dot{\boldsymbol{p}}, \tag{7.13}$$

where $\dot{\boldsymbol{p}}$ is the vector of output velocities defined as

$$\dot{\boldsymbol{p}} = \left(\dot{x} \ \ \dot{y} \ \ \dot{z} \right)^{\mathrm{T}}, \tag{7.14}$$

and $\dot{\boldsymbol{\rho}}$ is the vector of input velocities defined as

$$\dot{\boldsymbol{\rho}} = \left(\dot{z}_1 \ \ \dot{z}_2 \ \ \dot{z}_3 \right)^{\mathrm{T}}. \tag{7.15}$$

In Eq. (7.13), $\boldsymbol{A}$ and $\boldsymbol{B}$ are 3×3 matrices of the mechanism and can be expressed as follows:

$$\boldsymbol{A} = \begin{bmatrix} z - z_1 & 0 & 0 \\ 0 & z - z_2 & 0 \\ 0 & 0 & z - z_3 \end{bmatrix}, \quad \boldsymbol{B} = \begin{bmatrix} x - x_1 & y - y_1 & z - z_1 \\ x - x_2 & y - y_2 & z - z_2 \\ x - x_3 & y - y_3 & z - z_3 \end{bmatrix}. \tag{7.16}$$

If matrix $\boldsymbol{A}$ is nonsingular, the Jacobian matrix of the mechanism can be obtained thus:

$$\boldsymbol{J} = \boldsymbol{A}^{-1}\boldsymbol{B} = \begin{bmatrix} \frac{x - x_1}{z - z_1} & \frac{y - y_1}{z - z_1} & 1 \\ \frac{x - x_2}{z - z_2} & \frac{y - y_2}{z - z_2} & 1 \\ \frac{x - x_3}{z - z_3} & \frac{y - y_3}{z - z_3} & 1 \end{bmatrix}. \tag{7.17}$$

For the assembly mode shown in Fig. 7.2, the Jacobian matrix can be rewritten as

$$\boldsymbol{J} = \begin{bmatrix} \frac{x - R_2 \cos\varphi_1}{q_1} & \frac{y - R_2 \sin\varphi_1}{q_1} & 1 \\ \frac{x - R_2 \cos\varphi_2}{q_2} & \frac{y - R_2 \sin\varphi_2}{q_2} & 1 \\ \frac{x - R_2 \cos\varphi_3}{q_3} & \frac{y - R_2 \sin\varphi_3}{q_3} & 1 \end{bmatrix}, \tag{7.18}$$

in which $q_i = \sqrt{R_1^2 - (x - R_2 \cos\varphi_i)^2 - (y - R_2 \sin\varphi_i)^2}$ $(i = 1, 2, 3)$.

As provided in Chap. 5, singularity can be achieved from Jacobian matrices $\boldsymbol{A}$ and $\boldsymbol{B}$. When $|\boldsymbol{A}| = 0$ and $|\boldsymbol{B}| \neq 0$, from Eq. (7.18), we derive $z = z_i$ $(i = 1, 2, 3)$, indicating that the first kind of singularity occurs when any one of the three legs is in a plane parallel to the O-xy plane. When $|\boldsymbol{B}| = 0$ and $|\boldsymbol{A}| \neq 0$, the second kind of singularity arises, yielding $x = x_i$ or $y = y_i$. Physically, the three legs should be parallel to one another. Thus, singularity cannot occur when $R - r \neq 0$. $|\boldsymbol{B}| = 0$ and $|\boldsymbol{A}| = 0$ lead to the third kind of singularity. Singularity is a particular case in which the three legs are all in a plane parallel to the O-xy plane. Only $|R - r| = R_1$ results in the third kind of singularity.

Therefore, to assemble the mechanism and ensure that it works freely, $R > r$ and $R - r < R_1$ or $r > R$ and $r - R < R_1$ should hold. In this chapter, we focus on $R > r$ and $R - r < R_1$.

Equation (7.18) indicates that the Jacobian matrix is independent of the z value. That is, the LCI $1/\kappa$ values are the same when x and y are constant regardless of the z value. In the analysis, therefore, we may disregard z. To investigate the workspace, we need only be concerned about the workspace at the z-section.

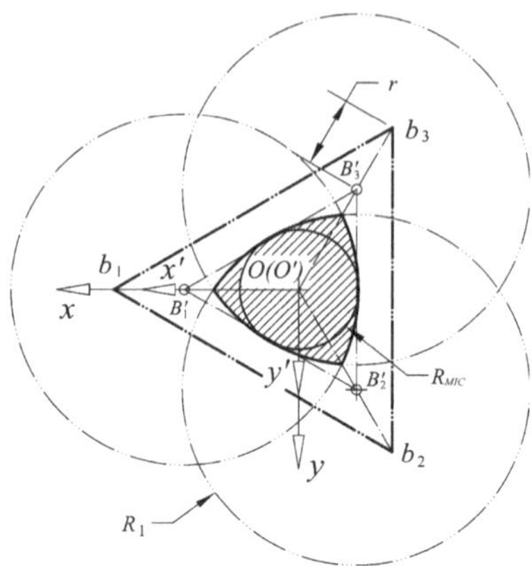

Fig. 7.5 Usable workspace in the *O-xy* plane

Disregarding the input, the maximal region that the mechanism can reach in the *O-xy* plane is determined by singularity, i.e., the first kind in which any one of the three legs $B_i P_i$ is parallel to the *O-xy* plane. The usable workspace in the *O-xy* plane, referred to as W_{xy} and shown as the shaded region in Fig. 7.5, is the intersection of three circles C_i, which can be expressed as

$$C_i : (x - x_i)^2 + (y - y_i)^2 = R_1^2, \quad i = 1, 2, 3. \tag{7.19}$$

If the volume of the usable workspace is denoted as V_{W-xy}, we can derive

$$V_{W-xy} = 3R_1^2 \tan^{-1}\left[\frac{\sqrt{3}\left(\sqrt{4R_1^2 - 3R_2^2} - R_2\right)}{3R_2 + \sqrt{4R_1^2 - 3R_2^2}}\right] - \frac{3\sqrt{3}}{4} R_2 \left(\sqrt{4R_1^2 - 3R_2^2} - R_2\right). \tag{7.20}$$

Figure 7.5 shows that a *maximal inscribed circle* (MIC) (Liu et al. 2004) exists within usable workspace W_{xy}. The MIC defines a workspace, which is referred to as the *maximal inscribed workspace* (MIW). The task workspace of the mechanism is usually a cylinder, whose section can be the MIC. The radius of the MIC can be expressed as follows:

$$R_{\text{MIC}} = R_1 - R_2. \tag{7.21}$$

The larger the usable workspace W_{xy}, the longer the MIC radius. Therefore, the MIC radius can be used to measure the volume size of W_{xy}.

Given that the GCW of the mechanism has a symmetrical structure, the distribution of the LCI should also be symmetrical. Therefore, an MIC should exist in the GCW. In this chapter, the MIC is defined as the *good-condition maximal inscribed circle* (GCMIC), and the MIW bounded by such an MIC is referred to as

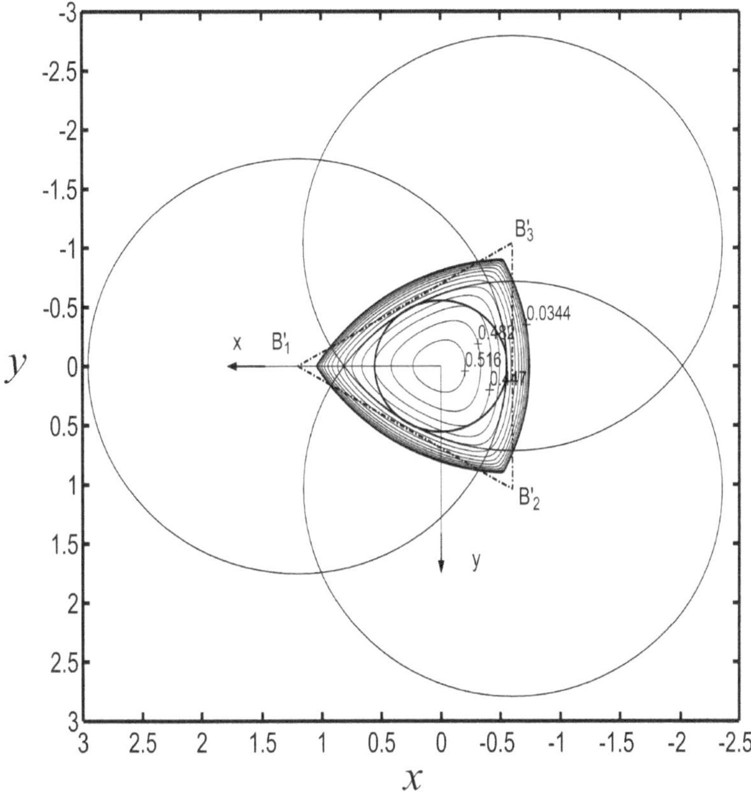

Fig. 7.6 Investigation of the LCI on the usable workspace

the *good-condition maximal inscribed workspace* (GCMIW). The task workspace is usually a cylinder; hence, the GCMIW is the section of the task workspace. The optimal design of the mechanism can be implemented with respect to the GCMIW.

Figure 7.6 illustrates the distribution of the LCI within usable workspace W_{xy}. The distribution is symmetrical about three lines, namely, $y = 0$, $y = \sqrt{3}x$, and $y = -\sqrt{3}x$. The LCI is zero on the boundary of the workspace. The nearer the distance to the point of origin, the better the LCI. The LCI is by far the best one at the point of origin. Moreover, within a circle centered at the point of origin, the LCI at the three points nearest the boundary of usable workspace W_{xy} is worst. The three points are located at lines $y = 0$, $y = \sqrt{3}x$, and $y = -\sqrt{3}x$. Their distances to the point of origin are same. The distance is, in fact, the radius of the GCMIC. Thus, to identify the GCMIW of a given mechanism, we can numerically search for the value of x within domain $[-R_{\text{MIC}}, 0]$ with respect to a specified LCI by setting $y = 0$. If the expected x value for a specified LCI is denoted as x_{GC} ($x_{\text{GC}} \leq 0$), the GCMIC radius R_{GCMIC} can be written as

$$R_{\text{GCMIC}} = -x_{\text{GC}}. \tag{7.22}$$

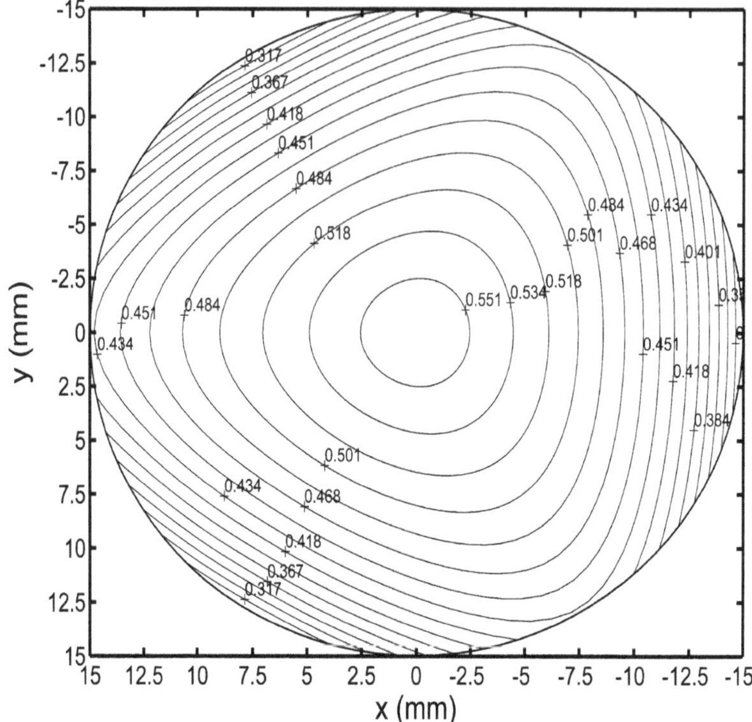

Fig. 7.7 The GCIW of the linear Tsai's mechanism when LCI ≥ 0.3

As an example, we investigate the mechanism with parameters $R_1 = 47.40$ mm, $R = 41.67$ mm, $r = 12$ mm, $z_{max} = 65.0727$ mm, and $z_{min} = 15.8541$ mm. $R_{GMIC} = 15$ mm when we allow LCI ≥ 0.3. The corresponding GCMIW is shown in Fig. 7.7.

7.1.3 *Stiffness*

Given that the manner by which a mechanism is applied cannot be predicted, we consider only the essential characteristics of a parallel mechanism. Thus, the analysis of stiffness in this chapter does not involve material parameters and the like.

Deformation occurs on the end-effector when an external force acts on it. The deformation is dependent on mechanism stiffness and external force. Mechanism stiffness affects the dynamics and position accuracy of the device, for which stiffness is an important performance index. In particular, the static stiffness (or rigidity) of the mechanism can be a primary consideration in the design of a parallel mechanism for certain applications (e.g., machine tools).

In the operational coordinate space, we define a stiffness matrix $\boldsymbol{K}$ that relates external force vector τ at the mobile platform to output displacement vector $\boldsymbol{D}$ of the mobile platform according to

$$\boldsymbol{D} = \boldsymbol{K}^{-1}\boldsymbol{\tau}, \tag{7.23}$$

where stiffness matrix $\boldsymbol{K}$ is expressed as

$$\boldsymbol{K} = \boldsymbol{J}^{\mathrm{T}}\boldsymbol{K}_p\boldsymbol{J}, \tag{7.24}$$

with

$$\boldsymbol{K}_p = \begin{bmatrix} k_{p1} & & \\ & \ddots & \\ & & k_{pn} \end{bmatrix}, \tag{7.25}$$

in which k_{pi} is a scalar that represents the stiffness of each of the actuators and n is the actuator number.

If $k_{pi} = 1$ $(i = 1, \ldots, n)$ and $\|\tau\|^2 = 1$, the maximum and minimum deformations can be obtained as follows by establishing the Lagrange equation (Liu 1999):

$$\|\boldsymbol{D}_{\max}\| = \sqrt{\max\left(|\lambda_{D\,i}|\right)} \text{ and } \|\boldsymbol{D}_{\min}\| = \sqrt{\min\left(|\lambda_{D\,i}|\right)}, \tag{7.26}$$

where $\lambda_{D\,i}$ $((i = 1, \ldots, n))$ are the eigenvalues of matrix $\left(\boldsymbol{K}^{-1}\right)^{\mathrm{T}}\boldsymbol{K}^{-1}$. $\|\boldsymbol{D}_{\max}\|$ and $\|\boldsymbol{D}_{\min}\|$ are the maximum and minimum deformations on the end-effector when both the external force vector and matrix $\boldsymbol{K}_p$ are unity. The maximum and minimum deformations form a deformation ellipsoid, whose axes lie in the directions of the eigenvectors of matrix $\left(\boldsymbol{K}^{-1}\right)^{\mathrm{T}}\boldsymbol{K}^{-1}$. The magnitudes of the ellipsoid are the maximum and minimum deformations given by Eq. (7.26). Usually, deformation $\|\boldsymbol{D}_{\max}\|$ can be used to evaluate mechanism stiffness, which is defined as the *local stiffness index* (LSI). The smaller the deformations, the better the stiffness.

Similarly, on the basis of Eq. (7.26), the *global stiffness index* (GSI), which can be used to evaluate the stiffness of a mechanism within the workspace, is defined as follows:

$$\eta_D = \frac{\int_W \|\boldsymbol{D}_{\max}\|\,\mathrm{d}W}{\int_W \mathrm{d}W}, \tag{7.27}$$

where for the mechanism studied here, W is a workspace (e.g., the GCW). Typically, η_D can be used as the criterion for design in a mechanism with respect to stiffness. Normally, we expect the index value to be as small as possible.

The distribution of the LSI within the GCMIW (LCI ≥ 0.3) of the linear Tsai's parallel mechanism with parameters $R_1 = 47.40$ mm, $R = 41.67$ mm, $r = 12$ mm,

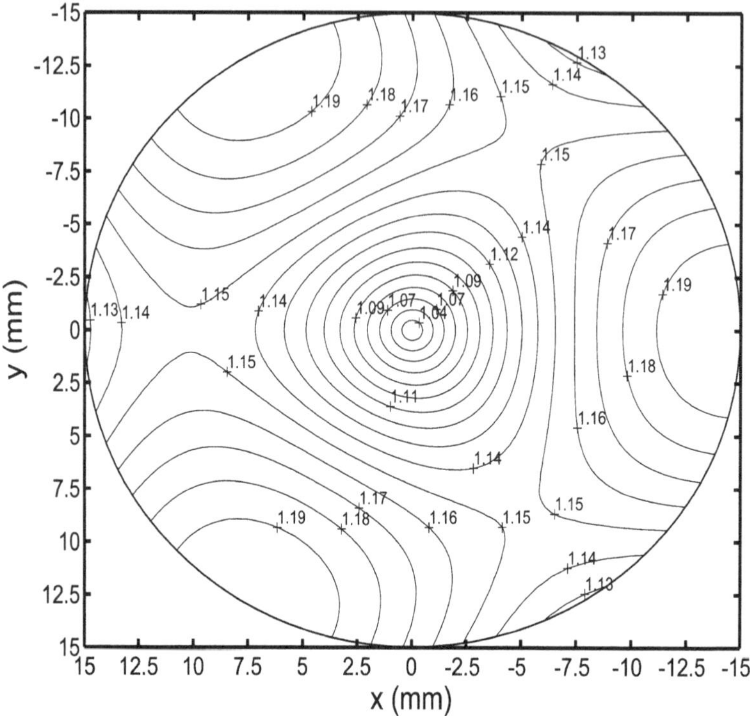

Fig. 7.8 Distribution of the LSI in the GCMIW of the linear Tsai's parallel mechanism

$z_{\max} = 65.0727$ mm, and $z_{\min} = 15.8541$ mm is shown in Fig. 7.8. The figure shows that the stiffness in the central area is high. Setting the workspace in Eq. (7.27) as the GCMIW (LCI ≥ 0.3) yields a GSI value of 1.1333.

7.2 Error/Accuracy

In this chapter, error is defined as the offset from the nominal pose of the reference point on the mobile platform of a mechanism; the offset occurs on account of the input errors. We consider the 3-$\underline{P}_V P_H$S parallel mechanism as an example.

Referring to Fig. 7.9, let E be a point on the mobile z-axis so that the distance between points E and o is l. After rotation at an angle θ about an a-axis (defined by an angle ϕ from the x-axis to the a-axis), position vector $\boldsymbol{e}$ of point E expressed in the base frame is

$$\boldsymbol{e} = \begin{bmatrix} x_E \ y_E \ z_E \end{bmatrix}^{\mathrm{T}} = \boldsymbol{R}_a(\theta) \begin{bmatrix} 0 \ 0 \ l \end{bmatrix}^{\mathrm{T}} + \boldsymbol{c}_{\Re}, \tag{7.28}$$

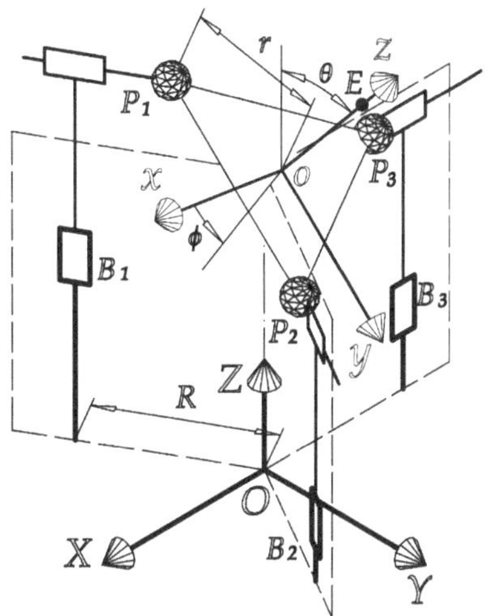

Fig. 7.9 Schematic of the 3-$\underline{P}_V P_H S$ parallel mechanism

where

$$\begin{cases} x_E = l \sin\phi \sin\theta - \frac{1}{2} r \cos 2\phi \,(1 - \cos\theta) \\ y_E = \frac{1}{2} r \sin 2\phi \,(1 - \cos\theta) - l \cos\phi \sin\theta \\ z_E = l \cos\theta + z \end{cases} . \tag{7.29}$$

The presence of input errors indicates that the mobile platform is undesirably offset from its nominal pose, as is point E. Suppose that at a given nominal pose $(\varphi,\ \theta,\ z)$, the nominal position of point E is at $E_N\,(x_E,\ y_E,\ z_E)$. The real position of point E is at $E_N^e\,(x'_E,\ y'_E,\ z'_E)$ because of the input errors. The real position of point E is near its nominal position; thus, to show the shape of the error region, defined by the possible real positions of the point for input range $[\rho_i - \varepsilon,\ \rho_i + \varepsilon]$ $(i = 1,\ 2,\ 3)$, we may project these positions onto the mobile platform plane, i.e., the o-xy plane, when at its nominal pose. The projection equation is

$$\begin{pmatrix} x''_E & y''_E & z''_E \end{pmatrix}^{\mathrm{T}} = \boldsymbol{R}_a^{-1}(\theta)\left[\begin{pmatrix} x'_E & y'_E & z'_E \end{pmatrix}^{\mathrm{T}} - \begin{pmatrix} x & y & z \end{pmatrix}^{\mathrm{T}}\right] \tag{7.30}$$

where $\begin{pmatrix} x''_E & y''_E \end{pmatrix}^{\mathrm{T}}$ is the position vector of the projection of point E_N^e in the o-xy plane and $(x\ \ y\ \ z)^{\mathrm{T}}$ denotes the platform nominal position vector. Figure 7.10 shows the error region for the 3-$\underline{P}_V P_H S$ parallel mechanism at $\phi = 50°$, $\theta = 20°$, $\varepsilon = 10\ \mu\text{m}$, $l = 1$ mm, and $z = 0$.

The distance from the nominal position to the real position is denoted as $\left|E_N E_N^e\right|$. The maximum distance, referred to as E_e, is defined as the maximum error at a given nominal pose.

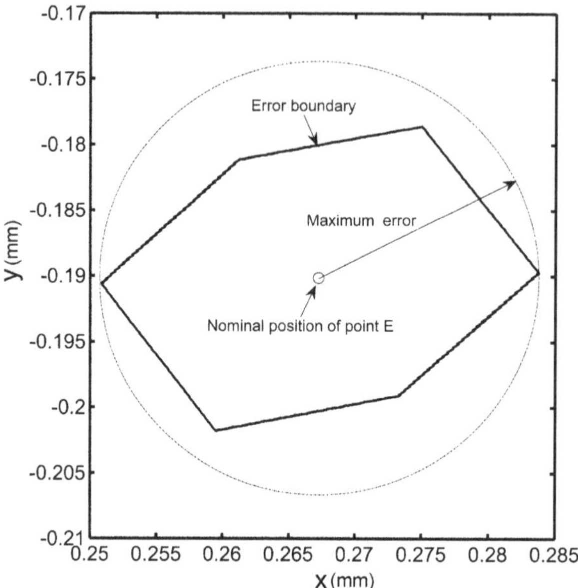

Fig. 7.10 Error region in the *o-xy* plane for a given pose of the 3-$\underline{P}_V P_H S$ parallel mechanism

To numerically obtain the value of E_e for a given nominal pose (ϕ, θ, z), we must first calculate the nominal position coordinates x_E, y_E, and z_E, as well as each corresponding input ρ_i ($i = 1, 2, 3$), using Eq. (7.29) and Eqs. (3.127)–(3.129)/(3.132), respectively. Subsequently, we sweep ϕ between $\phi - \phi_P$ and $\phi + \phi_P$, θ between $\theta - \theta_P$ and $\theta + \theta_P$, and z between $z - \varepsilon$ and $z + \varepsilon$ (where ϕ_P and θ_P are sufficiently large). For each pose, we identify corresponding input $\rho_{i\varepsilon}$ using Eqs. (3.127)–(3.129)/(3.132). Only the poses for which all corresponding $\rho_{i\varepsilon}$ belong to $[\rho_i - \varepsilon, \ \rho_i + \varepsilon]$ are retained. For each retained pose, corresponding position coordinates x'_E, y'_E, and z'_E are calculated, and distance $E_N E_N^e$ is computed. Finally, the maximal value of this distance, which is E_e, is retained. The maximum error of the parallel mechanism is clearly independent of the z value, and the error is proportional to parameter l.

Figure 7.11a illustrates the relationship between maximum error E_e and orientation ($\phi - \theta$) of the mobile platform of the 3-$\underline{P}_V P_H S$ parallel mechanism when $\varepsilon = 10$ μm and $l = 1$ mm. This relationship shows that the larger the tilt angle θ, the larger the error of point E, and that the maximum error is always larger than maximum input error ε. Figure 7.11b depicts the relationship between maximum error E_e and length parameter l for a specified orientation ($\phi = 50°$, $\theta = 20°$). This relationship indicates that the larger the length l, the larger the error of a certain orientation.

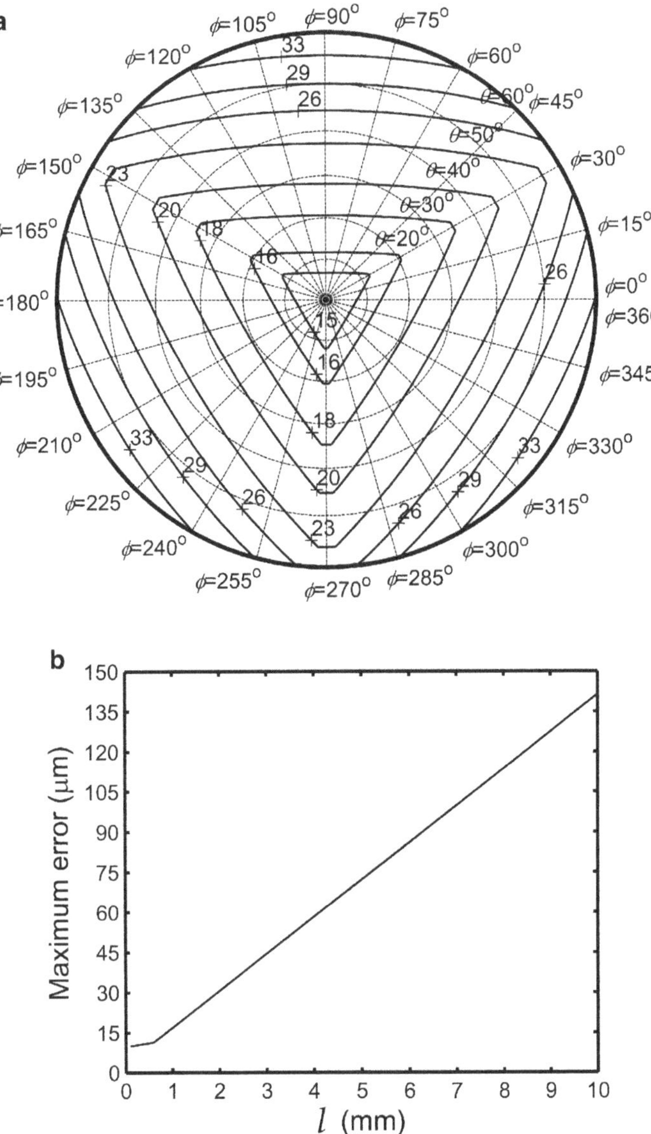

Fig. 7.11 Maximum error (in μm) versus platform orientation and length l for a 3-$\underline{P}_V P_H S$ parallel mechanism. (**a**) Maximum error for different orientations when $l = 1$ mm. (**b**) Maximum error for different l when $\phi = 50°$ and $\theta = 20°$

7.3 Motion/Force Transmissibility and Its Index

7.3.1 Problems with the LCI

Note that the LCI is a Jacobian matrix-based index. Similar to the Jacobian matrix, the LCI value is dependent on the frame coordinates. Its value accordingly changes when the frame varies, thereby causing problems in the analysis and design of a mechanism. Thus, even though the LCI can be used to evaluate the dexterity and closeness of a pose to the singularity of a mechanism, it cannot be assigned a definite value to indicate how far it is from the singularity. Merlet (2006) recently reviewed the LCI and GCI and discussed their severe inconsistencies when applied to parallel mechanisms with combined translational and rotational DOFs. The author concluded that these indices should not be used in parallel mechanisms with mixed types of DOFs (translations and rotations). Generally, defining a mathematical distance to the singularity of a parallel mechanism is impossible. No unified condition on the specified LCI exists in the identification of a GCW because we cannot determine whether the significance of the LCI is better or not.

Let us take the planar 5R parallel mechanism as an example. Figure 7.1 shows the distribution of the LCI within the workspace of a 5R parallel mechanism with parameters.

7.3.2 One Solution

The four-bar mechanism has been studied for a very long time. The *transmission angle* is an important index for the design of such a mechanism, as pointed out by Alt (1932). The author defined the concept using the forces that tend to move the driven link and tend to apply pressure to the driven link bearings as a simple index to evaluate the force transmission characteristics of a mechanism. The quality of motion/force transmission in a mechanism at its design stage can be evaluated using the transmission angle as an index. Therefore, the transmission angle is an index that enables the assessment of the quality of motion/force transmission. It facilitates decision making on the best option from among a family of possible mechanisms for the most effective force transmission (Balli and Chand 2002). A good transmission angle is the solution to most of the problems encountered in mechanisms. For example, Alt (1932) used the transmission angle to isolate better chains for various linkage applications. Eschenbach and Tesar (1971) pointed out that for many mechanical applications, a good transmission angle can guarantee the performance of linkage at higher speeds without unfavorable vibrations. However, it is not a cure-all for every design problem. The study of Hall (1961) shows that the most effective force transmission is realized when the transmission angle is equal to 90°. The output motion becomes less sensitive to the manufacturing tolerance of link lengths, clearance between joints, and change in dimensions because of thermal expansion. Mechanisms with transmission angles that deviate excessively from 90° exhibit poor operational characteristics, such as noise and jerking at high speeds (Hartenberg and Denavit 1964; Tao 1964); at 0°, self-locking takes place. Thus, the transmission angle of a mechanism is generally a very good indication of the quality of its motion, accuracy of its performance, and expected noise output and costs (Kimbrella 1991). A large transmission angle usually endows a mechanism with reasonable mechanical advantages and high-quality motion transmission. The study of link mechanisms shows that the transmission angle is significant not only as an indicator of good force and motion transmission but also as a prime factor in linkage sensitivity to small design parameter errors. The smaller the transmission angle, the more sensitive the linkage (Sutherland and Siddall 1974). Balli and Chand (2002) summarized the effects of transmission angle on velocity, input crank angle, friction, mechanical advantage, velocity, tolerance, clearance, and performance sensitivity. Many researchers have reached the conclusion that at an excessively small transmission angle, mechanical advantage diminishes; even a very small amount of friction causes the mechanism to jam. For high speed, high accuracy, and high quality of motion transmission, the most widely accepted design limits for the transmission angle are $(45°, 135°)$ (Tao 1964) and $(40°, 140°)$ (Alt 1932). Additionally, the transmission angle does not consider dynamic forces that stem from velocity and acceleration. Consequently, it is widely used in the kinematic synthesis stage, during which the lengths and mass properties of the links are still unknown to the designer (Soylu 1993); these parameters are unknown at this stage because a kinematically determined transmission angle does not reflect the action of gravity or dynamic forces.

7.3.2.1 Transmission Angle

Although we may not realize it, we are highly familiar with the transmission angle. In everyday life, we frequently attempt to move a somewhat constrained object, i.e., one that is attached to another component but cannot be moved freely: the handle of a crank, a curtain on a rail, and a sliding door. In all of these cases, the object cannot be moved even after pressure is exerted on it. Let us take the handle of a crank as an example. As shown in Fig. 7.12, the crank is attached to the base with a constant counterclockwise moment $\boldsymbol{M}$. Moving it necessitates the application of a right-hand force $\boldsymbol{F}$ at the end of the crank. When the direction of the force is constant, the crank is more or less easily rotated, depending on the position of the end point. This principle also explains why riding a bicycle is sometimes comfortable and at other times laborious. This phenomenon is related to the transmission angle. The direction of motion of a crank is always perpendicular to the crank; thus, the small angle between the force and crank is defined as the transmission angle, denoted as μ. When the force is normal to the crank, i.e., it has the same direction of motion as the crank, the force transmission is most effective; when the force is perpendicular to the direction of motion, the transmission is highly inefficient.

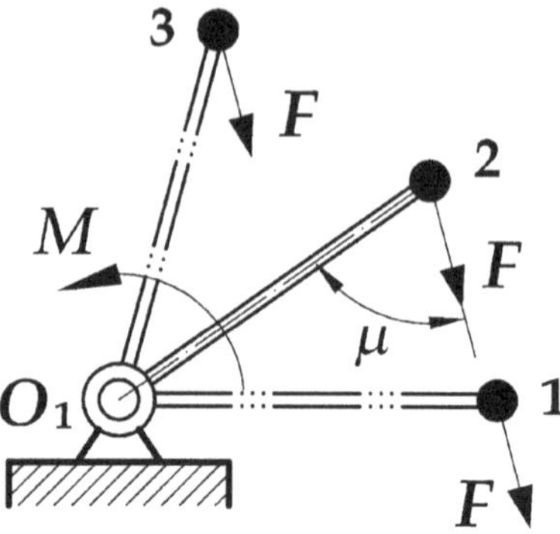

Fig. 7.12 Crank handle

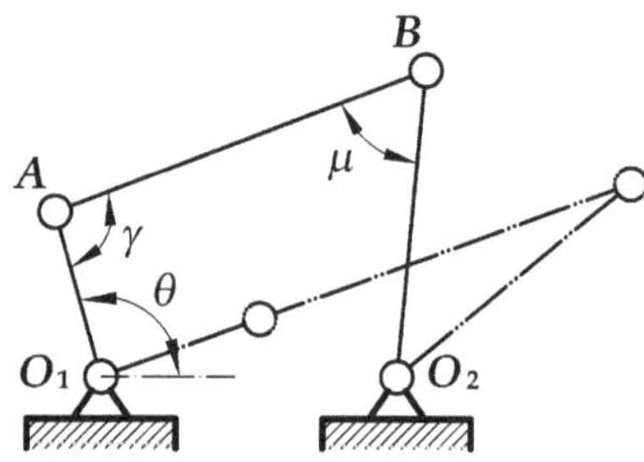

Fig. 7.13 Transmission angle

Considering the planar four-bar mechanism shown in Fig. 7.13, when O_1A is the input link, the force applied to output link BO_2 is transmitted through coupler link AB. For sufficiently slow motions (negligible inertia forces), the force on the coupler link is pure tension or compression (negligible bending action) and is directed along AB. For a given force in the coupler link, the torque transmitted to the output bar (about point O_2) is at its maximum when angle μ between coupler bar AB and output bar BO_2 is 90°. Therefore, angle ABO_2 is the transmission angle. Hartenberg and Denavit (1964) define the transmission angle as the small angle between the direction of the velocity difference vector of the driving link and the direction of the absolute velocity vector of the output link, both taken at the point of connection. Other definitions have been proposed [see Alt (1932) and Hain (1967), for instance], but all are somehow related to a joint variable(s) of the mechanism.

When the transmission angle deviates significantly from 90°, the torque on the output bar decreases and may be insufficient for overcoming friction in the system. For this reason, deviation angle $\alpha = |90° - \mu|$ should not be too large.

Meanwhile, when angle γ between input link O_1A and coupler link AB is 0° or 180° (Fig. 7.13), output point B does not move regardless of the input. Hence, the motion of the input cannot be effectively transmitted to the output; that is, the output point loses its DOF. Therefore, deviation angle $\beta = |90° - \gamma|$ should not be too large, either. In this chapter, angle μ is defined as the *forward transmission*

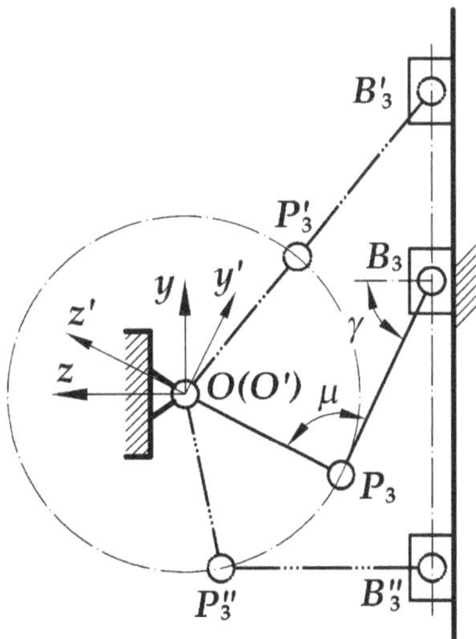

Fig. 7.14 Slider-crank mechanism

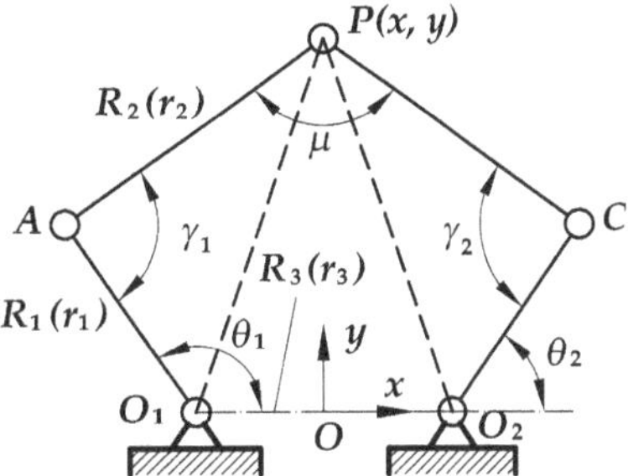

Fig. 7.15 5R parallel mechanism

angle, and angle γ is referred to as the *inverse transmission angle*. When BO_2 is an active link, angle O_1AB is the forward transmission angle and angle ABO_2 is the inverse transmission angle.

The slider-crank mechanism shown in Fig. 7.14 is taken as an example, in which angle $O'P_3B_3$ is the forward transmission angle. The angle between the coupler link and the normal to the straight-line path of the slider is the inverse transmission angle when the slider is actuated. For the 2-DOF 5R parallel mechanism shown in Fig. 7.15, the forward transmission angle is defined as the angle between couplers *AP* and *PC* (Ting and Tsai 1985); the inverse transmission angle is the angle between the input link and coupler. There are two inverse transmission angles for the mechanism. Using the concept of the transmission angle, Stanley (1961) investigated the design of a 2-DOF five-bar mechanism. A synthesis method that features transmission angle control for the 5R variable topology mechanism has also recently been proposed (Balli and Chand 2004). The forward and inverse transmission angles of a PRRRP mechanism are defined as shown in Fig. 7.16. The definitions follow those of the slider-crank and 5R mechanisms.

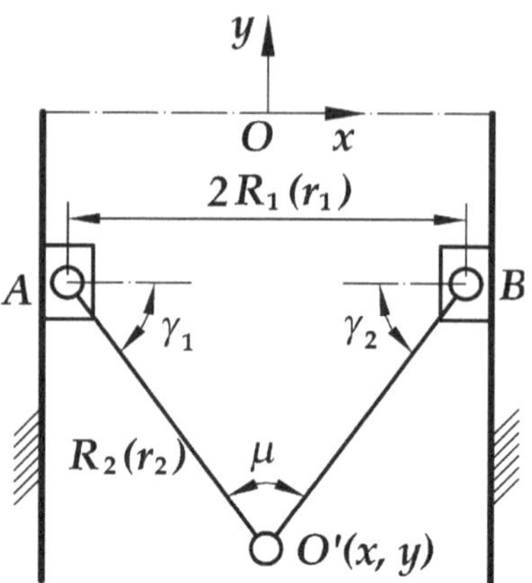

Fig. 7.16 PRRRP parallel mechanism

7.3.2.2 Relationship Between Transmission Angle and Singularity

The transmission angle is an important index that can evaluate the quality of motion/force transmission. Therefore, this concept has been used in the design of almost all types of four-bar mechanisms; some planar five-, six-, and seven-bar mechanisms (Philipp and Freudenstein 1965); spatial four-link mechanisms (Söylemez and Freudenstein 1982); and the 2-DOF spatial RSSRP mechanism (Alizade et al. 1975). The pressure angle, which is the complementary angle of the transmission angle, was also used in the design of a 6-RSS parallel mechanism (Takeda et al. 1997).

Singularity is one of the most important problems encountered in mechanisms, especially in parallel mechanisms. It causes loss of controllability and degradation of natural stiffness, prompting the extensive analysis of parallel mechanisms (Gosselin and Angeles 1990; Park and Kim 1999; Liu et al. 2003). In singularity, the condition number of the Jacobian matrix is normally null.

The *transmission angle* and singularity are closely related. Taking the four-bar mechanism shown in Fig. 7.13 as an example, when $\mu = 0°$ or $180°$, the mechanism is at a "dead point"—the second kind of singularity, according to the classification in Gosselin and Angeles (1990). Meanwhile, $\gamma = 0°$ or $180°$ corresponds to the first kind of singularity, in which the mechanism has a self-jamming characteristic. For the slider-crank mechanism shown in Fig. 7.14, configuration $O'P_3'B_3'$, in which $\mu = 180°$, is the second kind of singularity, whereas configuration $O'P_3''B_3''$, wherein, is the first type. The singularity of the 5R parallel mechanism shown in Fig. 7.15 was discussed systematically by Liu et al. (2006) and in Sect. 5.1 of the current work. We can see that $\gamma = 0°$ or $180°$ and $\mu = 0°$ or $180°$ represent the first and second kinds of singularities of the mechanism, respectively. The same holds true for the PRRRP mechanism in Fig. 7.16. Therefore, *forward* and *inverse transmission angles* μ and γ are closely related to the two kinds of singularities of the parallel mechanism. The simultaneous occurrence of $\gamma = 0°$ or $180°$ and $\mu = 0°$

or 180° easily leads to the conclusion that the mechanism will exhibit the third kind of singularity. This case is not only architecture dependent but also dimension dependent.

The study of planar four-bar mechanisms has a very long history—longer than that of serial robots and even more extensive than that of parallel mechanisms. The Jacobian matrix, manipulability, and condition number are concepts that have been proposed and used in serial robots. Study results were, in general, directly transposed to parallel mechanisms from the very beginning. Merlet (2006) has recently shown that the use of the condition number in the Jacobian matrix of a mixed parallel mechanism should be viewed critically when applied in optimal design. Even without the Jacobian matrix, the transmission angle can be used to evaluate the quality of motion transmission and velocity, as well as performance sensitivity to mechanical errors, singularity, tolerances, and clearances. A parallel mechanism is a type of mechanism with multiple closed-loop chains. It should inherit the design of planar closed-loop four-bar mechanisms, but not the design of serial robots. We believe that the transmission angle is a preferable parameter for accurately describing the motion/force transmission of mechanisms with closed chains and should be used in the optimal design of these mechanisms. However, the motion/force transmissibility of a spatial parallel mechanism cannot be measured simply by considering an angle. Identifying a method that can be applied to both planar and spatial parallel mechanisms is therefore necessary.

Much effort has been devoted to the analysis of the motion/force transmission of mechanisms. Several indices, including the transmission angle, pressure angle, and transmission factors, were accordingly proposed to evaluate its quality. Yuan et al. (1971) defined the virtual coefficient (Ball 1990) between the *transmission wrench* (TWS) and *output twists* as a transmission factor to evaluate the transmission performance of spatial mechanisms. To normalize the transmission factor, Sutherland and Roth (1973) proposed the transmission index using only the geometric properties of the mechanisms as bases. Tsai and Lee (1994) introduced the concept of the *generalized transmission wrench* to characterize the transmission properties of mechanisms. On the basis of this concept, they defined the measurements of transmissibility and manipulability and investigated the transmission performance of the variable lead screw mechanism. Takeda et al. (1997) developed a transmission index for parallel mechanisms on the basis of the minimum value of the pressure angle cosine at the connection point of the leg with the mobile platform of a single-DOF parallel mechanism. This minimum value is obtained by fixing all the input except one. They used the transmission index to evaluate the force transmission performance of the 6-RSS parallel mechanism. Lin and Chang (2002), Lin et al. (2003) proposed the *force transmission index* (FTI) for general single-DOF linkage mechanisms and developed the *mean force transmission index* (MFTI) as an extended definition of the FTI. The MFTI can be used as a quantitative measure of the force transmissibility performance of planar parallel mechanisms. Chen and Angeles (2007) proposed a generalized transmission index for spatial mechanisms; this index is applicable to single-loop spatial linkages with fixed output and single or multiple DOFs. The authors defined the transmission quality to evaluate the global performance of a single-loop mechanism. These analysis methods and indices for the force transmission of planar and spatial mechanisms motivate us to further study a generalized method for analyzing and evaluating the motion/force transmissibility of fully parallel mechanisms.

7.3.3 Mathematical Foundation

Screw theory is an efficient mathematical tool for studying spatial mechanisms. It was proven an efficient tool for solving the first-order analysis of closed chains more than two decades ago. Furthermore, no limits are imposed on the required order of analysis when screw theory is used (Lipkin 2005). This theory has been applied to kinematic and dynamic analyses (Gallardo et al. 2003) as well as in the type synthesis (Huang and Li 2002) of parallel mechanisms. In the current chapter, screw theory is employed as a mathematical resource for the analysis of the motion/force transmission of parallel mechanisms.

7.3.3.1 Twist and Wrench

In general, a unit screw can be represented in the form of Plücker coordinates:

$$\$=(\boldsymbol{s}\,;\boldsymbol{s}^0)=(\boldsymbol{s}\,;\boldsymbol{r}\times\boldsymbol{s}+h\boldsymbol{s})=(L,M,N\,;P^*,Q^*,R^*), \tag{7.31}$$

where

$\boldsymbol{s}$: the primal part, i.e., the unit vector along the axis of the screw, $\boldsymbol{s} = (L, M, N)$, and $L^2 + M^2 + N^2 = 1$

$\boldsymbol{s}^0$: the dual part, and $\boldsymbol{s}^0 = (P^*, Q^*, R^*)$

$\boldsymbol{r}$: the position vector of one arbitrary point on the screw axis

$\boldsymbol{h}$: the pitch of the screw, and $h = \frac{\boldsymbol{s}\bullet\boldsymbol{s}^0}{\boldsymbol{s}\bullet\boldsymbol{s}} = LP^* + MQ^* + NR^*$

With respect to a reference frame, the instantaneous motion of a rigid body can be represented by a twist, which can be expressed by

$$\$_1=w\hat{\$}_1=w(\boldsymbol{\omega}\,;\boldsymbol{r}\times\boldsymbol{\omega}+h\boldsymbol{\omega})=(L_1,M_1,N_1\,;P_1^*,Q_1^*,R_1^*), \tag{7.32}$$

where $\boldsymbol{w}$ is the amplitude of twist $\$_1$, $\hat{\$}_1$ is called the unit twist or the normalized twist, and $\boldsymbol{\omega}$ represents the unit rotational velocity of the rigid body with respect to the origin of the reference frame. When $h = 0$, the twist can be rewritten as $\$_1 = w(\boldsymbol{\omega};\ \boldsymbol{r}\times\boldsymbol{\omega})$, which can represent the instantaneous motion of a revolute joint. The twist with $h = \infty$ can be rewritten as $\$_1 = (0;\ \boldsymbol{v})$, which represents a translation; $\boldsymbol{v}$ is the translational velocity.

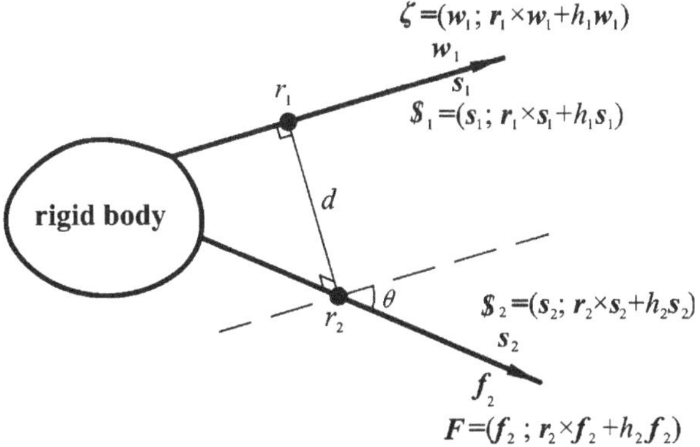

Fig. 7.17 Twist and wrench of a rigid body

Similarly, a wrench exerted on the rigid body can be expressed by a screw as follows:

$$\$_2 = f\$_2 = f(\boldsymbol{f};\, \boldsymbol{r}\times\boldsymbol{f} + h\boldsymbol{f}) = (L_2, M_2, N_2;\, P_2^*, Q_2^*, R_2^*), \tag{7.33}$$

where f is the amplitude of wrench $\$_2$, $\$_2$ is called the unit wrench or normalized wrench, and $\boldsymbol{f}$ represents the unit pure force vector. When $h = 0$, the wrench can be rewritten as $\$_2 = f(\boldsymbol{f};\ \boldsymbol{r}\times\boldsymbol{f})$, which represents pure force. The wrench with $h = \infty$ can be rewritten as $\$_2 = \tau(0;\ \boldsymbol{s})$, which represents pure torque, where τ is the amplitude of the torque, and $\boldsymbol{s}$ represents the unit vector of the torque with respect to the origin of the reference frame.

7.3.3.2 Reciprocal Product and Power Coefficient

Screws $\$_1$ and $\$_2$ are presented by

$$\$_1 = \rho_1\$_1 = \rho_1(\boldsymbol{s}_1\,;\boldsymbol{s}^{01}) = \rho_1(\boldsymbol{s}_1\,;\boldsymbol{r}_1\times\boldsymbol{s}_1 + h_1\boldsymbol{s}_1) = (L_1, M_1, N_1\,;P_1^*, Q_1^*, R_1^*)$$

and

$$\$_2 = \rho_2\$_2 = \rho_2(\boldsymbol{s}_2\,;\boldsymbol{s}^{02}) = \rho_2(\boldsymbol{s}_2\,;\boldsymbol{r}_2\times\boldsymbol{s}_2 + h_2\boldsymbol{s}_2) = (L_2, M_2, N_2\,;P_2^*, Q_2^*, R_2^*),$$

in which screws $\$_1$ and $\$_2$ (Fig. 7.17) are their unit screws, respectively. The axes of $\$_1$ and $\$_2$ lie on lines s_1 and s_2. d represents the length of the common perpendicular of these two lines, and θ is the small angle between these two lines.The reciprocal product between $\$_1$ and $\$_2$ is defined as

$$\begin{aligned}\$_1 \circ \$_2 &= \rho_1\rho_2(\boldsymbol{s}_1 \cdot \boldsymbol{s}^{02} + \boldsymbol{s}_2 \cdot \boldsymbol{s}^{01}) \\ &= \rho_1\rho_2\boldsymbol{s}_1 \cdot (\boldsymbol{r}_2 \times \boldsymbol{s}_2 + h_2\boldsymbol{s}_2) + \rho_1\rho_2\boldsymbol{s}_2 \cdot (\boldsymbol{r}_1 \times \boldsymbol{s}_1 + h_1\boldsymbol{s}_1) \\ &= \rho_1\rho_2(h_1 + h_2)(\boldsymbol{s}_1 \cdot \boldsymbol{s}_2) + \rho_1\rho_2(\boldsymbol{r}_2 - \boldsymbol{r}_1)\cdot(\boldsymbol{s}_2 \times \boldsymbol{s}_1) \\ &= \rho_1\rho_2\ (h_1 + h_2)\cos\theta - d\sin\theta\end{aligned}, \tag{7.34}$$

or

$$\$_1 \circ \$_2 = L_1P_2^* + M_1Q_2^* + N_1R_2^* + L_2P_1^* + M_2Q_1^* + N_2R_1^*. \tag{7.35}$$

Thus, the reciprocal product between unit screws $\$_1$ and $\$_2$ is

$$\$_1 \circ \$_2 = (h_1 + h_2)\cos\theta - d\sin\theta. \tag{7.36}$$

If the rigid body is permitted to move only along unit screw $\$_1$, its corresponding unit twist should be $\boldsymbol{\zeta} = (\boldsymbol{w}_1;\ \boldsymbol{v}_1) = (\boldsymbol{w}_1;\ \boldsymbol{r}_1 \times \boldsymbol{w}_1 + h_1\boldsymbol{w}_1)$. The unit wrench exerted on the rigid body along unit screw $\$_2$ should be expressed by $\boldsymbol{F} = (\boldsymbol{f}_2;\ \boldsymbol{\tau}_2) = (\boldsymbol{f}_2;\ \boldsymbol{r}_2 \times \boldsymbol{f}_2 + h_2\boldsymbol{f}_2)$; thus, the instantaneous power between $\boldsymbol{F}$ and $\boldsymbol{\zeta}$ can be represented by

$$\begin{aligned}P &= \boldsymbol{F} \circ \boldsymbol{\zeta} \\ &= \boldsymbol{f}_2 \cdot \boldsymbol{v}_1 + \boldsymbol{\tau}_2 \cdot \boldsymbol{w}_1 \\ &= \boldsymbol{f}_2 \cdot (\boldsymbol{r}_1 \times \boldsymbol{w}_1 + h_1\boldsymbol{w}_1) + \boldsymbol{w}_1 \cdot (\boldsymbol{r}_2 \times \boldsymbol{f}_2 + h_2\boldsymbol{f}_2) \\ &= (h_1 + h_2)(\boldsymbol{w}_1 \cdot \boldsymbol{f}_2) + (\boldsymbol{r}_2 - \boldsymbol{r}_1) \cdot (\boldsymbol{f}_2 \times \boldsymbol{w}_1) \\ &= (h_1 + h_2)\cos\theta - d\sin\theta.\end{aligned} \tag{7.37}$$

Equations (7.36) and (7.37) show that the reciprocal product of $\boldsymbol{F}$ and $\boldsymbol{\zeta}$ (pertaining to their instantaneous power) is equivalent to the reciprocal product of unit screws $\$_1$ and $\$_2$. Therefore, when the reciprocal product of $\$_1$ and $\$_2$ equals zero, the reciprocal product of $\boldsymbol{F}$ and $\boldsymbol{\zeta}$ is also equal to zero, and unit wrench $\boldsymbol{F}$ applies no work to the rigid body.

In the report by Chen and Angeles (2007), a global transmission index for single-loop mechanisms is defined as

$$\rho = \frac{|\$_1 \circ \$_2|}{|\$_1 \circ \$_2|_{\max}} = \frac{|(h_1 + h_2)\cos\theta - d\sin\theta|}{\max_{h_1,h_2,d}\sqrt{(h_1 + h_2)^2 + d^2}}. \tag{7.38}$$

When the wrench $\boldsymbol{F}$ and the twist $\boldsymbol{\zeta}$ are given, their pitches h_1 and h_2 are considered constant; thus, their reciprocal product is relevant only to their relative positions and directions. The maximal magnitude of the reciprocal product of $\boldsymbol{F}$ and $\boldsymbol{\zeta}$ becomes

$$\left|\$_1 \circ \$_2\right|_{max} = \sqrt{(h_1 + h_2)^2 + d_{max}^2}, \tag{7.39}$$

where d_{max} represents the potential maximal length of the common perpendicular between the axes of $\boldsymbol{F}$ and $\boldsymbol{\zeta}$. Thus, Eq. (7.38) can be rewritten as

$$\rho = \frac{\left|\$_1 \circ \$_2\right|}{\left|\$_1 \circ \$_2\right|_{max}} = \frac{\left|(h_1 + h_2)\cos\theta - d\sin\theta\right|}{\sqrt{(h_1 + h_2)^2 + d_{max}^2}}. \tag{7.40}$$

The larger the value of ρ, the more effectively the power is transmitted from the wrench to the twist, and the better the motion/force transmission performance. Consequently, ρ is referred to as the *power coefficient.*

Some special cases exist:

Case 1: When h_1 is infinite, unit twist $\boldsymbol{\zeta}$ becomes a unit translation represented by $(0;\ \boldsymbol{v}_1)$. Hence, the power coefficient is

$$\rho = \frac{|\boldsymbol{f}_2 \cdot \boldsymbol{v}_1|}{|\boldsymbol{f}_2 \cdot \boldsymbol{v}_1|_{max}}, \tag{7.41}$$

where $|\boldsymbol{f}_2 \cdot \boldsymbol{v}_1|_{max}$ represents the potential maximum of the dot product of unit vectors $\boldsymbol{f}_2$ and $\boldsymbol{v}_1$.

Case 2: When h_2 is infinite, unit wrench $\boldsymbol{F}$ becomes a unit torque represented by $(0; \boldsymbol{\tau}_2)$. Therefore, the power coefficient is

$$\rho = \frac{|\boldsymbol{\tau}_2 \cdot \boldsymbol{w}_1|}{|\boldsymbol{\tau}_2 \cdot \boldsymbol{w}_1|_{max}}, \tag{7.42}$$

where $|\boldsymbol{\tau}_2 \cdot \boldsymbol{w}_1|_{max}$ represents the potential maximum of the dot product of unit vectors $\boldsymbol{\tau}_2$ and $\boldsymbol{w}_1$.

Case 3: When both h_1 and h_2 are infinite, unit twist $\boldsymbol{\zeta}$ becomes a unit translation represented by $(0;\ \boldsymbol{v}_1)$, and unit wrench $\boldsymbol{F}$ becomes a unit torque represented by $(0;\ \boldsymbol{\tau}_2)$.

Thus, the reciprocal product of $\boldsymbol{F}$ and $\boldsymbol{\zeta}$ is

$$P = \boldsymbol{F} \circ \boldsymbol{\zeta} = 0 \cdot \boldsymbol{v}_1 + \boldsymbol{\tau}_2 \cdot 0 = 0, \tag{7.43}$$

indicating that the power coefficient vanishes and $\boldsymbol{F}$ transmits no power to $\boldsymbol{\zeta}$.

Notably, Eq. (7.40) shows that the power coefficient is related only to the parameters of $\boldsymbol{F}$ and $\boldsymbol{\zeta}$. In other words, the power coefficient is unrelated to the selected frame coordinate; alternatively, we can say that it is frame-free. Moreover, the value of the power coefficient belongs to $[0,\ 1]$. For these reasons, this coefficient can be used in the performance evaluation of the motion/force transmission of parallel mechanisms.

7.3.4 *Input Transmission Index*

The function of a mechanism is to transmit motion/force from its input member to its output member. On one hand, the motions from the input should be transmitted to the output to realize the required motion of the output member; on the other hand, mechanisms should transmit force from the input to the output to balance the payload of the output member. In the process of transmission, the arising internal wrench, namely, the transmission wrench, can be expressed by the TWS.

In this chapter, we first assume that gravity and friction are disregarded. Furthermore, the parallel mechanism considered here is non-redundant, and only one actuator exists in each of its legs. Such a mechanism is usually referred to as a fully parallel mechanism, with each leg having its own TWS. Notably, the TWS is reciprocal to all the passive joint screws in the leg.

Given the function of a parallel mechanism (transmitting motion/force from its input members to its output members), the transmission performance with respect to both its input and output should be considered. Here, we first investigate the input transmission.

Each leg of a fully parallel mechanism has its own actuator, and each input is independent of the other. Hence, to evaluate the manipulability of the *i*th leg, the power coefficient between the TWS and input twist is defined as the input transmission index represented by

$$\lambda_i = \frac{\left|\$_{Ti} \circ \$_{Ii}\right|}{\left|\$_{Ti} \circ \$_{Ii}\right|_{\max}}, \tag{7.44}$$

where $\$_{Ti}$ and $\$_{Ii}$ represent the unit TWS and the unit input twist of the *i*th leg, respectively. Equation (7.44) shows that λ_i is frame-free, with values ranging from 0 to 1, i.e., $0 \leq \lambda_i \leq 1$.

The transmission wrenches and input joints vary for different legs. The input transmission indices of some typical legs used in parallel mechanisms are studied below.

7.3.4.1 Input Transmission Index in an $\underline{R}$SS Leg

Figure 7.18 shows an $\underline{R}$SS leg that contains one actuating revolute joint and two passive spherical joints S_1 and S_2. Parameters a and b are the lengths of the input link and coupler link, respectively.

With respect to coordinate system *o-xyz*, unit input twist $\$_I$ can be expressed by

$$\$_I = (\boldsymbol{w}_{12}; \boldsymbol{0}) = (1,\ 0,\ 0,\ 0,\ 0,\ 0). \tag{7.45}$$

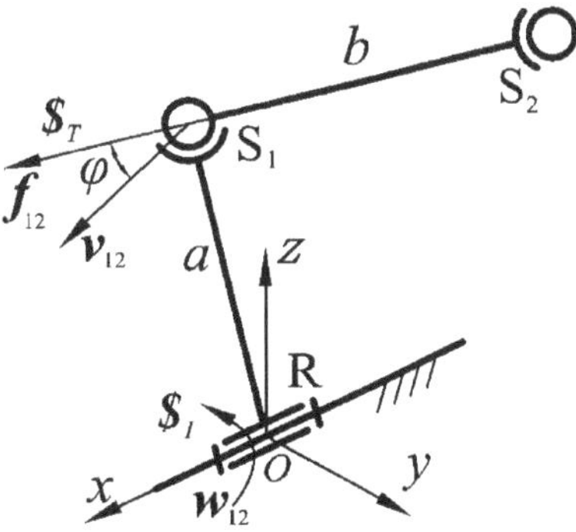

Fig. 7.18 RSS leg

The transmission wrench is a pure force along the coupler link; hence, the TWS can be represented by

$$\$_T = (f_{12};\, a \times f_{12}), \tag{7.46}$$

where $\boldsymbol{a}$ represents a vector along the input link and its norm is a. Thus, the input transmission index of the RSS leg can be obtained as follows:

$$\lambda_{12} = \frac{|\$_T \circ \$_I|}{|\$_T \circ \$_I|_{\max}} = \frac{|\,a \times f_{12} \;\cdot w_{12}|}{|\,a \times f_{12} \;\cdot w_{12}|_{\max}} = \frac{|f_{12} \cdot \;w_{12} \times a\,|}{|f_{12} \cdot \;w_{12} \times a\,|_{\max}} = \frac{|f_{12} \cdot v_{12}|}{|f_{12} \cdot v_{12}|_{\max}}, \tag{7.47}$$

where $\boldsymbol{v}_{12}$ represents the velocity of the center of the joint (spherical joint S_1) connected to the input link. Given that $\boldsymbol{f}_{12}$ and $\boldsymbol{w}_{12}$ are unit vectors, we obtain

$$|\boldsymbol{f}_{12} \cdot \boldsymbol{v}_{12}|_{\max} = |\boldsymbol{f}_{12} \cdot (\boldsymbol{w}_{12} \times \boldsymbol{a})|_{\max} = a. \tag{7.48}$$

Thus, Eq. (7.47) can be rewritten as

$$\lambda_{12} = \frac{|\boldsymbol{f}_{12} \cdot \boldsymbol{v}_{12}|}{|\boldsymbol{f}_{12} \cdot \boldsymbol{v}_{12}|_{\max}} = \frac{|\boldsymbol{f}_{12} \cdot \boldsymbol{v}_{12}|}{a} = |\cos\varphi|. \tag{7.49}$$

Equation (7.49) shows that the input transmission index of the RSS leg is independent of frame *o-xyz* and equal to the absolute cosine value of the angle between the velocity of the center of the joint connected to the input link and the unit vector along the coupler link. Moreover, this result also fits some similar legs, such as the RUS and planar RRR legs. In each of these legs, the TWS is a pure force along the coupler link, and the actuator is a revolute joint attached to the base.

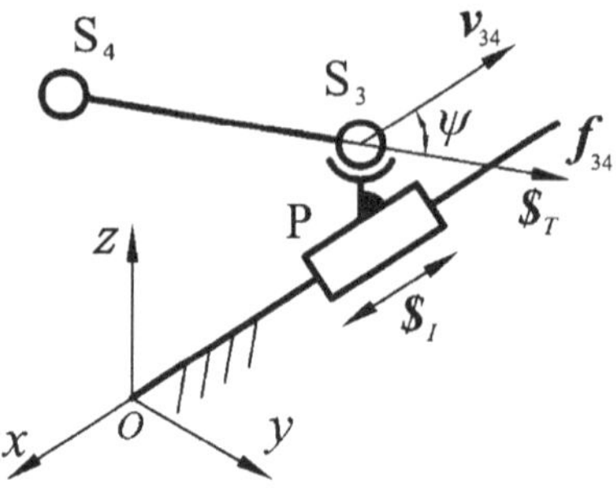

Fig. 7.19 PSS leg

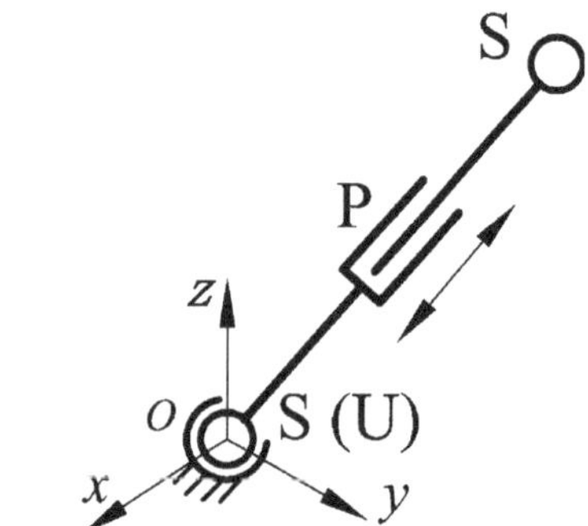

Fig. 7.20 SPS or UPS leg

7.3.4.2 Input Transmission Index in a Leg with a Prismatic Actuator

Figure 7.19 shows a PSS leg, which contains one actuating prismatic joint and two passive spherical joints S_3 and S_4. Similar to the transmission wrench of the RSS leg, that of the PSS leg is a pure force along the coupler link. The pitch of the input twist is infinite because the input twist is a translation. According to Eq. (7.41), therefore, the input transmission index of the PSS leg can be obtained as

$$\lambda_{34} = \frac{|\boldsymbol{f}_{34} \cdot \boldsymbol{v}_{34}|}{|\boldsymbol{f}_{34} \cdot \boldsymbol{v}_{34}|_{\max}} = |\boldsymbol{f}_{34} \cdot \boldsymbol{v}_{34}| = |\cos \psi| , \tag{7.50}$$

which is also independent of the coordinate system *o-xyz*.

Equation (7.50) shows that the input transmission index of the PSS leg is equal to the absolute cosine value of the angle between the translational direction of the prismatic joint and the unit vector along the coupler link. This result fits other similar legs, such as the PUS, PRS, and planar PRR legs. In each of these legs, the TWS is a pure force along the coupler link, and the actuator is a prismatic joint attached to the base.

Notably, an SPS or UPS leg (shown in Fig. 7.20), where the prismatic joint between two passive joints is actuated, is typically used in parallel mechanisms. In such a leg, the position of the prismatic actuator differs from that of the

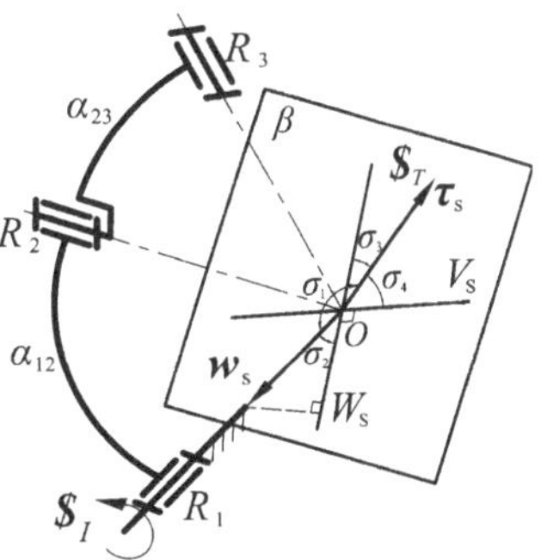

Fig. 7.21 Spherical RRR leg

PSS leg. However, the input transmission index of the SPS or UPS leg can still be obtained using the above-mentioned result. The translational direction of the prismatic actuator is along the coupler link; thus, the input transmission index of the SPS or UPS leg is constant and its value is equal to 1.

7.3.4.3 Input Transmission Index in a Spherical RRR Leg

As shown in Fig. 7.21, a spherical RRR leg contains three revolute joints, whose axes (OR_1, OR_2, and OR_3) intersect at a common point O. The revolute joint attached to the base is actuated. Parameter α_{12} represents the angle between axes OR_1 and OR_2, and parameter α_{23} stands for the angle between axes OR_2 and OR_3.

In such a spherical RRR leg, the transmission wrench is a pure torque, whose axis is perpendicular to plane OR_2R_3 and passes through point O. Thus, the pitch of the TWS is infinite. According to Eq. (7.42), the input transmission index of the spherical RRR leg can be obtained as follows:

$$\lambda_s = \frac{|\boldsymbol{\tau}_s \cdot \boldsymbol{w}_s|}{|\boldsymbol{\tau}_s \cdot \boldsymbol{w}_s|_{\max}}, \tag{7.51a}$$

where $\boldsymbol{\tau}_s$ and $\boldsymbol{w}_s$ represent the unit vectors along the axes of the TWS and revolute actuator, respectively. Given that the axis of the TWS is perpendicular to line OR_2, all the potential locations of vector $\boldsymbol{\tau}_s$ must lie in plane β, which is perpendicular to line OR_2 and passes through point O.

As shown in Fig. 7.21, the projection of vector $\boldsymbol{w}_s$ onto plane β lies on line OW_S. Line OV_S is perpendicular to line OW_S and lies in plane β; hence, OV_S is the normal line of plane OR_1R_2. The angle between unit vectors $\boldsymbol{\tau}_s$ and $\boldsymbol{w}_s$ is minimal when $\boldsymbol{\tau}_s$ is parallel to line OW_S. Thus, Eq. (7.51a) can be rewritten as

$$\lambda_s = \frac{|\boldsymbol{\tau}_s \cdot \boldsymbol{w}_s|}{|\boldsymbol{\tau}_s \cdot \boldsymbol{w}_s|_{\max}} = \frac{|\cos\sigma_1|}{|\cos\sigma_2|} = |\cos\sigma_3| = |\sin\sigma_4|, \tag{7.51b}$$

where

σ_1 is the angle between unit vectors $\boldsymbol{\tau}_s$ and $\boldsymbol{w}_s$
σ_2 is the small angle between lines OR_1 and OW_S
σ_3 is the small angle between vector $\boldsymbol{\tau}_s$ and line OW_S
σ_4 is the small angle between vector $\boldsymbol{\tau}_s$ and line OV_S, as well as the complement angle of σ_3

σ_4 is also the small angle between planes OR_1R_2 and OR_2R_3. Thus, Eq. (7.51b) can be rewritten as

$$\lambda_s = |\sin\sigma_4| = |\sin\angle R_1\text{-OR}_2\text{-R}_3|\,, \tag{7.51c}$$

where $\angle R_1$-OR$_2$-R$_3$ represents the angle between planes OR_1R_2 and OR_2R_3.

7.3.5 Output Transmission Index

Generally, if a fully parallel mechanism has n DOFs, n legs will be connected to its output member. To balance the external load exerted on the mobile platform, each leg contributes to the force transmission of the parallel mechanism. Thus, the output transmission index should be different from the input transmission index. Takeda et al. (1996) defined a transmission index using the pressure angle at the connection point of the leg with the mobile platform of a single-DOF parallel mechanism obtained by fixing all the input, except one. In this section, we further discuss this method, on which a proposed output transmission index is based.

The arbitrary motion of a rigid body can be generated by a screw motion that includes a rotation and translation along the same axis. Hence, similar to the instantaneous unit motion of an n-DOF parallel mechanism at a certain configuration, that of the mobile platform can be represented by a unit twist $\$_{Oi}$ $(i = 1, 2, \ldots, n)$ when fixing all of its input, except that in the ith leg. In this case, only the transmission wrench represented by $\$_{Ti}$ can contribute to the mobile platform, whereas all the other transmission wrenches apply no work to the mobile platform. Hereby, we may obtain

$$\$_{Tj} \circ \$_{Oi} = 0 \quad (j = 1, 2, \ldots, n, \quad j \neq i). \tag{7.52}$$

Using the method above, n output twists $\$_{Oi}$ $(i = 1, 2, \ldots, n)$ can be obtained, and their relationship can be described as follows.

Theorem 1 *As in an n-DOF motion-definite*[1] *fully parallel mechanism, when the mechanism is at a nonsingular configuration, unit twists* $\$_{O1}, \$_{O2}, \ldots, \$_{On}$ *are linearly independent.*

Proof Let us assume that $\$_{O1}, \$_{O2}, \ldots, \$_{On}$ are linearly dependent. Hence, each can be expressed by a linear combination of the others; for example,

$$\$_{On} = k_1 \$_{O1} + k_2 \$_{O2} + \cdots + k_{n-1} \$_{O(n-1)} . \tag{7.53}$$

Thus, we have

$$\$_{Tn} \circ \$_{On} = k_1 \left(\$_{Tn} \circ \$_{O1} \right) + k_2 \left(\$_{Tn} \circ \$_{O2} \right) + \cdots + k_{n-1} \left(\$_{Tn} \circ \$_{O(n-1)} \right) . \tag{7.54a}$$

Substituting Eq. (7.52) into Eq. (7.54a) yields

$$\$_{Tn} \circ \$_{On} = 0 , \tag{7.54b}$$

which means that the transmission wrench represented by $\$_{Tn}$ does not contribute to the output member. Similarly, the transmission wrenches represented by $\$_{Ti}$ $(i = 1, 2, \ldots, n-1)$ do not contribute to the mobile platform.

In this case, all the transmission wrenches in the n-DOF parallel mechanism transmit no power to the output member, indicating that the mechanism is at its singular configuration. This result is inconsistent with the condition that the mechanism is at a nonsingular configuration. Thus, the assumption that $\$_{O1}, \$_{O2}, \ldots, \$_{On}$ are linearly dependent does not hold. Hereby, the theorem is proved.

Consequently, an n-order screw system can be spanned by linearly independent unit screws $\$_{O1}, \$_{O2}, \ldots, \$_{On}$. At a certain configuration, an arbitrary motion of the output member can be represented by a twist $\$_{\forall}$ that belongs to the n-order screw system. Twist $\$_{\forall}$ can be expressed by a linear combination of $\$_{O1}, \$_{O2}, \ldots, \$_{On}$, as below:

$$\$_{\forall} = l_1 \$_{O1} + l_2 \$_{O2} + \cdots + l_n \$_{On} . \tag{7.55}$$

Therefore, the power transmitted to the output member can be expressed as

$$P = \sum_{i=1}^{n} \left| m_i \$_{Ti} \circ \$_{\forall} \right| , \tag{7.56a}$$

[1] When the input of the manipulator is given, if its output motion is definite, the manipulator is said to be motion definite; otherwise, the manipulator is considered motion indefinite.

where m_i is the module of the TWS in the ith leg. Using Eqs. (7.52) and (7.55) as bases, we can rewrite Eq. (7.56a) as

$$P=\sum_{i=1}^{n}\left|m_i\$_{Ti}\circ l_i\$_{Oi}\right|=\sum_{i=1}^{n}\left(\eta_i\cdot m_il_i\left|\$_{Ti}\circ\$_{Oi}\right|_{\max}\right), \tag{7.56b}$$

where η_i represents the power coefficient of the ith leg and can be expressed as

$$\eta_i=\frac{\left|\$_{Ti}\circ\$_{Oi}\right|}{\left|\$_{Ti}\circ\$_{Oi}\right|_{\max}}. \tag{7.57}$$

Equation (7.57) shows that η_i is independent of any coordinate system, and $0\leq\eta_i\leq 1$. Here, η_i is referred to as the output transmission index of the ith leg.

In Eq. (7.56b), m_i is determined only by the effective external load exerted on the output member, and l_i is determined only by the desired motion of the mobile platform. Thus, m_i and l_i should be considered external factors. The transmission performance of a mechanism is one of its essential characteristics and is unrelated to external conditions. Therefore, m_i and l_i should be disregarded in the evaluation of the transmission performance of parallel mechanisms. $|\$_{Ti}\circ\$_{Oi}|_{\max}$ is constant at a certain configuration; hence, η_i should be as large as possible to transmit more power to the output member. Furthermore, to obtain good transmissibility to the output under any combination of m_i and l_i, all the output transmission indices should be as large as possible. That is, the minimum of η_i should be as large as possible.

7.3.6 *LTI and GTI of Parallel Mechanisms*

In Sects. 7.3.4 and 7.3.5, the input and output transmission indices of the ith leg in a fully parallel mechanism are denoted as λ_i and η_i, respectively. In Sect. 7.3.5, we conclude that each output transmission index should be as large as possible to obtain good-transmission performance. Aside from the output transmission indices, every input transmission index should be sufficiently large to better transmit motion/force from the input; if one of the input transmission indices is very small or vanishes, the transmission of motion/force from the corresponding input to the output will be confronted with considerable difficulty. In this case, the mechanism is said to be singular or near its singularity.

For an integrated n-DOF parallel mechanism with n legs, therefore, when transmission indices λ_i and η_i $(i=1,2,\ldots,n)$ are both large enough, good motion/force transmission performance is said to occur; when any one of them is small, the mechanism has poor motion/force transmissibility. For these reasons, a transmission index is defined as

$$\chi=\min\left\{\lambda_i,\,\eta_i\right\},\quad (i=1,2,\ldots,n). \tag{7.58}$$

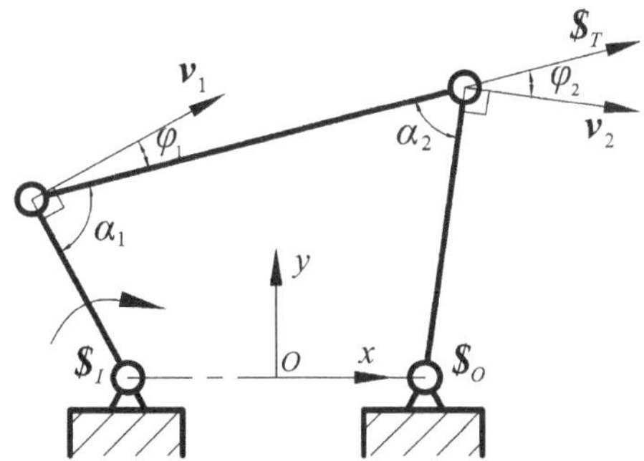

Fig. 7.22 Planar four-bar mechanism

At a different configuration, the value of the transmission index also differs; hence, χ is referred to as the LTI. A larger χ indicates better motion/force transmission performance of the parallel mechanism at the configuration.

As mentioned in Sect. 7.3.3, the power coefficient is unrelated to the selected frame coordinate or is frame-free. Hence, Eqs. (7.54) and (7.57) show that λ_i and η_i are frame-free indices. According to Eq. (7.58), LTI χ can be regarded as a frame-free index, which is most important for the optimal design of parallel mechanisms. Given that the values of λ_i and η_i range from 0 to 1, the value range of χ can be expressed by

$$0 \leq \chi \leq 1. \tag{7.59}$$

The LTI is relative to the pose of the mobile platform of a parallel mechanism. To globally evaluate the motion/force transmissibility over a workspace, the GTI is defined as the LTI that is integrated over the workspace and is divided by the volume of the workspace. Therefore, the GTI can be expressed as

$$\Gamma = \frac{\int_W \chi \, \mathrm{d}W}{\int_W \mathrm{d}W}, \tag{7.60}$$

in which W is the workspace of the mechanism.

7.3.7 Good-Transmission Workspace

To evaluate the transmission performance of mechanisms, an LTI limit should be defined. First, we investigate the application of the transmission index in a planar four-bar mechanism (Fig. 7.22). Screws $\$_I$ and $\$_O$ represent the unit input and output twists, respectively. Unit TWS $\$_I$ represents a unit pure force along the coupler link. According to the results in Sects. 7.3.4 and 7.3.5, the input and output transmission index can be obtained as follows:

$$\lambda = |\cos \varphi_1| = |\sin \alpha_1| \tag{7.60a}$$

and

$$\eta = |\cos\varphi_2| = |\sin\alpha_2|\,. \tag{7.60b}$$

Thus, the LTI is

$$\chi = \min\left\{\lambda,\, \eta\right\} = \min\left\{|\sin\alpha_1|\,,\, |\sin\alpha_2|\right\}. \tag{7.61}$$

The LTI of the planar parallel mechanism shown in Fig. 7.15 or 7.16 can be expressed as $\chi = \min\{|\sin\gamma_1|\,,\ |\sin\gamma_2|\,,\ |\sin\mu|\}$ (see the Appendix).

The most important index for evaluating the motion/force transmissibility of the planar four-bar mechanism is the transmission angle. Angles α_1 and α_2 are defined as the forward and inverse transmission angles (Wang et al. 2009) for the planar four-bar mechanism. When one of the transmission angles deviates excessively from 90°, the motion or force from the input is not effectively transmitted to the output, indicating poor motion/force transmission performance of the mechanism. At the same time, the value of the LTI may be small. When these transmission angles are equal to or close to 90°, the mechanism exhibits good motion/force transmission performance, and the LTI has a large value. Thus, similar to the transmission angle, the LTI can be a quality measurement for the motion/force transmission of a planar four-bar mechanism.

For high-speed and high-quality motion transmission, the most widely accepted design limits for the transmission angle are [45°, 135°] (Tao 1964). For the same purpose, the LTI should be no smaller than $\sin 45° \approx 0.7$. In this study, 0.7 is set as the LTI limit. When the LTI ≥ 0.7, the parallel mechanism is considered to have good motion/force transmission at the corresponding configuration; at an LTI < 0.7, the parallel mechanism is at a configuration with poor motion/force transmission performance. The set of poses at which the LTI is no smaller than 0.7 is therefore defined as the *good-transmission workspace* (GTW) of the mechanism. Within the GTW, the parallel mechanism has good motion/force transmissibility at every configuration and is far from its singularity.

7.3.8 Examples

To introduce the analytical method and proposed frame-free index in detail, the motion/force transmission performance analysis of two parallel mechanisms is presented.

7.3.8.1 3-DOF Spherical Parallel Mechanism

Figure 7.23 shows a 3-DOF spherical parallel mechanism (SPM), whose mobile platform is connected to the base by three identical spherical $\underline{R}$RR legs. This

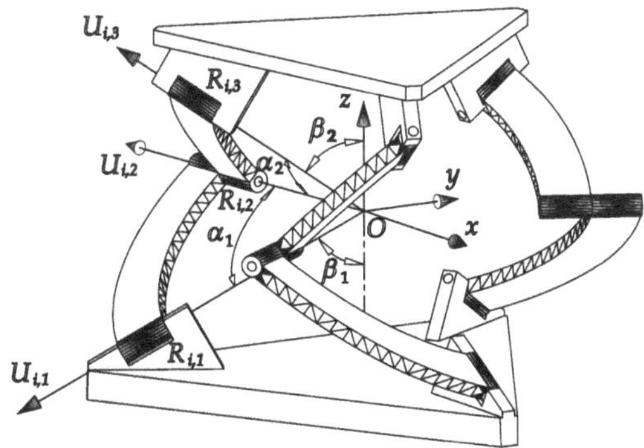

Fig. 7.23 3-DOF spherical parallel mechanism

spherical parallel mechanism is usually applied to a wrist robot or a camera-orienting device (Söylemez and Freudenstein 1982) because of its high rotational workspace, high speed, and high load capacity. A special feature of this parallel mechanism is that all the axes of the nine revolute joints intersect at origin O of global coordinate system O-xyz. In the ith leg, the axial directions of three revolute joints $R_{i,1}$, $R_{i,2}$, and $R_{i,3}$ are denoted by three unit vectors $\boldsymbol{U}_{i,1}$, $\boldsymbol{U}_{i,2}$ and $\boldsymbol{U}_{i,3}$, respectively. The revolute joint attached to the base in each leg is actuated.

The symmetrical SPM has three rotational DOFs; thus, the orientation of the mobile platform with respect to coordinate system O-xyz can be described by *Tilt-and-Torsion*(T&T) angles (ϕ, θ, σ) (Alizade et al. 1975), where ϕ is the *azimuth*, θ is the *tilt* angle, and σ denotes the *torsion* angle. When the T&T angles of the mobile platform are (0°, 0°, 0°), the SPM is considered located at its initial orientation, and unit vectors $\boldsymbol{U}_{i,1}$ and $\boldsymbol{U}_{i,3}$ lie on the same vertical plane. In Fig. 7.23, angles α_1, α_2, β_1, and β_2 denote the four geometric parameters of the 3-DOF SPM.

As indicated in the analysis in Sect. 7.3.4.3, the transmission wrench of the ith leg is a pure torque, whose axis is perpendicular to plane $OR_{i,2}R_{i,3}$ and passes through origin O. Thus, the unit TWS of the ith leg can be expressed by

$$\$_{Ti} = (0;\ \boldsymbol{U}_{i,2} \times \boldsymbol{U}_{i,3} / |\boldsymbol{U}_{i,2} \times \boldsymbol{U}_{i,3}|). \tag{7.62}$$

According to Eq. (7.51c), the input transmission index of the ith leg can be represented by

$$\lambda_i = |\sin \angle R_{i,1}\text{-}OR_{i,2}\text{-}R_{i,3}|, \tag{7.63}$$

where $\angle R_{i,1}\text{-}OR_{i,2}\text{-}R_{i,3}$ represents the angle between planes $OR_{i,1}R_{i,2}$ and $OR_{i,2}R_{i,3}$.

When the second and third input revolute joints ($R_{2,1}$ and $R_{3,1}$) are fixed and the first ($R_{1,1}$) is driven, the transmission wrenches represented by $\$_{T2}$ and $\$_{T3}$ become two constraint wrenches for the output member, and only the transmission wrench represented by $\$_{T1}$ can contribute to the mobile platform. Thus, the mobile platform in this case can rotate only around an axis $OR_{1,4}$, which can be denoted by a unit vector $\boldsymbol{U}_{1,4}$, i.e.,

$$\boldsymbol{U}_{1,4} = \frac{(\boldsymbol{U}_{2,2} \times \boldsymbol{U}_{2,3}) \times (\boldsymbol{U}_{3,2} \times \boldsymbol{U}_{3,3})}{|(\boldsymbol{U}_{2,2} \times \boldsymbol{U}_{2,3}) \times (\boldsymbol{U}_{3,2} \times \boldsymbol{U}_{3,3})|}. \tag{7.64a}$$

Similarly, when only input revolute joint $R_{2,1}$ or $R_{3,1}$ is driven, the corresponding rotational axes $OR_{2,4}$ or $OR_{3,4}$ can be denoted by unit vectors

$$\boldsymbol{U}_{2,4} = \frac{(\boldsymbol{U}_{1,2} \times \boldsymbol{U}_{1,3}) \times (\boldsymbol{U}_{3,2} \times \boldsymbol{U}_{3,3})}{|(\mathbf{U}_{1,2} \times \boldsymbol{U}_{1,3}) \times (\boldsymbol{U}_{3,2} \times \boldsymbol{U}_{3,3})|} \tag{7.64b}$$

and

$$\boldsymbol{U}_{3,4} = \frac{(\boldsymbol{U}_{1,2} \times \boldsymbol{U}_{1,3}) \times (\boldsymbol{U}_{2,2} \times \boldsymbol{U}_{2,3})}{|(\boldsymbol{U}_{1,2} \times \boldsymbol{U}_{1,3}) \times (\boldsymbol{U}_{2,2} \times \boldsymbol{U}_{2,3})|}. \tag{7.64c}$$

Thus, similar to how the input transmission index is obtained, the output transmission index of the ith leg can be derived as follows:

$$\eta_i = |\sin \angle R_{i,2}\text{-}OR_{i,3}\text{-}R_{i,4}|\,, \tag{7.65}$$

where $\angle R_{i,2}\text{-}OR_{i,3}\text{-}R_{i,4}$ represents the angle between planes $OR_{i,2}R_{i,3}$ and $OR_{i,3}R_{i,4}$.

According to Eqs. (7.58), (7.63), and (7.65), therefore, the LTI of the 3-DOF SPM can be obtained as

$$\chi = \min\{|\sin \angle R_{i,1}\text{-}OR_{i,2}\text{-}R_{i,3}|\,,\ |\sin \angle R_{i,2}\text{-}OR_{i,3}\text{-}R_{i,4}|\}\,, \quad (i = 1, 2, 3). \tag{7.66}$$

For a specific 3-DOF spherical parallel mechanism with geometric parameters $\alpha_1 = \alpha_2 = 90°$, $\beta_1 = 30°$, and $\beta_2 = 60°$, Figs. 7.24a, b illustrate the distribution of the LTI within its constant-torsion orientational workspace. This distribution is achieved by fixing torsion angle $\sigma = 0°$ and $\sigma = 30°$. The area enveloped by the line (LTI = 0.7) is the GTW of the 3-DOF SPM.

Takeda et al. (1995) defined a transmission index considering only transmissibility to the output for the 3-DOF SPM. The index was obtained as the minimum cosine value of the angle between unit vector $\mathbf{U}_{i,4}$ and the axis of the unit transmission torque, represented by $\$_{Ti}$. Therefore, the output transmission index in Eq. (7.65) differs from the transmission index introduced by Takeda et al. (1995). Moreover, the LTI in Eq. (7.66) considers both transmissibility to the output and manipulability from the input.

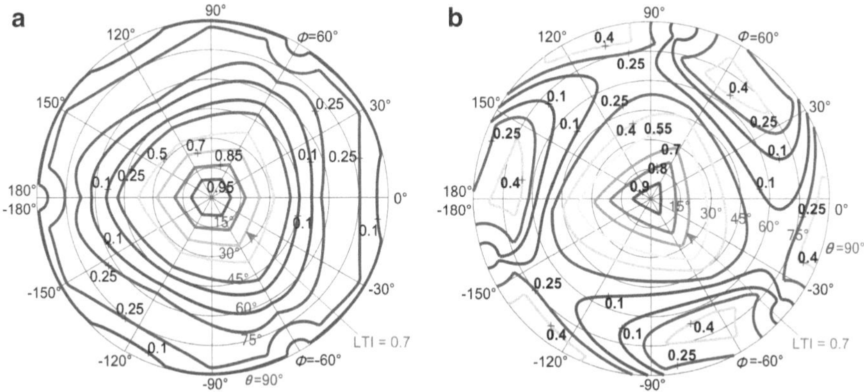

Fig. 7.24 Distribution of the LTI within the constant-torsion orientational workspace of the 3-DOF SPM with parameters $\alpha_1 = \alpha_2 = 90°$, $\beta_1 = 30°$, and $\beta_2 = 60°$: (**a**) $\sigma = 0°$; (**b**) $\sigma = 30$

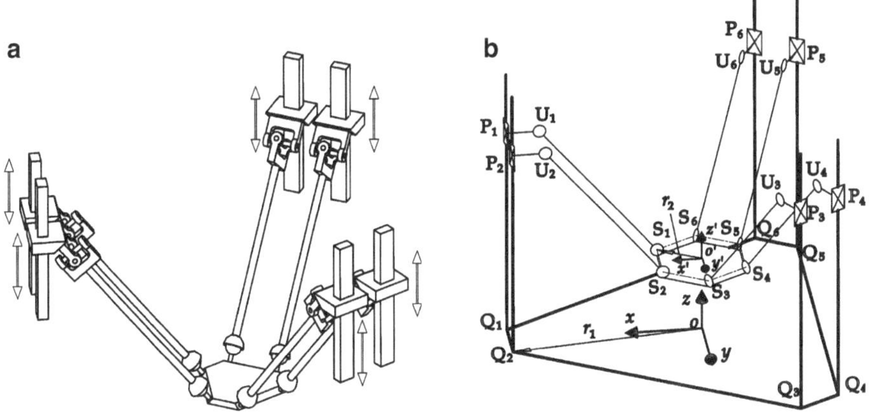

Fig. 7.25 6-PUS parallel mechanism: (**a**) kinematic structure; (**b**) kinematic scheme

7.3.8.2 Spatial 6-DOF 6-PUS Parallel Mechanism

Figure 7.25a shows the kinematic structure of a spatial 6-DOF parallel mechanism called the 6-PUS parallel mechanism. Its kinematic scheme is illustrated in Fig. 7.25b. The mobile platform is connected to the base through six identical PUS legs, each including an active prismatic joint, passive universal joint, and passive spherical joint by turns. Line P_iQ_i $(i = 1, 2, \ldots, 6)$, which denotes the translation axis of the ith prismatic actuator, is parallel to the z-axis and perpendicular to the base. U_i $(i = 1, 2, \ldots, 6)$ represents the center of the universal joint connected to the ith prismatic actuator. The center of the spherical joint attached to the circular mobile platform is represented by S_i $(i = 1, 2, \ldots, 6)$.

With respect to global coordinate system **R**:O-xyz attached to the base, the position of center O' of the mobile platform can be described by $(x_{O'}, y_{O'}, z_{O'})$. The orientation of the mobile platform can be denoted by T&T angles $(\phi,\ \theta,\ \sigma)$. The geometric parameters of the 6-$\underline{\text{P}}$US parallel mechanism can be expressed by

$$|\mathrm{OQ}_i| = r_1 \quad (i = 1, 2, \ldots, 6), \tag{7.67a}$$

$$\left|\mathrm{O'S}_i\right| = r_2 \quad (i = 1, 2, \ldots, 6), \tag{7.67b}$$

$$|\mathrm{S}_i\mathrm{S}_{i+1}| = |\mathrm{Q}_i\mathrm{Q}_{i+1}| = r_3 \quad (i = 1, 3, 5), \tag{7.67c}$$

$$|\mathrm{U}_i\mathrm{S}_i| = r_4 \quad (i = 1, 2, \ldots, 6). \tag{7.67d}$$

The analysis result in Sect. 7.3.4.2 indicates that the transmission wrench of the ith leg is a pure force along line $\mathrm{U}_i\mathrm{S}_i$. With respect to global coordinate system **R**:O-xyz, the unit TWS of the ith leg can be expressed by

$$\$_{Ti} = \left(\overrightarrow{\mathrm{U}_i\mathrm{S}_i}\big/\left|\overrightarrow{\mathrm{U}_i\mathrm{S}_i}\right|;\ \ \overrightarrow{\mathrm{OS}_i}\times\overrightarrow{\mathrm{U}_i\mathrm{S}_i}\big/\left|\overrightarrow{\mathrm{U}_i\mathrm{S}_i}\right|\right) = \left(m_{i,1}, m_{i,2}, m_{i,3};\ \ m_{i,4}, m_{i,5}, m_{i,6}\right), \quad (i=1,2,\ldots,6), \tag{7.68}$$

where $\overrightarrow{\mathrm{U}_i\mathrm{S}_i}$ represents the vector from point U_i to point S_i and $\overrightarrow{\mathrm{OS}_i}$ represents the vector from point O to point S_i. The six-unit TWS can be obtained if the configuration of the mobile platform is given. According to Eq. (7.50), the input transmission index of the ith leg can be represented by

$$\lambda_i = |\cos\psi_i|, \quad (i = 1, 2, \ldots, 6), \tag{7.69}$$

where ψ_i represents the angle between lines $\mathrm{P}_i\mathrm{Q}_i$ and $\mathrm{U}_i\mathrm{S}_i$.

When only prismatic joint P_1 is driven and the other prismatic joints are fixed, the transmission wrenches represented by $\$_{Ti}$ $(i = 2, 3, \ldots, 6)$ become five constraint wrenches for the mobile platform. Only the transmission wrench represented by $\$_{Ti}$ can contribute to the mobile platform. In this case, the 6-DOF parallel mechanism becomes a single-DOF mechanism, and the unit instantaneous motion of the mobile platform can be represented by a unit twist $\$_{O1}$, i.e.,

$$\$_{O1} = \left(k_{1,1}, k_{1,2}, k_{1,3};\ \ k_{1,4}, k_{1,5}, k_{1,6}\right). \tag{7.70}$$

Considering that $\$_{O1}$ is a unit screw, we may obtain

$$k_{1,1}^2 + k_{1,2}^2 + k_{1,3}^2 = 1. \tag{7.71}$$

Because there are six undetermined elements in Eq. (7.70), the other five conditions are needed to determine unit twist $\$_{O1}$. According to Eq. (7.52), five equations can be obtained as follows:

$$\$_{Tj} \circ \$_{O1} = 0 \quad (j = 2, 3, \ldots, 6), \tag{7.72a}$$

that is,

$$m_{j,4}k_{1,1} + m_{j,5}k_{1,2} + m_{j,6}k_{1,3} + m_{j,1}k_{1,4} + m_{j,2}k_{1,5} + m_{j,3}k_{1,6} = 0$$
$$(j = 2, 3, \ldots, 6). \tag{7.72b}$$

Therefore, unit twist $\$_{O1}$ can be determined using Eqs. (7.71) and (7.72b) as bases. In the same manner, unit twists $\$_{O1}$ $(j = 2, 3, \ldots, 6)$ can be obtained.

According to Eqs. (7.40) and (7.57), the output transmission index of the first leg can be expressed as

$$\eta_1 = \frac{|\$_{T1} \circ \$_{O1}|}{|\$_{T1} \circ \$_{O1}|_{\max}} = \frac{|(h_{T1} + h_{O1})\cos\theta - d\sin\theta|}{\sqrt{(h_{T1} + h_{O1})^2 + d_{\max}^2}}, \tag{7.73}$$

where $d_{\max}$ denotes the potential maximal length of the common perpendicular between the axes of unit screws $\$_{T1}$ and $\$_{O1}$; h_{T1} and h_{O1} represent the pitches of unit TWS $\$_{T1}$ and unit twist $\$_{O1}$, respectively. On the basis of Eqs. (8.67) and (7.70), the numerator in Eq. (7.73) can be obtained as

$$|\$_{T1} \circ \$_{O1}| = |(h_{T1} + h_{O1})\cos\theta - d\sin\theta| = |m_{1,4}k_{1,1} + m_{1,5}k_{1,2} + m_{1,6}k_{1,3} + m_{1,1}k_{1,4} + m_{1,2}k_{1,5} + m_{1,3}k_{1,6}|. \tag{7.74a}$$

The transmission wrench in each leg is a pure force; therefore, the pitch of the TWS vanishes, i.e.,

$$h_{T1} = 0. \tag{7.74b}$$

The pitch of unit twist $\$_{O1}$ can be obtained as follows:

$$h_{O1} = k_{1,1}k_{1,4} + k_{1,2}k_{1,5} + k_{1,3}k_{1,6}. \tag{7.74c}$$

As shown in Fig. 7.26, the center of the spherical joint (S_1) is the intersection point of the axis of $\$_{T1}$. The mobile platform and its position relative to the axis of $\$_{O1}$ are constant. Line segment *CD* represents the distance between the axes of unit screws $\$_{T1}$ and $\$_{O1}$. The position and direction of $\$_{T1}$ relative to $\$_{O1}$ changes; thus, the position of line segment *CD* also differs. Considering that the axis of $\$_{T1}$ always passes through point S_1, length d of line segment *CD* obtains its maximal value $d_{\max}$ when point C coincides with point S_1. Thus, $d_{\max}$ is equal to the distance from the

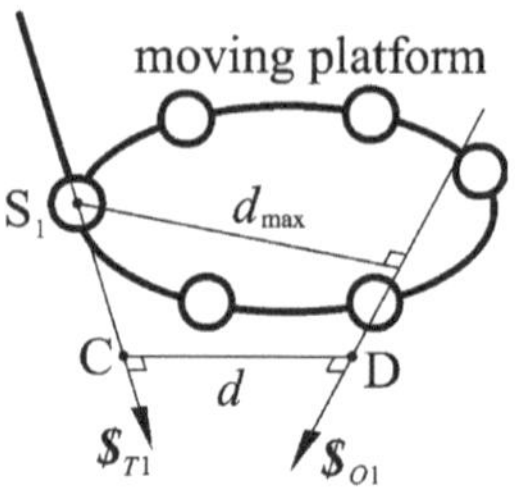

Fig. 7.26 Position and direction of $\$_{T1}$ relative to $\$_{O1}$

center of the spherical joint (S_1) to the axis of $\$_{O1}$. Using the formula of point-line distance, we can obtain $d_{\max}$ as follows:

$$d_{\max} = |\boldsymbol{s}_1 \times \boldsymbol{l}|, \tag{7.74d}$$

where $\boldsymbol{s}_1$ represents the unit vector along the axis of $\$_{O1}$, i.e., $\boldsymbol{s}_1 = (k_{1,1},\ k_{1,2},\ k_{1,3})$, and $\boldsymbol{l}$ denotes the vector from the center of the spherical joint (S_1) to an arbitrary point on the axis of $\$_{O1}$. The axis of $\$_{O1}$ can be expressed by the following equation:

$$\boldsymbol{r} \times \boldsymbol{s}_1 = \boldsymbol{s}_1^0 - h_{O1}\boldsymbol{s}_1, \tag{7.75}$$

where $\boldsymbol{r} = (x, y, z)$ denotes a vector from origin O to a point on the axis of $\$_{O1}$ and $\boldsymbol{s}_1^0 = (k_{1,4},\ k_{1,5},\ k_{1,6})$.

Therefore, substituting the results of Eqs. (7.74a), (7.74b), (7.74c), and (7.74d) into those of Eq. (7.73) yields the output transmission index of the first leg. Similarly, all the other output transmission indices ($\eta_i,\quad i = 2, 3, \ldots, 6$) can be obtained. On the basis of Eq. (7.58), therefore, the LTI of the 6-PUS parallel mechanism can be obtained by $\chi = \min\{\lambda_i,\ \eta_i\},\quad (i = 1, 2, \ldots, n)$.

Given that the 6-PUS parallel mechanism (Fig. 7.25) has the same transmission performance along the z-axis when $x_{O'}$, $y_{O'}$, ϕ, θ, and σ are constant, we consider only the distribution of the LTI in a plane parallel to the o-xy plane. If the geometric parameters of the mechanism are $r_1 = 15$, $r_2 = 50$, $r_3 = 25$, and $r_4 = 70$, the distributions of the LTI (Fig. 7.27) can be determined when fixing the orientation of the mobile platform as two groups of T&T angles $(0°, 0°, 0°)$ and $(30°, 20°, 10°)$. In the atlas, the area enveloped by line LTI = 0.7 is the GTW of the 6-PUS parallel mechanism.

7.4 Closeness to Singularities

A parallel manipulator always loses control at a singular configuration, and in its neighborhood, it should therefore work far from singularities. However, how far it should be is still a challenging problem. In this section, some performance indices are introduced to measure the closeness between a pose and a singular configuration.

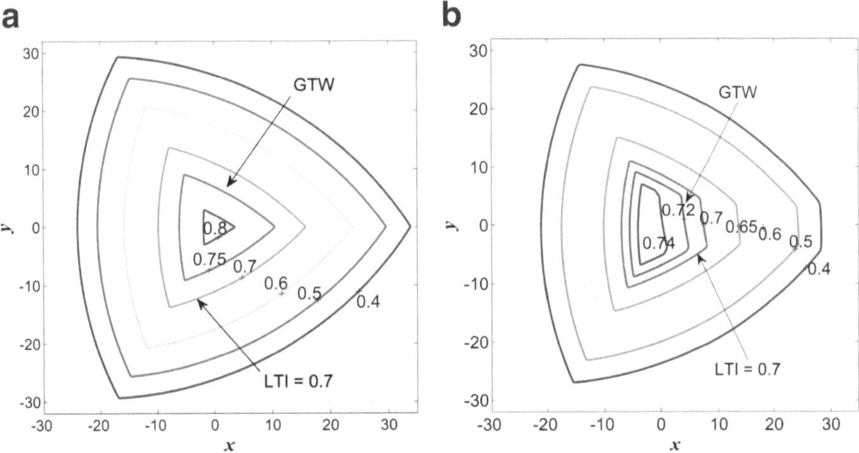

Fig. 7.27 The LTI distribution of the 6-PUS parallel mechanism with parameters $r_1 = 15$, $r_2 = 50$, $r_3 = 25$, and $r_4 = 70$, in which the orientation angles are fixed as (**a**) 0°, 0°, 0°; (**b**) 30°, 20°, 10°

7.4.1 Performance Indices

In Sect. 7.3.5, the *power coefficient* was defined to evaluate the effectiveness of motion/force transmission between a wrench and a twist. It can be represented by

$$\rho = \frac{|\$_1 \circ \$_2|}{|\$_1 \circ \$_2|_{\max}} = \frac{|(h_1 + h_2)\cos\theta - d\sin\theta|}{\sqrt{(h_1 + h_2)^2 + d_{\max}^2}} \tag{7.76}$$

Notably, the definition of $d_{\max}$ is as follows: there should be a point of application of TWS with respect to OTS, and the axis of TWS is always passing through the point; supposing that the axis of TWS could rotate along the point, the potential maximal length of the common normal of the axes of TWS and OTS is then equal to the length from the point to the axis of OTS; thus, the length from the point to the axis of OTS is $d_{\max}$. Since $d_{\max}$ has a relation with the configuration of the manipulator, it is a local parameter. The concept of $d_{\max}$ is more or less the same as the characteristic length proposed by (Chen and Angeles 2007).

For an n-DOF non-redundant parallel manipulator, there exist n transmission wrench screws that have a one-to-one correspondence with n actuators. The power coefficients between the ith TWS and its corresponding ITS and OTS can be represented by

$$\lambda_i = \frac{|\$_{Ti} \circ \$_{Ii}|}{|\$_{Ti} \circ \$_{Ii}|_{\max}} \tag{7.77a}$$

and

$$\eta_i = \frac{\left|\$_{Ti} \circ \$_{Oi}\right|}{\left|\$_{Ti} \circ \$_{Oi}\right|_{\max}} \tag{7.77b}$$

respectively. Since there are always two or more transmission wrench screws in a parallel manipulator, the minimums of λ_i and η_i $(i = 1, 2, \ldots, n)$ are defined to evaluate the input and output transmission performance of the entire manipulator; they can be represented by

$$\gamma_I = \min_i \{\lambda_i\} = \min_i \left\{ \frac{\left|\$_{Ti} \circ \$_{Ii}\right|}{\left|\$_{Ti} \circ \$_{Ii}\right|_{\max}} \right\} \quad (i = 1, 2, \ldots, n) \tag{7.78}$$

$$\gamma_O = \min_i \{\eta_i\} = \min_i \left\{ \frac{\left|\$_{Ti} \circ \$_{Oi}\right|}{\left|\$_{Ti} \circ \$_{Oi}\right|_{\max}} \right\} \quad (i = 1, 2, \ldots, n) \tag{7.79}$$

Here, γ_I and γ_O are referred to as the *input transmission index* (ITI) and *output transmission index* (OTI), respectively. They will be applied to measure the closeness to input and output transmission singularities, respectively.

As to the closeness to constraint singularity, it has not been investigated so far. A new index is hereby proposed to measure the closeness to constraint singularity. According to the previous assumption, there are n transmission wrench screws and q $(q \geq 6 - n)$ constraint wrench screws in an n-DOF $(n < 6)$ parallel manipulator. If we take the n transmission wrench screws and $(5 - n)$ constraint wrench screws together to represent five constraint wrenches, the manipulator can be considered as a single-DOF one, and the twist screw ($\$_{Oj}$) of the moving platform can be determined. Then we can obtain the power coefficients between $\$_{Oj}$ and the other $[q-(5-n)]$ constraint wrench screws. In order to counterbalance the external wrench along the axis of $\$_{Oj}$, there should be at least one power coefficient that is larger than zero. In other words, if the maximum of the power coefficients is larger than zero, there will exist at least one constraint wrench to counterbalance the external wrench along the axis of $\$_{Oj}$. Thus, an index with respect to $\$_{Oj}$ is defined as

$$\varepsilon_j = \max_k \left\{ \frac{\left|\$_{Ck} \circ \$_{Oj}\right|}{\left|\$_{Ck} \circ \$_{Oj}\right|_{\max}} \right\} \quad (k = 1, 2, \ldots, [q-(5-n)]) \tag{7.80}$$

Note that the number of combinations of $(5 - n)$ constraint wrench screws selecting from q constraint wrench screws is equal to

$$\chi = C_q^{5-n} = \frac{q \cdot (q-1) \cdots (5-n+1)}{(5-n)!} \tag{7.81}$$

and the number of $\$_{Oj}$ is also equal to χ, then we get $j = 1, 2, \ldots, \chi$.

If any one of ε_j equals zero, there will exist the twist screw $\$_{Oj}$, along whose axis external wrench cannot be counterbalanced; this means that the constraint singularity occurs. If all of ε_j are larger than zero, all the external wrenches along the axes of $\$_{Oj}$ $(j = 1, 2, \ldots, \chi)$ can be counterbalanced, and there will be no constraint singularity. In other words, if the minimum of ε_j is larger than zero, no constraint singularity will occur. Thus, the index that can be used to measure the closeness to constraint singularity is defined as

$$\gamma_C = \min_j \left\{ \varepsilon_j \right\} = \min_j \left\{ \max_k \left\{ \frac{\left| \$_{Ck} \circ \$_{Oj} \right|}{\left| \$_{Ck} \circ \$_{Oj} \right|_{\max}} \right\} \right\} \quad (j = 1, 2, \ldots, \chi;\ k = 1, 2, \ldots, [q - (5-n)]) \tag{7.82}$$

which is referred to as the *constraint transmission index* (CTI). A larger value of CTI indicates a larger "distance" from the configuration to constraint singularity.

Furthermore, an index to measure the closeness to all types of singularities in this study is defined as

$$\gamma = \min \{\gamma_I, \gamma_O, \gamma_C\}. \tag{7.83}$$

Since γ is related to the local configuration of the manipulator, it is referred to as the *local singularity index* (LSI).

The power coefficient has no relation to the selected frame coordinate, and its value range is [0, 1]; thus, all the indices defined here, i.e., ITI, OTI, CTI, and LSI, are frame-free, and their values have a range of [0, 1].

7.4.2 Examples

In order to introduce the applications of the proposed indices in measuring closeness to singularities, the distributions of the indices in the workspace of three parallel manipulators are presented in this section. Owing to the limitation of space, the determinations of ITI, OTI, CTI, and LSI will not be presented, and we will directly show the distributions of index values in the workspace of the manipulator.

7.4.2.1 The Planar 3-RRR Parallel Manipulator

A planar three-DOF 3-RRR parallel manipulator is shown in Fig. 7.28, where the moving platform is connected to the base by means of three legs, each of

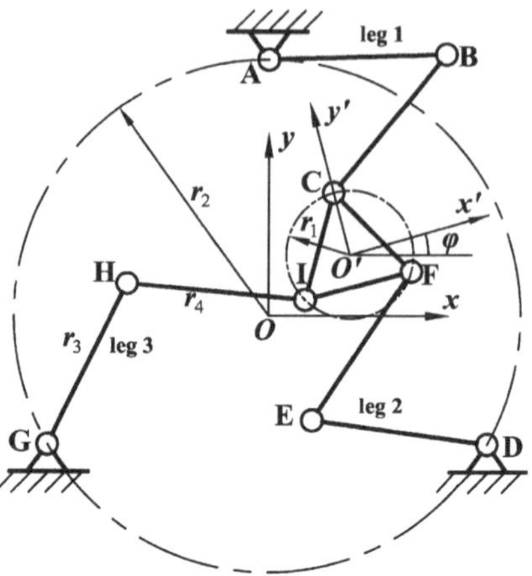

Fig. 7.28 A planar 3-RRR parallel manipulator

which includes three revolute joints and two bars. The position of the center O' of the moving platform can be described by $(x_{O'};\ y_{O'})$ with respect to the global coordinate system O-xy, and the orientation of the moving platform can be described by the angle φ measured counterclockwise from the x-axis to the x'-axis of the local coordinate system which is attached to the moving platform. The three revolute joints A, D, and G fixed on the base are actuated, and they are equally distributed at a nominal angle of 120° on a circle whose center is the origin O and radius is r_2. The moving platform has the shape of a equilateral triangle, the circumcircle of which has a radius of r_1. The parameter r_3 is the length of the input links AB, DE, and GH, and r_4 the length of the coupler links BC, EF, and HI.

Considering that there is no constraint singularity in the planar 3-RRR manipulator, the CTI will not be investigated here, and the LSI is defined as the minimum of ITI and OTI. Giving the geometric parameters as $r_1 = 0.2$, $r_2 = 2$, and $r_3 = r_4 = 1.4$, Fig. 7.29 illustrates the distribution of ITI, OTI, and LSI within its constant-orientation workspace by fixing the orientation angle $\varphi = 0$. The locus marked with "input transmission singular locus" is the boundary of the constant-orientation workspace. It can be seen from Fig. 7.29a, c that the values of ITI and LSI increase as the configuration gets far from the input transmission singular locus. Since the output transmission singularity does not occur in the constant-orientation workspace of the manipulator, the distribution of OTI in Fig. 7.29b cannot clearly show the "distance" from a configuration to output transmission singular locus.

For the planar 3-RRR manipulator, when the three lines BC, EF, and HI intersect at a point, the output transmission singularity occurs. Fixing $x_{O'} = y_{O'} = 0$, the manipulator reaches the output transmission singular configuration for $\varphi = -44°$. This is consistent with the result in Fig. 7.30.

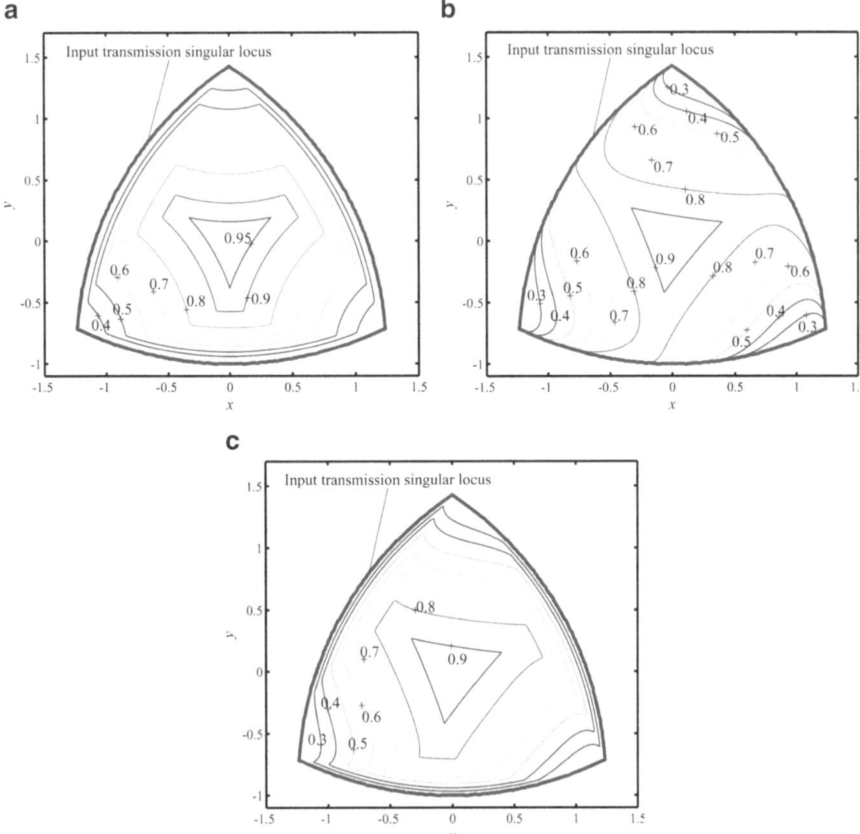

Fig. 7.29 Distributions of index values in the constant-orientation workspace of the planar 3-RRR manipulator with $\varphi = 0$: (**a**) ITI; (**b**) OTI; (**c**) LSI

7.4.2.2 A Three-DOF Articulated Tool Head

Figure 7.31a shows a 3-DOF articulated tool head with parallel kinematic chains. The spindle is fixed in the moving platform. The said platform connects to the base by means of three identical PRS legs, each of which contains a P pair, an R pair, and an S pair in turn. The three P pairs are actuated. The three S pairs attached to the moving platform are equally spaced at a nominal angle of 120°. The slideways of P pairs are correspondingly fixed to three columns that are also spaced at a nominal angle of 120° from one another. Thus, the structure of the tool head is actually a 3-PRS parallel manipulator, and its kinematic scheme is shown in Fig. 7.31b.

It is well known that the 3-PRS parallel manipulator has three independent DOFs, i.e., two rotational DOFs and one translational DOF. Since all the slideways of P pairs are vertical, any performance is independent of the position workspace along the z-axis. Thus, performance of the manipulator needs to be investigated only in the

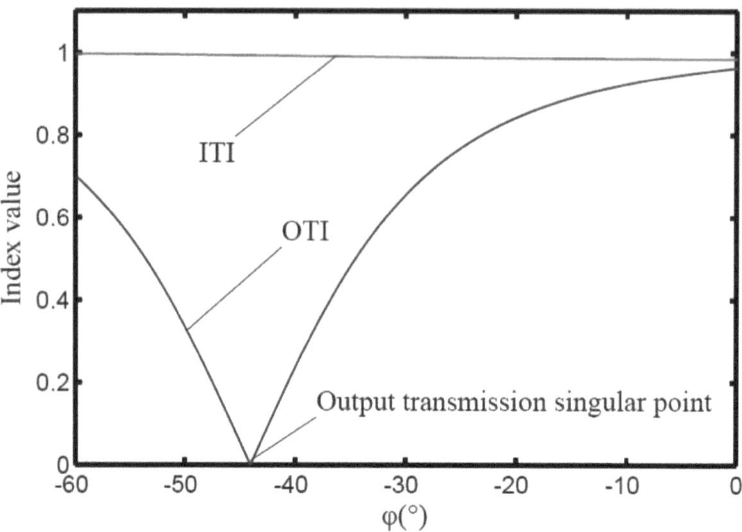

Fig. 7.30 Relationships between angle φ and the value of OTI by fixing $x_{O'} = y_{O'} = 0$

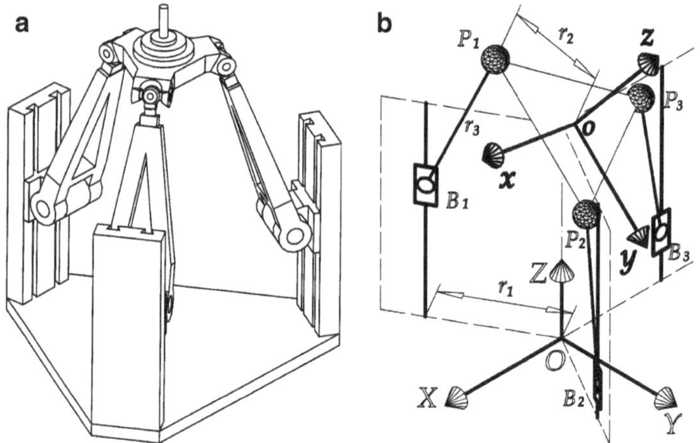

Fig. 7.31 A 3-axis articulated tool head: (**a**) kinematic structure; (**b**) kinematic scheme

orientation workspace. Bonev (2002) introduced the *Tilt-and-Torsion* (T&T) angles (φ, θ, ψ) to represent the orientation of a platform, where φ is the azimuth, θ is the tilt angle, and ψ is the torsion angle; he also pointed out that the 3-PRS parallel manipulator is a zero-torsion mechanism, and (φ, θ, 0°) can represent the orientation of the moving platform.

By giving the geometric parameters as $r_1 = 200$, $r_2 = 130$, and $r_3 = 220$, the distributions of values of ITI, OTI, and CTI in the orientation workspace of the

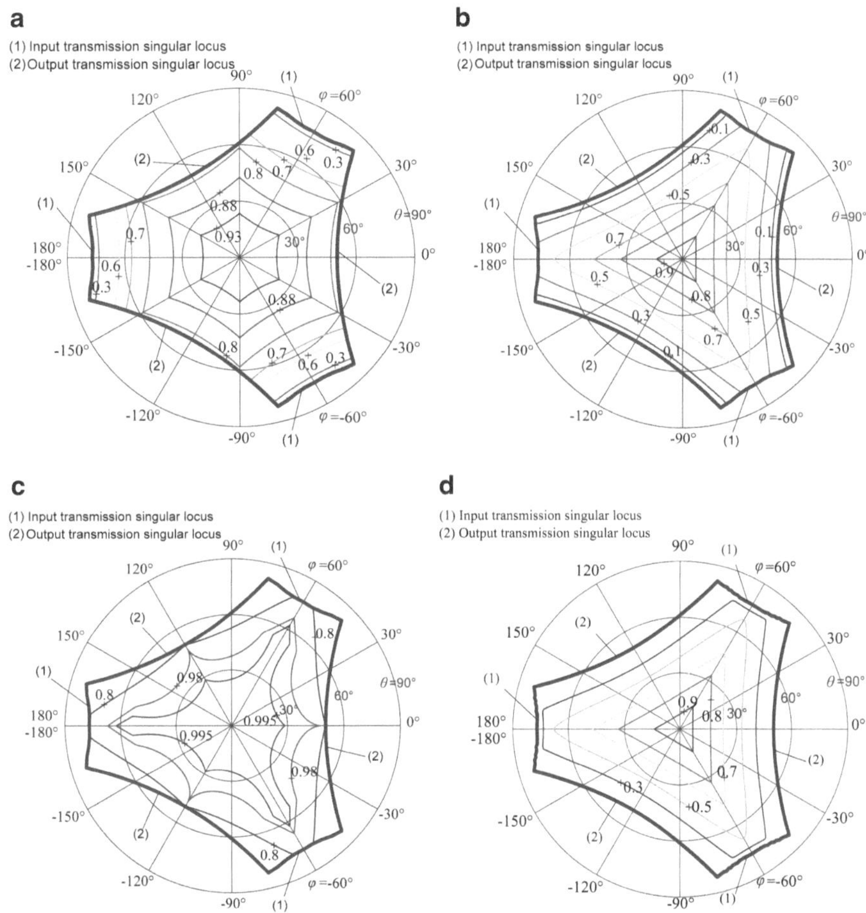

Fig. 7.32 Distributions of index values in the orientation workspace of the 3-PRS manipulator: (**a**) ITI; (**b**) OTI; (**c**) CTI; (**d**) LSI

manipulator are illustrated by Fig. 7.32, where the area included by the input and output transmission singular loci is a singularity-free orientation workspace. It can be seen from Fig. 7.32a that the ITI value increases as the orientation is far away from the input transmission singular configurations. Figure 7.32b shows the OTI has a larger value with a larger "distance" to the output transmission singular locus. From Fig. 7.32c, it can be seen that the values of CTI are very large, which means that the 3-PRS manipulator is far from constraint singularity in the singularity-free orientation workspace. It can be seen from Fig. 7.32d that the ISI value increases as the orientation is far away from the singular locus.

Zlatanov et al. (2002) found that the constraint singularity occurs in the 3-RPS parallel manipulator when the moving platform is upside down. In other words, when the moving platform tilt at 180°, the 3-RPS manipulator has a constraint

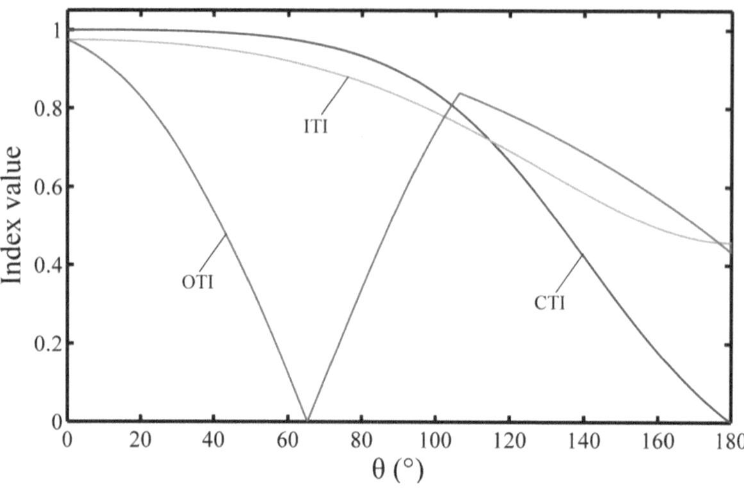

Fig. 7.33 Relationships between tilt angle (θ) and the values of ITI, OTI, and CTI by fixing $\varphi = 0$

singularity. This result also fits for the 3-PRS parallel manipulator. Figure 7.33 shows the relationships between tilt angle (θ) and the values of ITI, OTI, and CTI by fixing $\varphi = 0°$. According to the curve marked with "CTI," when $\theta = 180°$, the value of CTI is equal to zero, which means that the constraint singularity occurs. Also, it can be seen from the curve marked with "OTI" that the value of OTI vanishes at $\theta = 65.5°$, which means that when the moving platform tilt at 65.5°, the coupler link B_3P_3 and the moving platform are coplanar, and the output transmission singularity occurs. Moreover, the values of ITI, OTI, and CTI have adequate distributions within [0, 1]; the larger the index value is, the further the pose is away from the corresponding singularity.

It can be seen from Figs. 7.31 and 7.33 in this chapter that the values of ITI, OTI, CTI, and LSI have adequate distributions within [0, 1] in the singularity-free workspace of all the three manipulators; in other words, the values of ITI, OTI, CTI, and LSI can be considered at the same level for different types of parallel manipulators. The larger value the index has, the further the pose is away from the corresponding singularity. Thus, ITI, OTI, and CTI can be used to measure the closeness to input transmission, output transmission, and constraint singularities for different types of parallel manipulators, respectively; as the minimum of ITI, OTI, and CTI, LSI can be used to measure the closeness to all the proposed singularities in this study.

Since ITI, OTI, CTI, and LSI are frame-free, a certain index value indicates the same "distance" from a pose to the corresponding singular configuration. It should be noted that the "distance" does not mean the physical distance or length from a point to the singular locus in the workspace. For example, the physical distance from a point to the singular locus in the constant-orientation workspace (Fig. 7.29) is not comparable with that in the orientation workspace (Fig. 7.32). For different types of parallel manipulators, the "distance" means how good the motion/force transmission performance of the manipulator is. A larger "distance" to singularity indicates better motion/force transmission performance.

Moreover, the limit of the indices needs to be defined to judge whether the motion/force transmission performance is good or bad. In Ref. by Wang et al. (2009), for the purpose of high speed, high accuracy, and high quality of motion transmission, 0.7 is defined as the limit of the *local transmission index* (LTI). When the LTI is greater than 0.7, the manipulator is considered good at motion/force transmission; if the LTI is smaller than 0.7, the manipulator is said to have bad motion/force transmission performance. 0.7 may also be defined as the limit of ITI, OTI, CTI, and LSI. However, 0.7 is not the only one limit of the indices. The limit of the indices depends on the design requirement of practical applications. Once the value of limit is identified under some practical design requirements, it will be suitable for all types of non-redundant parallel manipulators. A uniform "metric" can therefore be defined to measure the closeness to singularities of different types of non-redundant parallel manipulators.

Appendix: LTI of the Planar 5R and PRRRP Parallel Mechanisms

The planar 5R parallel mechanism is shown in Fig. 7.15. The transmission wrench of each leg is the pure force along the coupler link of the corresponding leg. For example, the transmission wrench of leg 1 is a pure force along link *AP*. With respect to local coordinate system *O-xy*, the unit transmission wrenches of legs 1 and 2 can be represented by

$$\$_{T1} = (\boldsymbol{f}_1 ; \boldsymbol{f}_1 \times \boldsymbol{b}) = (\cos(\gamma_1 + \theta_1), \quad \sin(\gamma_1 + \theta_1), \quad 0 ; 0, \quad 0, \quad r_2 \sin\gamma_1 - r_1 \sin(\gamma_1 + \theta_1))$$

and

$$\$_{T2} = (\boldsymbol{f}_2 ; \boldsymbol{f}_2 \times \boldsymbol{d}) = (-\cos(\alpha_2 - \theta_2), \quad \sin(\alpha_2 - \theta_2), \quad 0 ; 0, \quad 0, \quad r_2 \sin\alpha_2 + r_1 \sin(\alpha_2 - \theta_2)),$$

where $\boldsymbol{f}_1$ and $\boldsymbol{f}_2$ represent the unit vectors along coupler links *AP* and *CP*, respectively; $\boldsymbol{b}$ and $\boldsymbol{d}$ stand for the vectors from origin *O* to the center of joints *A* and *C*, respectively.

Revolute joints O_1 and O_2 are the input joints; thus, the two input twists can be represented by

$$\$_{I1} = (0, \quad 0, \quad 1 ; 0, \quad r_1, \quad 0)$$

and

$$\$_{I2} = (0, \ 0, \ 1; 0, \ -r_1, \ 0)$$

respectively. According to Eq. (7.35), the reciprocal product between $\$_{T1}$ and $\$_{I1}$ can be obtained as

$$\$_{T1} \circ \$_{I1} = r_2 \sin \gamma_1$$

Thus, on the basis of Eq. (7.49), the input transmission index of leg 1 can be obtained as follows:

$$\lambda_1 = \frac{|\$_{T1} \circ \$_{I1}|}{|\$_{T1} \circ \$_{I1}|_{\max}} = \frac{|r_2 \sin \gamma_1|}{|r_2 \sin \gamma_1|_{\max}} = \frac{|r_2 \sin \gamma_1|}{r_2} = |\sin \gamma_1| .$$

When input revolute joint C is fixed and joint A is actuated, the transmission force of leg 2 becomes a constraint force for the end-effector. In this case, end-effector P can move only along the direction perpendicular to coupler link DP. Given that the position vector of joint D can be expressed by

$$\overrightarrow{OD} = \begin{bmatrix} D_x \\ D_y \\ D_z \end{bmatrix} = \begin{bmatrix} -r_1 + r_2 \cos\theta_1 - r_3 \cos(\theta_1 + \gamma_1) + r_3 \cos(\theta_1 + \gamma_1 + \beta) \\ r_2 \sin\theta_1 - r_3 \sin(\theta_1 + \gamma_1) + r_3 \sin(\theta_1 + \gamma_1 + \beta) \\ 0 \end{bmatrix},$$

the output twist can be represented by

$$\$_{O1} = (0, \ 0, \ 1; D_y, \ -D_x, \ 0).$$

According to Eq. (7.52), therefore, the reciprocal product between $\$_{T1}$ and $\$_{O1}$ can be obtained as

$$\$_{T1} \circ \$_{O1} = r_3 \sin \mu .$$

Using Eq. (7.57) as basis, we can derive the output transmission index of leg 1 thus:

$$\eta_1 = \frac{|\$_{T1} \circ \$_{O1}|}{|\$_{T1} \circ \$_{O1}|_{\max}} = \frac{|r_3 \sin \mu|}{|r_3 \sin \mu|_{\max}} = \frac{|r_3 \sin \mu|}{r_3} = |\sin \mu| .$$

Similar to the input and output transmission indices of leg 1, those of leg 2 can be obtained as

$$\lambda_2 = |\sin \gamma_2|$$

and

$$\eta_2 = |\sin \mu|,$$

respectively.

Thus, the LTI of the planar 5R parallel manipulator can be expressed by

$$\gamma = \min\{\lambda_1, \lambda_2, \eta_1, \eta_2\} = \min\{|\sin\gamma_1|, |\sin\gamma_2|, |\sin\mu|\}.$$

Consequently, μ is defined as the forward transmission angle, and γ_1 and γ_2 are referred to as the inverse transmission angles of the planar 5R parallel manipulator.

The planar PRRRP parallel mechanism is shown in Fig. 7.16. The transmission wrench of each leg is the pure force along the coupler link of the corresponding leg. For example, the transmission wrench of leg 1 is a pure force along link *AO*'. With respect to global coordinate system *O-xy*, the unit transmission wrenches of legs 1 and 2 can be represented by

$$\$_{T1} = (\boldsymbol{f}_1;\ \boldsymbol{f}_1 \times \boldsymbol{a}) = (\cos\gamma_1,\ \ -\sin\gamma_1,\ \ 0;\ 0,\ \ 0,\ \ d_1)$$

and

$$\$_{T2} = (\boldsymbol{f}_2;\ \boldsymbol{f}_2 \times \boldsymbol{b}) = (\cos\gamma_2,\ \ \sin\gamma_2,\ \ 0;\ 0,\ \ 0,\ \ d_2),$$

where $\boldsymbol{f}_1$ and $\boldsymbol{f}_2$ represent the unit vectors along coupler links *AO*' and *BO*', respectively; $\boldsymbol{a}$ and $\boldsymbol{b}$ stand for the vectors from origin *O* to the center of revolute joints *A* and *B*, respectively; and d_1 and d_2 stand for the distances from origin *O* to coupler links *AO*' and *BO*', respectively.

The two prismatic joints are the input joints; hence, the two input twists can be represented by

$$\$_{I1} = \$_{I2} = (0,\ \ 0,\ \ 0;\ 0,\ \ 1,\ \ 0).$$

According to Eq. (7.35), the reciprocal product between $\$_{T1}$ and $\$_{I1}$ can be obtained as

$$\$_{T1} \circ \$_{I1} = -\sin\gamma_1.$$

Thus, on the basis of Eq. (7.49), the input transmission index of leg 1 can be obtained as follows:

$$\lambda_1 = \frac{|\$_{T1} \circ \$_{I1}|}{|\$_{T1} \circ \$_{I1}|_{\max}} = \frac{|-\sin\gamma_1|}{|-\sin\gamma_1|_{\max}} = |\sin\gamma_1|.$$

When the input prismatic joint in leg 2 is fixed and that in leg 1 is actuated, the transmission force of leg 2 becomes a constraint force for the end-effector. In this case, end-effector *P* can move only along the direction perpendicular to coupler link *BO*', and the output twist can be represented by

$$\$_{O1} = \left(0, \quad 0, \quad 0;\ \sin\gamma_2, \quad -\cos\gamma_2, \quad 0\right).$$

According to Eq. (7.52), the reciprocal product between $\$_{T1}$ and $\$_{O1}$ can be obtained as

$$\$_{T1} \circ \$_{O1} = \sin(\gamma_1 + \gamma_2)$$

Using Eq. (7.49) as basis, we can obtain the output transmission index of leg 1 as follows:

$$\eta_1 = \frac{\left|\$_{T1} \circ \$_{O1}\right|}{\left|\$_{T1} \circ \$_{O1}\right|_{\max}} = \frac{\left|\sin(\gamma_1 + \gamma_2)\right|}{\left|\sin(\gamma_1 + \gamma_2)\right|_{\max}} = \left|\sin(\gamma_1 + \gamma_2)\right| = \left|\sin\mu\right|.$$

Similar to the input and output transmission indices of leg 1, those of leg 2 can be obtained as

$$\lambda_2 = |\sin\gamma_2|$$

and

$$\eta_2 = |\sin\mu|,$$

respectively.

Thus, the LTI of the planar PRRRP parallel manipulator can be expressed by

$$\gamma = \min\{\lambda_1,\ \lambda_2,\ \eta_1,\ \eta_2\} = \min\{|\sin\gamma_1|,\ |\sin\gamma_2|,\ |\sin\mu|\}.$$

For the planar PRRRP parallel manipulator, therefore, μ is defined as the forward transmission angle, and γ_1 and γ_2 are referred to as the inverse transmission angles.

References

Alizade RI, Mohan Rao AV, Sandor GN (1975) Optimum synthesis of two degree of freedom planar and spatial function generating mechanism using the penalty function approach. J Eng Ind Trans ASME 97(2):629–634

Alt VH (1932) Der uberstragungswinkel und seine bedeutung fur dar konstruieren periodischer getriebe. Werksstattstechnik 26(4):61–65

Ball RS (1990) A treatise on the theory of screws. Cambridge University Press, Cambridge

Balli SS, Chand S (2002) Transmission angle in mechanisms. Mech Mach Theory 37:175–195

Balli SS, Chand S (2004) Synthesis of a five-bar mechanism of variable topology type with transmission angle control. J Mech Des 126:128–134

Bonev IA (2002) Geometric analysis of parallel mechanisms. Ph.D. thesis, Laval University, Quebec

Chen C, Angeles J (2007) Generalized transmission index and transmission quality for spatial linkages. Mech Mach Theory 42:1225–1237

Eschenbach PW, Tesar D (1971) Link length bounds on the four bar chain. J Eng Ind Trans ASME 93:287–293

Gallardo J, Rico JM, Frisoli A et al (2003) Dynamics of parallel manipulators by means of screw theory. Mech Mach Theory 38:1113–1131

Gosselin CM, Angeles J (1989) The optimum kinematic design of a spherical three-degree- of-freedom parallel manipulator. J Mech Trans Autom Des 111(2):202–207

Gosselin CM, Angeles J (1990) Singularity analysis of closed loop kinematic chains. IEEE Trans Robot Autom 6(3):281–290

Gosselin CM, Angeles J (1991) A global performance index for the kinematic optimization of robotic manipulators. Trans ASME J Mech Des 113:220–226

Hain K (1967) Applied kinematics. McGraw-Hill, New York

Hall AS (1961) Kinematics and linkage design. Prentice-Hall, Englewood Cliffs, pp 41

Hartenberg RS, Denavit J (1964) Kinematic synthesis of linkages. McGraw-Hill, New York, pp 46–47

Huang Z, Li QC (2002) General methodology for the type synthesis of lower-mobility symmetrical parallel mechanisms and several novel mechanisms. Int J Robot Res 21(2):131–145

Kim HS, Tsai L-W (2003) Design optimization of a Cartesian parallel manipulator. J Mech Des 125(1):43–51

Kimbrella JT (1991) Kinematics analysis and synthesis. McGraw-Hill, New York, pp 14–15

Lin CC, Chang WT (2002) The force transmissivity index of planar linkage mechanisms. Mech Mach Theory 37(12):1465–1485

Lin CC, Chang WT, Lee JJ (2003) Force transmissibility performance of parallel manipulators. J Robot Syst 20:659–670

Lipkin H (2005) Time derivatives of screws with applications to dynamic and stiffness. Mech Mach Theory 40:259–273

Liu GF, Lou YJ, Li ZX (2003) Singularities of parallel manipulators: a geometric treatment. IEEE Trans Robot Autom 19(4):579–594

Liu X-J (1999) The relationships between the performance criteria and link lengths of the parallel manipulators and their design theory. Ph.D thesis, Yanshan University, Qinhuangdao (in Chinese)

Liu X-J, Wang J, Oh K-K, Kim J (2004) A new approach to the design of a DELTA robot with a desired workspace. J Intell Robot Syst 39(2):209–225

Liu X-J, Wang J, Pritschow G (2006) Kinematics, singularity and workspace of planar 5R symmetrical parallel mechanisms. Mech Mach Theory 41(2):145–169

Merlet J-P (2006) Jacobian, manipulability, condition number, and accuracy of parallel robots. ASME J Mech Des 128:199–206

Park FC, Kim JW (1999) Singularity analysis of closed kinematic chains. J Mech Des 121:32–38

Philipp MRE, Freudenstein F (1965) Synthesis of two-degree-of-freedom linkages—a feasibility study of numerical methods of synthesis of bivariate function generators. J Mech 1:9–21

Ryu J, Cha J (2001) Optimal architecture design of parallel manipulators for best accuracy. In: Proceedings of the 2001 IEEE/RSJ international conference on intelligent robots and systems, IEEE press, Piscataway, N.J., Maui, Hawwaii, pp 1281–1286

Söylemez E, Freudenstein F (1982) Transmission optimization of spatial 4-link mechanisms. Mech Mach Theory 17:263–283

Soylu R (1993) Analytical synthesis of mechanism part-I, transmission angle synthesis. Mech Mach Theory 28:825–833

Stanley R (1961) Five-bar loop synthesis. Mach Des 33(21):189–195
Strang G (1976) Linear algebra and its application. Academic Press, New York
Sutherland G, Roth B (1973) A transmission index for spatial mechanisms. ASME J Eng Ind 95(2):589–597
Sutherland GH, Siddall JN (1974) Dimensional synthesis of linkage by multifactor optimization. Mech Mach Theory 9:81–95
Takeda Y, Funabashi H, Ichimaru H (1997) Development of spatial in-parallel actuated manipulators with six degrees of freedom with high motion transmissibility. JSME Int J Series C 40:299–308
Takeda Y, Funabashi H, Sasaki Y (1995) Motion transmissibility of in-parallel actuated manipulators. JSME Int J Series C 38:749–755
Takeda Y, Funabashi H, Sasaki Y (1996) Development of a spherical in-parallel actuated mechanism with three degrees of freedom with large working space and high motion transmissibility. JSME Int J Series C 39:541–548
Tao DC (1964) Applied linkage synthesis. Addison-Wesley, Reading, pp 7–12
Ting KL, Tsai GH (1985) Mobility and synthesis of five-bar programmable linkages. In: Proceedings of 9th OSU applied mechanisms conference, Kanas City, MD, pp III-1–III-8, ASME press, New York
Tsai LW, Stamper R (1996) A parallel manipulator with only translational degrees of freedom. In: Proceedings of the ASME design engineering technical conference, Irvine, CA, 96-DETC-MECH–1152, ASME press, New York
Tsai MJ, Lee HW (1994) Generalized evaluation for the transmission performance of mechanisms. Mech Mach Theory 29:607–618
Voglewede PA, Ebert-Uphoff I (2004) Measuring 'closeness' to singularities for parallel manipulators. In: Proceedings of IEEE international conference on robotics and automation, New Orleans, IEEE press, Piscataway, N.J., pp 4539–4544
Wang J, Liu XJ, Wu C (2009) Optimal design of a new spatial 3-DOF parallel robot with respect to a frame-free index. Sci China Series E-Technol Sci 52:986–999
Yuan MSC, Freudenstwin F, Woo LS (1971) Kinematic analysis of spatial mechanism by means of screw coordinates. Part 2-Analysis of spatial mechanisms. J Eng Ind Trans ASME 91:67–73
Zlatanov D, Bonev IA, Gosselin CM (2002) Constraint singularities of parallel mechanisms. In: Proceedings of the 2002 IEEE international conference on robotics and automation, Washington, DC, IEEE press, Piscataway, N.J., pp 496–502

Chapter 8
Dimensional Synthesis of Parallel Mechanisms

Abstract This chapter presents a general geometric-based methodology for dimensional synthesis and optimal design of parallel mechanisms. The proposed method is different from those frequently used methods implemented by establishing the objective functions with specified constraints and then searching for the result using an optimization algorithm; it is indeed a geometrical method by aid of performance charts or atlas, which is highly useful in illustrating the relationship between a performance index provided in Chap. 7 and the design parameters involved in a visual way. The proposed method is illustrated by several parallel mechanisms including a planar 5R parallel mechanism and several 3-DOF spatial parallel mechanisms.

Keywords Dimensional synthesis • Optimal design • Performance chart • Parallel mechanism • Design parameter

The objective of dimensional synthesis is to determine the values of each geometric parameter of a mechanism by taking into account the desired performance. Each geometric parameter may have any value between zero and infinity, and the performance criteria are usually antagonistic, making dimensional synthesis one of the most challenging issues in the field. For parallel mechanisms, the most frequently used methods are establishing the objective functions with specified constraints and then searching for the result using an optimization algorithm. This approach is referred to as the cost-function method, which may provide a mathematical optimal result. Such a result may occur just as likely as does an impractical one. Additionally, the method fails to illustrate the relationship between an index and the design parameters involved. The user cannot determine how optimal the result is. Furthermore, if the objective function changes, a designer will have to initiate the optimization from scratch. For such reasons, the synthesis procedure cannot be accepted as a unified methodology.

Performance charts are applied extensively in industrial design and used in most design manuals. The final objective of dimensional synthesis is to identify

X.-J. Liu and J. Wang, *Parallel Kinematics: Type, Kinematics, and Optimal Design*, Springer Tracts in Mechanical Engineering, DOI 10.1007/978-3-642-36929-2_8,

optimal geometric parameters. To this end, using performance charts may be one of the most effective ways to show the relationship between a performance index and geometric parameters. In accordance with this concept, defining a space that can embody all possible mechanisms (with different geometric parameters) is of primary importance.

8.1 Parameter Design Space

8.1.1 Establishing the Parameter Design Space

A parallel mechanism usually consists of at least two kinematic chains. Consequently, several geometric parameters are involved in any parallel mechanism. Suppose that a mechanism has n characteristic linear parameters, each denoted as L_i $(1 \leq i \leq n)$. These n parameters are all related to the performance indices (kinematics, workspace, singularity, etc.) of the mechanism. They are therefore defined as design parameters, which are dimensional synthesized objects. The input parameters are not defined as the design parameters because they are eventually excluded in the expression of an index.

Given that n parameters L_i may have any value between zero and infinity, graphically illustrating the relationship between a performance index and design parameters is impossible. Therefore, the parameter infinity is the most troublesome problem in the kinematic design of a parallel mechanism when using performance charts. The problems that arise can be summarized as follows: (a) the manner by which the design parameter number is reduced; (b) procedures involved in reasonably specifying the bounds of each parameter; (c) the manner by which a *parameter design space* (PDS) is defined, in which the optimal kinematic design can be implemented logically; and (d) whether such a design space exists and how the relationship between mechanisms with finite and infinite parameters is addressed.

Solving these problems first necessitates the identification of a technique for logically identifying the limit of each linear parameter and at the same time retaining the performance similarity of the mechanism. The parameter normalization technique is one solution to these problems. The key problem in parameter normalization is the selection of a normalization factor. Yang (1987) proposed a normalization method to analyze the performance of all four-bar mechanisms. To address the problem of nonhomogeneous physical units in the Jacobian matrix, the concepts of "characteristic length (CL)" and "natural length (NL)" were introduced (Tandirci et al. 1992; Ma and Angeles 1991; Angeles 1995). The normalization methods were established on the factor that is the average of related linear parameters. The CL and NL were used to normalize the dimensional elements in the Jacobian matrix, but these could not generate infinite elements because they were not the average of all the dimensional elements. Finiteness, which is very important in the problem indicated in this chapter, is not necessary for such normalization. Nevertheless, the idea, especially that of Yang's generation of the normalization factor, remains beneficial to the present study. We extend the idea to a general case. Suppose n characteristic parameters exist in a mechanism, and these parameters are denoted as L_i $(i = 1, 2, \ldots, n)$. Let

$$D = \frac{\sum_{i=1}^{n} L_i}{d}, \tag{8.1}$$

where d can be any positive number. Parameter D is defined as the normalization factor of the mechanism. We can obtain n non-dimensional parameters l_i using

$$l_i = \frac{L_i}{D}. \tag{8.2}$$

Therefore,

$$\sum_{i=1}^{n} l_i = d, \tag{8.3}$$

which not only reduces the parameter number from n to $n-1$ but also yields the bound of each normalized parameter l_i, i.e.,

$$l_n = d - \sum_{i=1}^{n-1} l_i \tag{8.4}$$

and

$$0 \leq l_i \leq d. \tag{8.5}$$

In some cases, other conditions are imposed on these parameters because parameters l_i $(i = 1, 2, \ldots, n)$ should be used to set up the mechanism and ensure that it works. Together with these conditions, Eqs. (8.3) and (8.5) define an $(n-1)$-dimensional finite space. Such a space is referred to as the PDS.

Equations (8.3) and (8.5) show that the limit of the PDS depends on the value of d. For the convenience of prescribing the bound of every normalized parameter and expressing the PDS in a limited space, we usually assign parameter d with an integer, typically the number 1 or n, even though it can be any positive number. The selection of parameter d merely determines the size of the PDS, but does not affect its shape and the final result. When $d = 1$, D is the sum of all the characteristic parameters. When $d = n$, D is the average of all the parameters. Regardless of what parameter d is, a mechanism can be transformed from one with dimension into one with no dimension using Eqs. (8.1), (8.2), (8.3), (8.4), and (8.5). Most important, this technique also changes an n-dimensional problem to an $(n-1)$-dimensional one and, at the same time, defines the limits of each normalized parameter. This technique is one of the most important contributions to mechanisms and is referred to as the *parameter-finiteness normalization method* (PFNM).

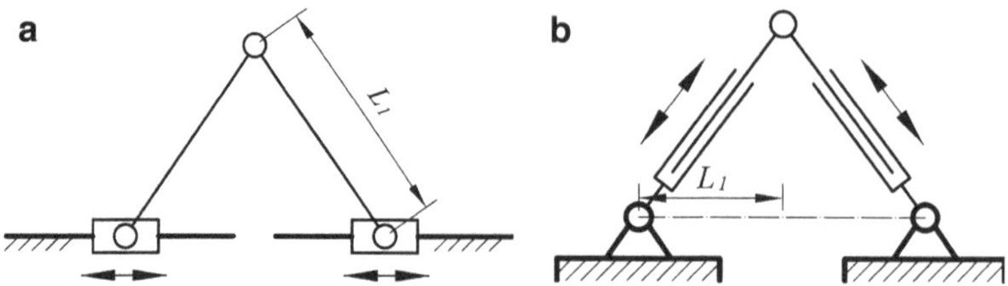

Fig. 8.1 Mechanisms with one characteristic parameter

The size of the mechanism with Dl_i is similar to that of the mechanism with l_i. With different D, varied kinds of mechanisms exist. The mechanisms with Dl_i (with different D) are defined as the *similarity mechanisms* (SMs), and the mechanism with l_i is defined as the *basic similarity mechanism* (BSM). All of the SMs have the same ratio among characteristic parameters $Dl_i x$. The ratio is also the same as that between l_i. The selection of parameter d does not affect the ratio. For example, considering a mechanism with $L_1 = 6$ mm and $L_2 = 4$ mm, when $d = 1$, then $l_1/l_2 = 0.6/0.4 = 1.5$. When $d = 5$, then $l_1/l_2 = 3/2 = 1.5$. Therefore, the selection of parameter d does not affect the application of the PFNM in the analysis and design of mechanisms, nor does it influence the analytical results.

Characteristic parameters $L_i = D\,l_i$ of a mechanism make up an n-dimensional space. This space is infinite because each of the parameters can have any value between zero and infinity. Therefore, dimensional mechanism space $\Pi = \left[C_1^n,\ C_2^n,\ C_3^n,\ \ldots\right]$ is an infinite space and *combination* $C^n = (L_1,\ L_2, \ldots, L_n)$ can be found in this infinite space. All parameters l_i are limited; thus, normalized mechanism space $\pi = \left[c_1^n,\ c_2^n,\ c_3^n,\ \ldots\right]$, which is actually the PDS, is finite. C*ombination* $c^n = (l_1,\ l_2, \ldots,\ l_n)$ is an element in this finite space. Therefore, the PFNM establishes the relationship between the elements in finite and infinite spaces. Every combination in infinite space Π has a unique counterpart in finite space π. However, one element in the finite space corresponds to an infinite number of elements in the infinite space. The defined SMs are in space Π, whereas the BSM exists in space π, i.e., the PDS.

The parameter number affects the difficulty of the kinematic design of a mechanism. Using the PFNM as basis, we provide some examples with different parameter numbers. In the following examples, $d = 1$.

(a) $n = 1$

The case $n = 1$ means that only one characteristic parameter L_1 exists in the mechanism. The P̲RRRP̲ 2-DOF and RP̲RP̲R 2-DOF parallel mechanisms in Fig. 8.1a, b are such mechanisms. The normalization factor is L_1 itself. In Eq. (8.2), $l_1 = 1$, indicating that the PDS is a point.

(b) $n = 2$

In this case, there are two characteristic parameters L_1 and L_2, namely, the P̲RRRP̲ 2-DOF parallel mechanism actuated vertically by linear actuators and the

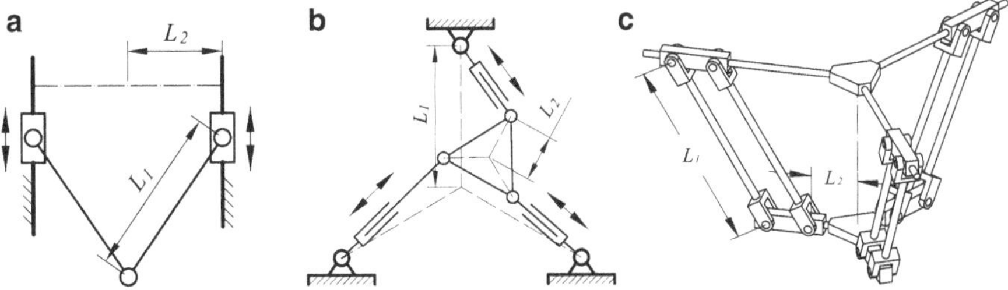

Fig. 8.2 Mechanisms with two characteristic parameters

3-RPR parallel mechanism, shown in Fig. 8.2a, b, respectively. The Star-like robot in Fig. 8.2c, 3-RPS parallel mechanism, and linear Tsai's mechanism (Fig. 7.2) also have two characteristic parameters. The parameters of the Star-like robot are the length of each leg and radius of the mobile platform. Those of the linear Tsai's mechanism are the length of each leg, as well as the margin between the radius of the base and that of the mobile platform. The normalization factor of these mechanisms is $D = L_1 + L_2$. Because $l_1 + l_2 = 1$ for the normalized parameters, the PDS is actually a closed line section. Particularly, condition $l_1 \geq l_2$, i.e., $l_1 \geq 0.5$, is imposed on the mechanism in Fig. 8.2a. The PDS of the mechanism is a closed interval $l_1 \in [0.5,\ \ 1]$.

(c) $n = 3$

Parallel mechanisms, such as the 5R planar 2-DOF (Fig. 8.3a) and 3-PRS (Fig. 8.3b) mechanisms, as well as the DELTA robot, have three parameters. Other 3-DOF translational parallel mechanisms with revolute actuators (e.g., Tsai's mechanism (Fig. 1.23)) are also such types of mechanisms. Similar to the 3-PRS mechanism, most non-translational parallel mechanisms that are actuated vertically by linear actuator shave three characteristic parameters. Examples of these mechanisms are the 6-PUS 6-DOF parallel mechanism and *HALF* mechanism in Fig. 8.3c, d, respectively.

For these mechanisms, $l_1 + l_2 + l_3 = 1$. The normalized parameters should be specified as $0 < l_1, l_2, l_3 < 1$ and $l_2 \geq |l_3 - l_1|$. An additional condition, $l_1 > l_3$, should be imposed on the 3-PRS, 6-PUS, and *HALF* mechanisms. In any case, the PDS for any one of these mechanisms is a closed planar space. For example, the PDS for the 3-PRS parallel mechanism is depicted as isosceles triangle ABC in Fig. 8.4.

(d) $n = 4$

Most parallel mechanisms with revolute actuator shave four characteristic parameters. Such mechanisms include the 6-RRRS mechanism (Hunt 1983), HEXA mechanism (Pierrot et al. 1991), 3-RRR parallel mechanism (Fig. 8.5a), CaPaMan mechanism (Ottaviano and Ceccarelli 2002; Fig. 8.5b), t321-HEXA mechanism (Bruyninckx 1997), TURIN mechanism (Sorli, et al. 1997), and *HALF* mechanism with revolute actuators (Liu et al. 2005).

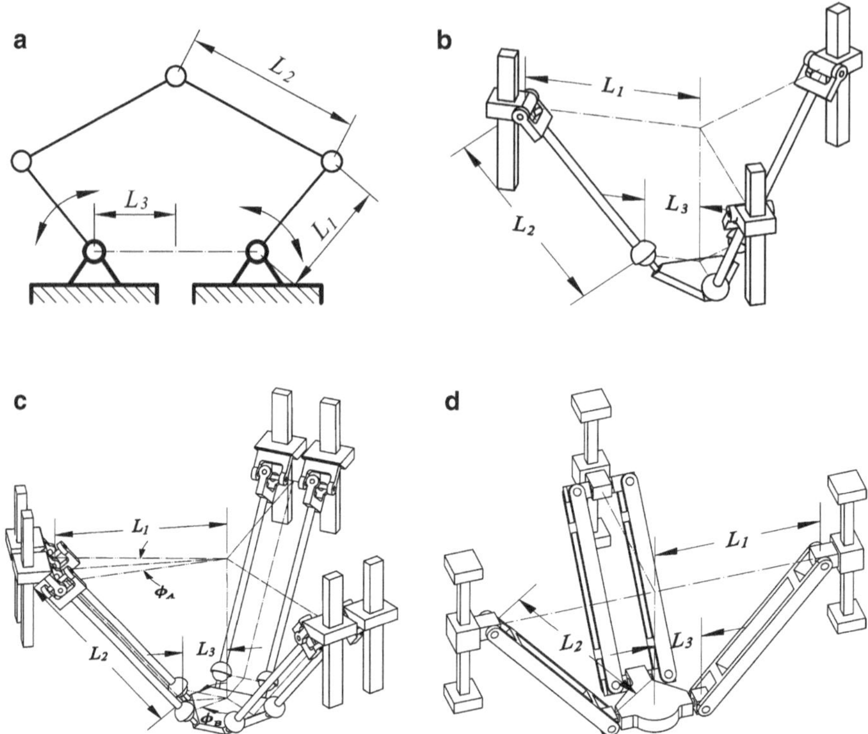

Fig. 8.3 Mechanisms with three characteristic parameters

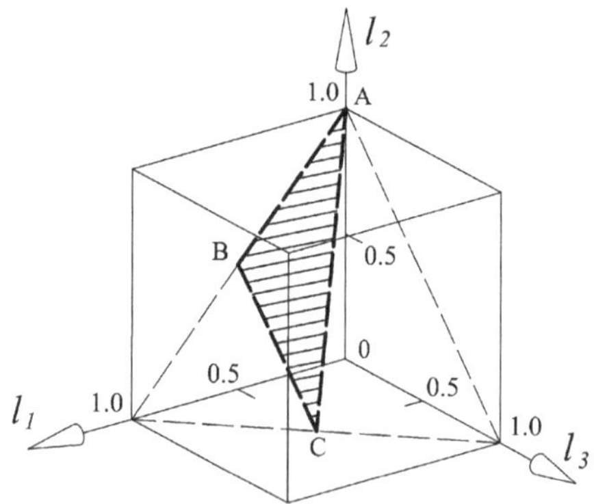

Fig. 8.4 PDS of the 3-PRS parallel mechanism in Fig. 8.3b

When $n = 4$, $l_1 + l_2 + l_3 + l_4 = 1$, which defines a unit cube. Other parameter conditions always exist; thus, the PDS of such a mechanism is usually a 3-dimensional polyhedron. For example, to guarantee that the 3-RRR parallel mechanism in Fig. 8.5a can be assembled, parameter conditions $l_2 + l_3 + l_4 \geq l_1$

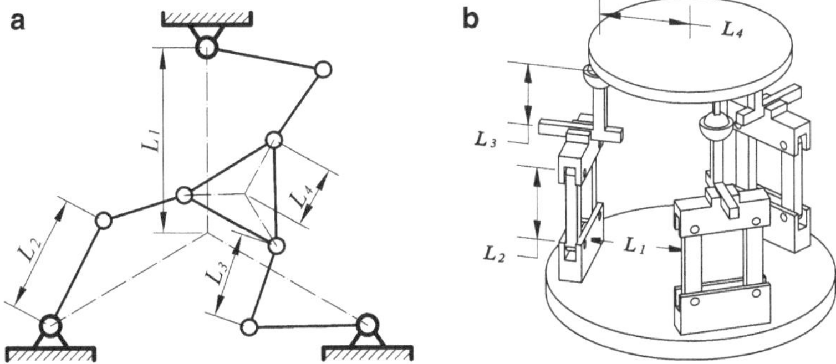

Fig. 8.5 Mechanisms with four characteristic parameters

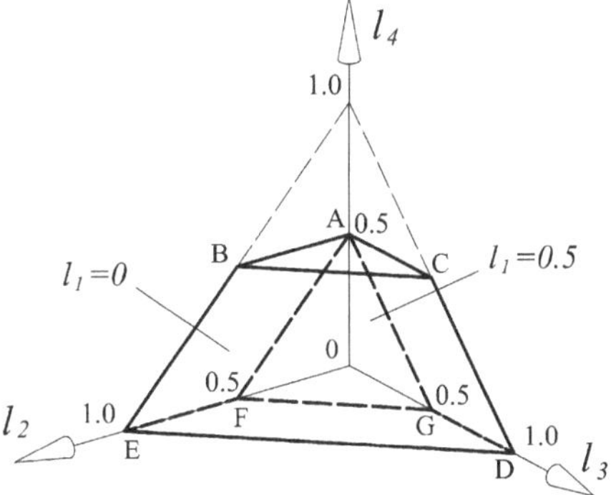

Fig. 8.6 PDS of the 3-RRR parallel mechanism in Fig. 8.5a

$(l_1 \leq 0.5)$ and $l_2 + l_3 + l_1 \geq l_4$ $(l_4 \leq 0.5)$ must be satisfied. Therefore, the PDS of the mechanism is actually polyhedron ABCDEFG in Fig. 8.6.

(e) $n \geq 5$

Most fully parallel mechanisms with k DOFs consist of k legs, each made up of three joints and two moving links. Parallel mechanisms such as the DELTA robot, Tsai's mechanism, Star-like robot, CaPaMan mechanism, and *HALF* mechanism contain simple mechanisms in their legs. In this study, the combination of one characteristic link and one joint is regarded as the simple mechanism. In addition, parallel mechanisms usually have a fully symmetrical kinematic structure, indicating that the parameters of one leg are the same as those of the other legs in terms of number and length. In general, therefore, the characteristic parameter number of a fully parallel mechanism cannot exceed 4.

The case $n \geq 5$ arises when a parallel mechanism is asymmetrical. For example, a 5R 2-DOF dissymmetrical parallel mechanism has five characteristic parameters. The parameter number of a planar 3-RRR dissymmetrical parallel mechanism can be up to 12. Regardless of how many parameters exist, the number of parameters can be reduced from n to $n-1$, and the $n-1$ normalized parameters can be limited using the PFNM. However, the PDS cannot be expressed in a 3-dimensional space because it exists in a space with dimensions GE 4.

8.1.2 Relationship Between a BSM and Its SMs

As presented previously, the PFNM not only reduces the parameter number but also logically specifies the bounds of each normalized parameter. The normalization makes the establishment of a design space within a limited space possible. If the technique can be used in parallel mechanisms, especially mechanisms with fewer than five parameters, the analysis can be implemented and the optimal design issues can be considered completely and precisely. To these ends, the inherent relationship between a BSM and its corresponding SMs should be clarified.

As shown in Eq. (8.2), a BSM and any one of its SMs have the relationship $L_i = Dl_i$ in terms of size. The planar workspace of an SM should be D^2-times that of its BSM. The factor is D^3 for spatial workspaces. A BSM and any one of its SMs are also similar in performance. We provide two examples to illustrate the relationships in detail.

The first example is the linear Tsai's mechanism (Fig. 7.2). The inverse kinematics, Jacobian matrix, and workspace analysis are discussed in Sect. 7.1.2.2. For a prescribed position $(x,\ y,\ z)$ of point O', the input of points B_k $(k = 1, 2, 3)$ for the working mode in Fig. 7.2 can be described by $z_k = \sqrt{L_2^2 - (x - x_k)^2 - (y - y_k)^2} + z$. Its Jacobian matrix is given in Eq. (7.18).

As workspaces, the LCI and GCI are the criteria used in most designs. These indices are employed here to compare a mechanism with its normalized mechanism.

Suppose that $R_i = L_i$. Replacing parameters L_i in Eq. (7.19) with $D\,l_i$ enables us to easily conclude that the maximal workspace of an SM is D^2-times that of its BSM. Equation (7.18) shows that the Jacobian matrix of the SM of the mechanism at point $(x,\ y)$ is identical to that of the BSM at point $(x\,/D,\ y\,/D)$. As a result, the corresponding LCI values are the same. Therefore, the GCMIWs of an SM and its BSM also exhibit the D^2-times relationship. In the end, the GCI values in their corresponding GCMIWs are also the same. For example, when the specified LCI is 0.3, the GCMIWs of an SM with parameters $L_1 = 8$ mm and $L_2 = 12$ mm, as well as its BSM with $l_1 = 0.4$ and $l_2 = 0.6$ $(D = 20$ mm), are 37.1764 mm^2 and 0.0929 (37.1764/20^2), respectively. The GCI values over their GCMIWs are both 0.5386, and their LCI distributions are shown in Fig. 8.7, respectively. The distributions are similar to each other.

The first example shows a comparison of positional workspaces. The second concerns the orientational workspace of a mechanism. The angle is defined as the figure formed by two lines diverging from a common point or by two planes diverging from a common line. Thus, the angle is usually measured by the ratio of

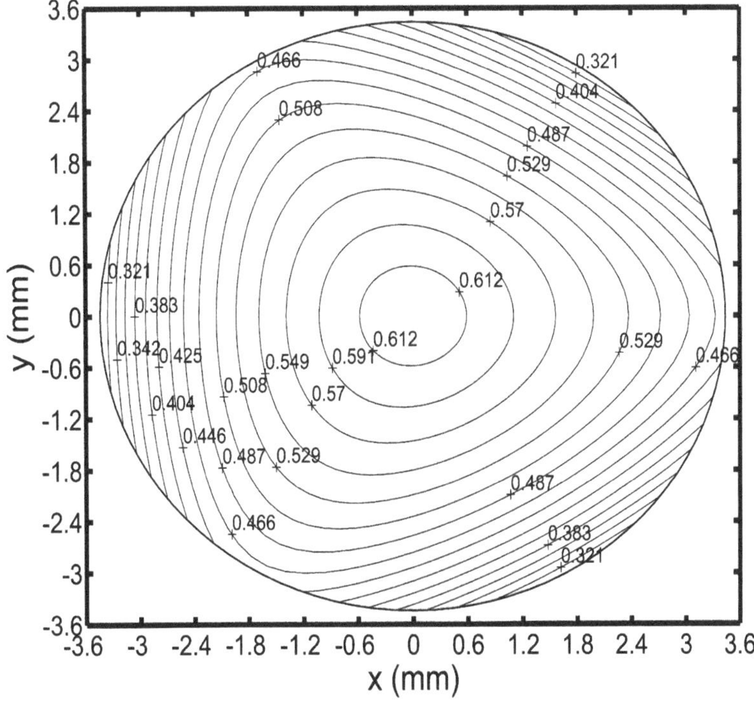

Fig. 8.7 Distribution of the LCI in the GCW of the linear Tsai's mechanism for the SM

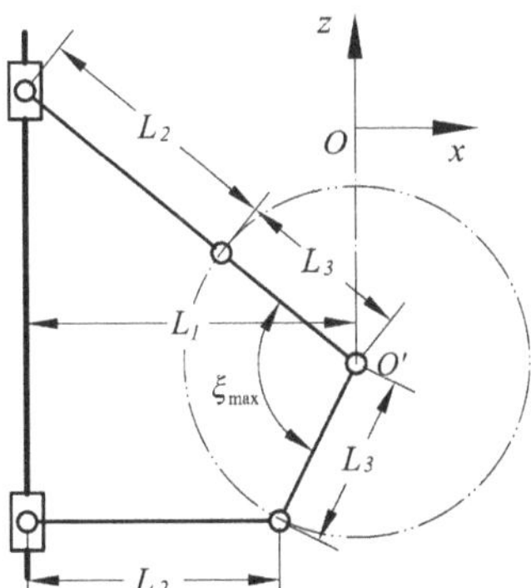

Fig. 8.8 Orientation capability of the mechanism in Fig. 3.15

two linear parameters. For such reason, normalization factor D cannot generate any change in the orientation capability of a mechanism. For example, Fig. 8.8 shows the maximal orientation capability that the mechanism in Fig. 3.15 can reach at one point. The capability at point O' can be calculated by

$$\xi_{\max} = \frac{180^\circ}{\pi}\left[\cos^{-1}\left(\frac{L_1}{L_3 + L_2}\right) + \cos^{-1}\left(\frac{L_1 - L_2}{L_3}\right)\right], \tag{8.6}$$

which shows that replacing L_i with L_i/D (i.e., l_i) does not change value $\xi_{\max}$. Hence, regardless of what normalization factor D is, the orientational workspaces of a BSM and its SMs are the same.

Therefore, an SM and its BSM are similar not only in size but also in performance, indicating that determining the performance of a BSM enables the realization of the performance of all of its SMs. Having checked workspaces LCI and GCI, a BSM and its SMs are also similar in terms of other performance indices such as the LTI, GTI, and GTW. Therefore, if a BSM is optimal, any one of its SMs will also be optimal. This relationship is highly important in developing a methodology for optimal kinematic design.

8.2 Performance Chart

A performance chart or atlas is highly useful in illustrating the relationship between an index and the design parameters involved. With a chart, we can visually and globally determine the index for a mechanism with different parameter values.

8.2.1 *Jacobian Matrix-Based Performance Chart*

8.2.1.1 Performance Chart of Planar 5R Parallel Mechanisms

Parameter Design Space

Because of its symmetrical structure, the 5R parallel mechanism leaves three parameters, namely, R_1, R_2, and R_3 (Fig. 3.1). Any one of these parameters can have a value between zero and infinity. This is the largest difficulty encountered in developing a design space that can embody all the mechanisms (with different link lengths) within a finite space. For this reason, the physical link size of the mechanisms should be eliminated. Let

$$D = \frac{(R_1 + R_2 + R_3)}{3}. \tag{8.7}$$

We can obtain three non-dimensional parameters r_i using

$$r_1 = \frac{R_1}{D}, r_2 = \frac{R_2}{D}, r_3 = \frac{R_3}{D}, \tag{8.8}$$

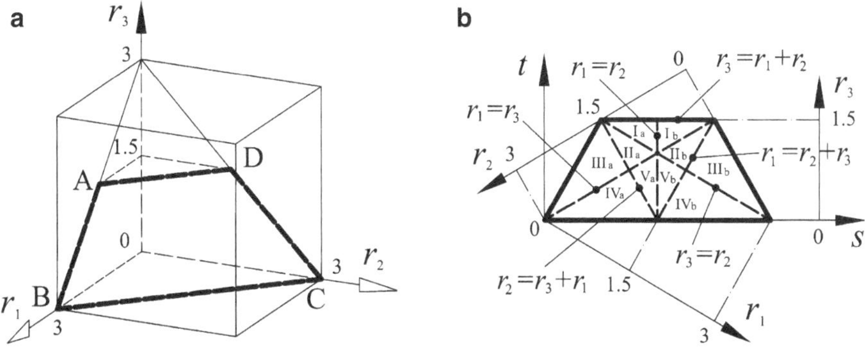

Fig. 8.9 Design space of the 5R parallel mechanism

which then yields

$$r_1 + r_2 + r_3 = 3. \tag{8.9}$$

As shown in Eq. (8.9), non-dimensional parameters r_1, r_2, and r_3 can theoretically take on any value between 0 and 3. For the 5R parallel mechanism studied here, the analysis of the workspace and singularity shows that r_1 and r_2 cannot be 0, and $r_1 + r_2$ cannot be less than r_3. Otherwise, they result in the failure of mechanism assembly. Therefore, the three parameters should be

$$0 < r_1,\ r_2 < 3, \quad \text{and} \quad 0 \leq r_3 \leq 1.5. \tag{8.10}$$

Using Eqs. (8.9) and (8.10) as bases, we can establish a design space; in Fig. 8.9a, trapezoid $ABCD$ is the design space and is restricted by r_1, r_2, and r_3. Therefore, the design space can be depicted in another form (Fig. 8.9b), referred to as the planar-closed configuration of the design space.

For convenience, two orthogonal coordinates s and t are used to express r_1, r_2, and r_3. Thus, using

$$\begin{cases} s = 2r_1 / \sqrt{3} + r_3 / \sqrt{3} \\ t = r_3 \end{cases}, \tag{8.11}$$

coordinates r_1, r_2, and r_3 can be transformed into s and t. Equation (8.11) is useful in constructing a performance atlas.

To study the performance of the parallel mechanism in detail, we use five lines, $r_1 = r_3$, $r_2 = r_3$, $r_2 = r_1 + r_3$, $r_1 = r_3 + r_2$, and $r_1 = r_2$, to divide the design space into ten subregions, namely, I_a, I_b, II_a, II_b, III_a, III_b, IV_a, IV_b, V_a, and V_b (Fig. 8.9b). The first four lines indicate four types of change point mechanisms. $r_1 = r_2$ is used because when $r_1 = r_2$, the boundaries of the theoretical workspace are only two

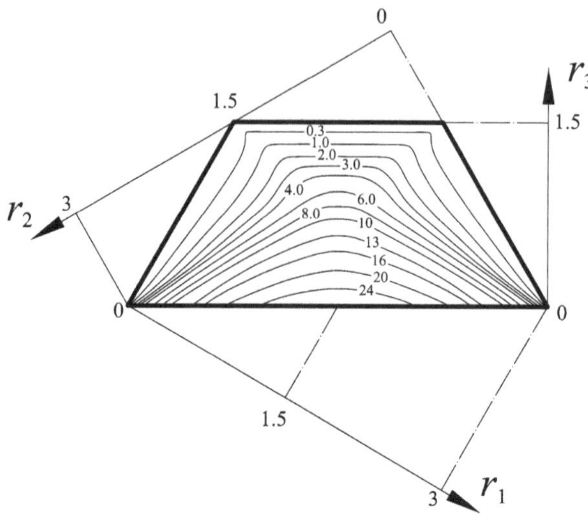

Fig. 8.10 Atlas of the theoretical workspace area

circles C_{1o} and C_{2o}, given by Eqs. (5.3) and (5.5), respectively. Such a design space facilitates the comprehensive investigation of the performance characteristics of all 5R parallel mechanisms with possible combinations of R_1 (r_1), R_2 (r_2), and R_3 (r_3).

Atlas of the Theoretical Workspace Area

The theoretical workspace can be determined using Eqs. (5.3), (5.4), (5.5), and (5.6) as bases. The workspace area can be calculated following the equations provided by Gao et al. (1996). The atlas for the theoretical workspace area is shown in Fig. 8.10, which indicates that:

- The theoretical workspace area is inversely proportional to parameter r_3.
- The area atlas is symmetrical with respect to $r_1 = r_2$, which means that the area of a mechanism with $r_1 = a$, $r_2 = b$ $(a, b < 3)$, and $r_3 = 3 - a - b$ is identical to that of a mechanism with $r_1 = b$, $r_2 = a$ $(a, b < 3)$, and $r_3 = 3 - a - b$.
- When $r_3 = 1.5$ (or $r_1 + r_2 = r_3$), the area is 0.
- The area reaches its maximum value when $r_1 = r_2 = 1.5$ and $r_3 = 0$. The maximum value is 9π.

Distribution Characteristics of the Theoretical Workspace Shape in the Design Space

Meanwhile, the workspace of a typical mechanism in each of the ten subregions can also be obtained. The distribution characteristics of the workspace shape can be

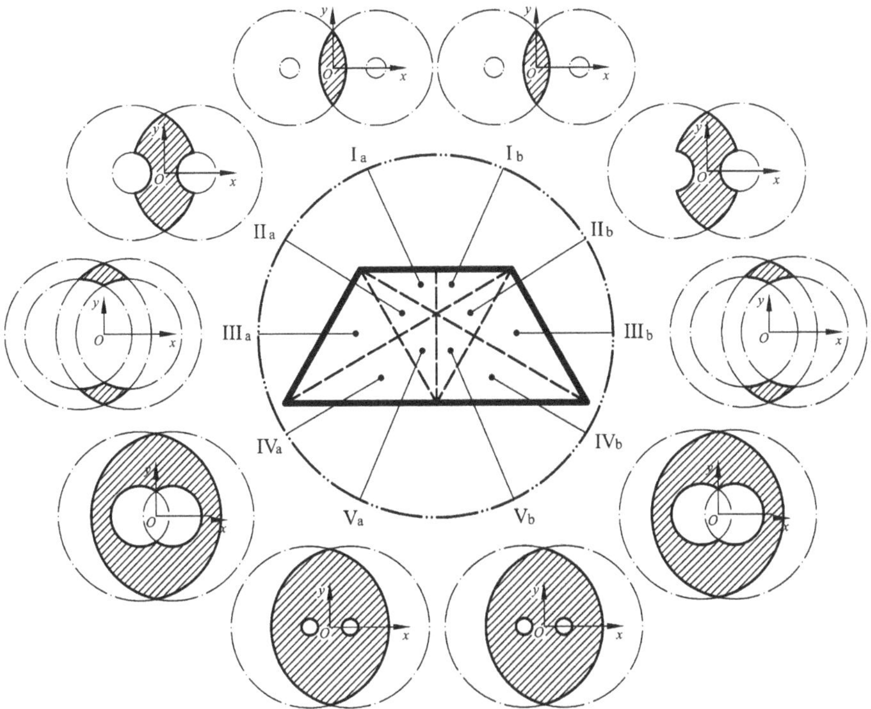

Fig. 8.11 Distribution of the theoretical workspace shape in the design space

summarized on the basis of these workspaces. The distribution in the design space is shown in Fig. 8.11; it is symmetrical with respect to line $r_1 = r_2$. This symmetry means that the theoretical workspace shape of a mechanism with $r_1 = a$, $r_2 = b$ $(a, b < 3)$, and $r_3 = 3-a-b$ is the same as that of a mechanism with $r_1 = b, r_2 = a$ $(a, b < 3)$, and $r_3 = 3 - a - b$. This characteristic is as it should be because the radii of boundary circles C_{1o}, C_{2o}, C_{1i}, and C_{2i} (given in Eqs. 5.3), (5.4), (5.5), and (5.6)) of the theoretical workspace are related to $r_1 + r_2$ and $|r_1 - r_2|$. Hence, there are only five types of theoretical workspace shapes for the 5R parallel mechanism. Moreover, each theoretical workspace is symmetrical about the x- and y-axes.

Distribution Characteristics of Singular Loci in the Design Space

The analysis of singularity shows that two kinds of singular loci exist. The first kind is the inverse singular locus. The loci are C_{1o}, C_{1i}, C_{2o}, and C_{2i} given by Eqs. (5.3), (5.4), (5.5), and (5.6) and are the boundaries of the theoretical workspace. The second type is the forward singular locus, which presents two cases: one is that when points B_1 and B_2 are coincident. The loci are two circles given by Eq. (5.7), denoted as $C_{\text{Coin-U}}$ and $C_{\text{Coin-D}}$. The second case occurs when B_1PB_2 is

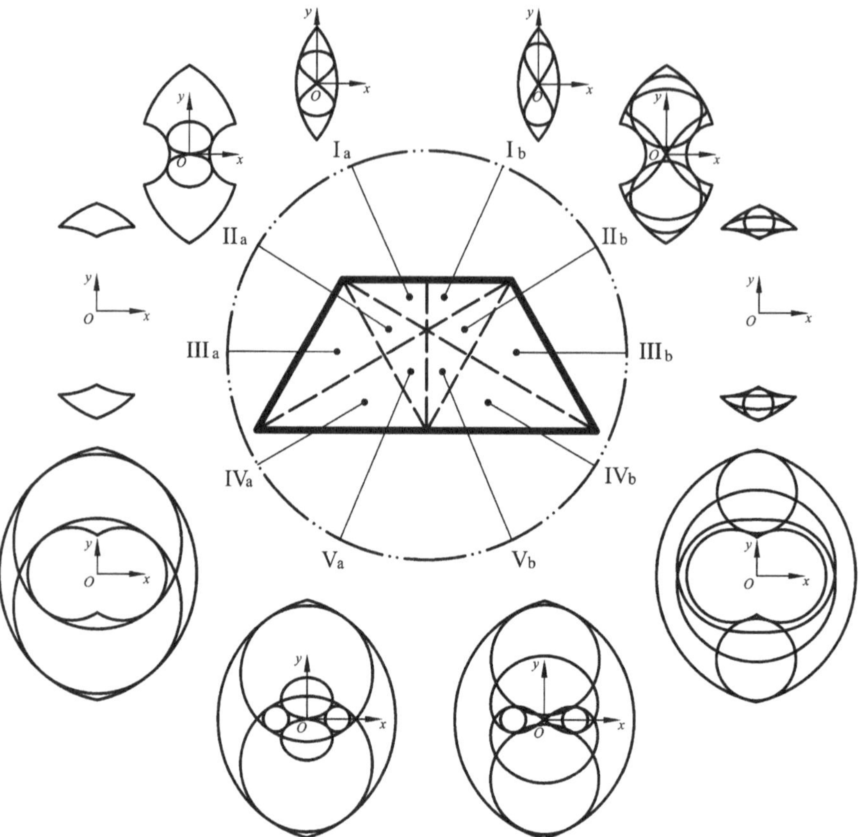

Fig. 8.12 Distribution of singular loci in the design space

completely extended. The locus is presented by Eq. (5.9) or (5.10), denoted as C_{Col}. The singular loci for the mechanism with $r_1 = 1.2$, $r_2 = 1.0$, and $r_3 = 0.8$ are shown in Fig. 6.1.

The singular loci for the mechanism with different link lengths differ. They are illustrated in Fig. 8.12, which indicates the following:

- For a specified mechanism, the distribution of singular loci within the workspace is symmetrical about the x- and y-axes.
- The distribution in the design space is not symmetrical with respect to $r_1 = r_2$. This indicates that the singular loci of a mechanism with $r_1 = a$, $r_2 = b$ $(a, b < 3)$, and $r_3 = 3 - a - b$ are not the same as those of a mechanism with $r_1 = b$, $r_2 = a$ $(a, b < 3)$, and $r_3 = 3 - a - b$.
- For the mechanisms in subregions $\mathrm{I_a}$, $\mathrm{I_b}$, $\mathrm{II_a}$, and $\mathrm{III_a}$ where $r_1 < r_3$, no singular loci $C_{\mathrm{Coin-U}}$ and $C_{\mathrm{Coin-D}}$ exist, i.e., no singular configuration exhibits coincidence between points B_1 and B_2 within their workspaces.

- For the mechanisms in subregions III_a and Iva where $r_1 + r_3 < r_2$, no singular locus C_{Col} is found in their workspaces. For these mechanisms, $B_1 P B_2$ cannot be completely extended.
- The mechanisms in subregion III_a, where $r_1 < r_3$ and $r_1 + r_3 < r_2$, do not exhibit the uncertainty singularity.

The workspace where a real mechanism works is different from the theoretical workspace. The region should contain no internal singular loci. From the distribution of the theoretical workspace area and shape, we can see that the workspace area and shape of the mechanism with parameters $r_1 = 0.7$, $r_2 = 1.3$, and $r_3 = 1.0$ are identical to those of the mechanism with parameters $r_1 = 1.3$, $r_2 = 0.7$, and $r_3 = 1.0$. However, the analysis of singularity shows that the singular loci for these two mechanisms are different. This difference directly results in varied usable workspaces. As indicated in Figs. 8.10 and 8.11, any design result cannot be provided for a practical mechanism.

In Sect. 6.1, the singular loci for planar 5R parallel mechanisms with different working modes can be identified. We are concerned only with the mechanism that has the working mode "$+ -$". Figure 8.13 shows the singular loci of all the mechanisms with the aforementioned working mode in the design space.

The singular loci in Fig. 8.13 can be used to determine the usable workspace of a 5R parallel mechanism characterized by both the working mode "$+ -$" and up-configuration. The workspace of a mechanism should be continuous. As shown in Fig. 8.13, the singular loci divide the theoretical workspace into several regions. The regions for the mechanisms in all the subregions are separated. A practical mechanism cannot access the singular loci from one region to another. Nevertheless, Fig. 8.13 shows that the two actuated links of the mechanism in subregions IV_a, IV_b, V_a, and V_b are both crank links. A mechanism in these four subregions can be selected to identify a mechanism with two actuated crank links and a surrounding workspace.

Distribution of the Usable Workspace Shape in the Design Space

The usable workspace shape for the 5R parallel mechanism is classified in the design space, as shown in Fig. 8.14. The distribution indicates the following:

- In Fig. 8.14, each of the hatched regions is the usable workspace of the mechanism, characterized by both the "$+ -$" working mode and up-configuration.
- Each usable workspace is symmetrical about the y-axis.
- The distribution in the design space is not symmetrical about $r_1 = r_2$. The usable workspace shape of a mechanism with $r_1 = a$, $r_2 = b$ $(a, b < 3)$, and $r_3 = 3 - a - b$ is no longer the same as that of a mechanism with $r_1 = b$, $r_2 = a$ $(a, b < 3)$, and $r_3 = 3 - a - b$.
- The usable workspace of a mechanism in subregion IV_a is half of its theoretical workspace.
- The flattest part in the workspace is always located at the region where $y = 0$.

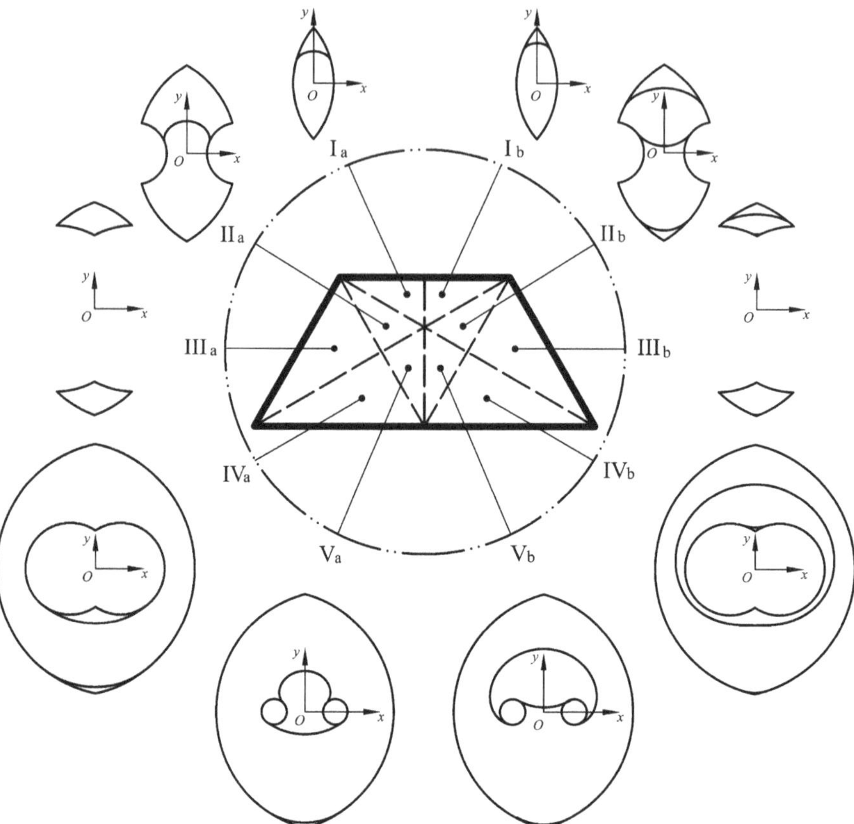

Fig. 8.13 Distribution of singular loci in the design space of the mechanism with the working mode "+ −"

Maximal Inscribed Circle and Maximal Inscribed Workspace

In the design process, we usually select the thickest part in the usable workspace as a search objective to generate a desired workspace that is as large as possible and as far as possible from singularity. As shown in Fig. 8.10, the thickest part is always symmetrical about the y-axis. It can be characterized by a circle, which is tangential to the singular loci. Here, the circle is referred to as the MIC, defined as the circle that is located at the y-axis and is tangential to the usable workspace boundary curves. According to this definition, the MIC of a planar 5R parallel mechanism can be described as

$$x^2 + (y - y_{\mathrm{MIC}})^2 = r_{\mathrm{MIC}}^2, \tag{8.12}$$

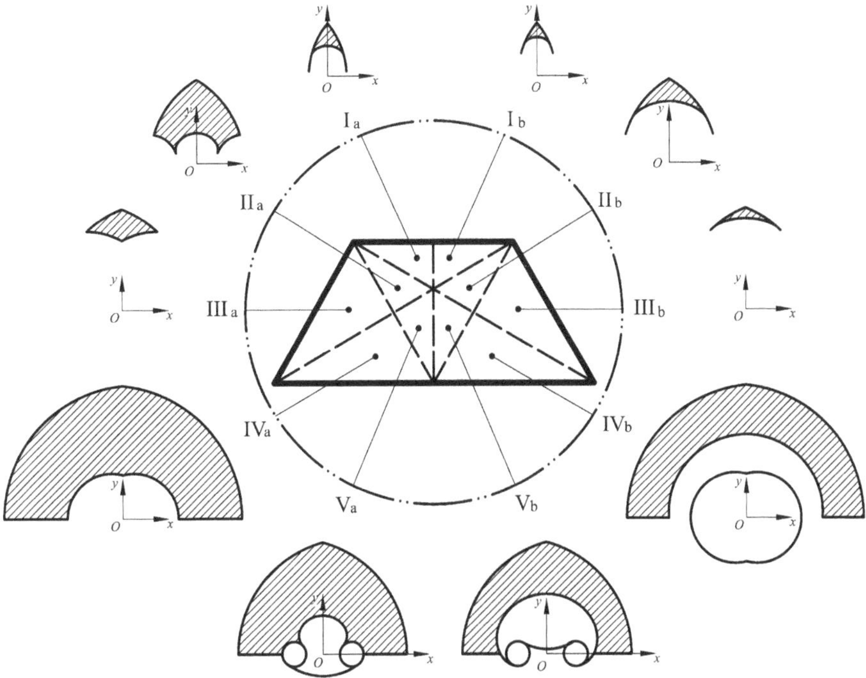

Fig. 8.14 Usable workspace shape in the design space

where r_{MIC} is the radius and $(0,\ y_{\mathrm{MIC}})$ is the center. The analysis of the usable workspace shows that for the mechanisms in subregions $\mathrm{III_a}$ and Iva where $r_1+r_3<r_2$, the MIC should be tangential to circles C_{1o}, C_{1i}, C_{2o}, and C_{2i} given by Eqs. (5.3), (5.4), (5.5), and (5.6). In this case, the radius and y_{MIC} of the MIC can be given as

$$r_{\mathrm{MIC}}=(r_1+r_2-|r_1-r_2|)/2 \quad \text{and} \quad y_{\mathrm{MIC}}=\sqrt{(r_1+r_2+|r_1-r_2|)^2/4-r_3^2}. \tag{8.13}$$

For the mechanisms in the other subregions where $r_1+r_3>r_2$ in the design space, the MIC should be tangential to C_{1o}, C_{2o}, and C_{Col}. Furthermore, the MIC is the circle that passes through the intersection point between C_{Col} and the y-axis. It is tangential to C_{1o} and C_{2o}. Its radius and y_{MIC} can be expressed as follows:

$$r_{\mathrm{MIC}}=|y_{\mathrm{MIC}}|-y_{\mathrm{Col}} \quad \text{and} \quad y_{\mathrm{MIC}}=\frac{\left[(r_1+r_2+y_{\mathrm{Col}})^2-r_3^2\right]}{2\,(r_1+r_2+y_{\mathrm{Col}})}, \tag{8.14}$$

where $y_{\mathrm{Col}}=\sqrt{r_1^2-(r_2-r_3)^2}$.

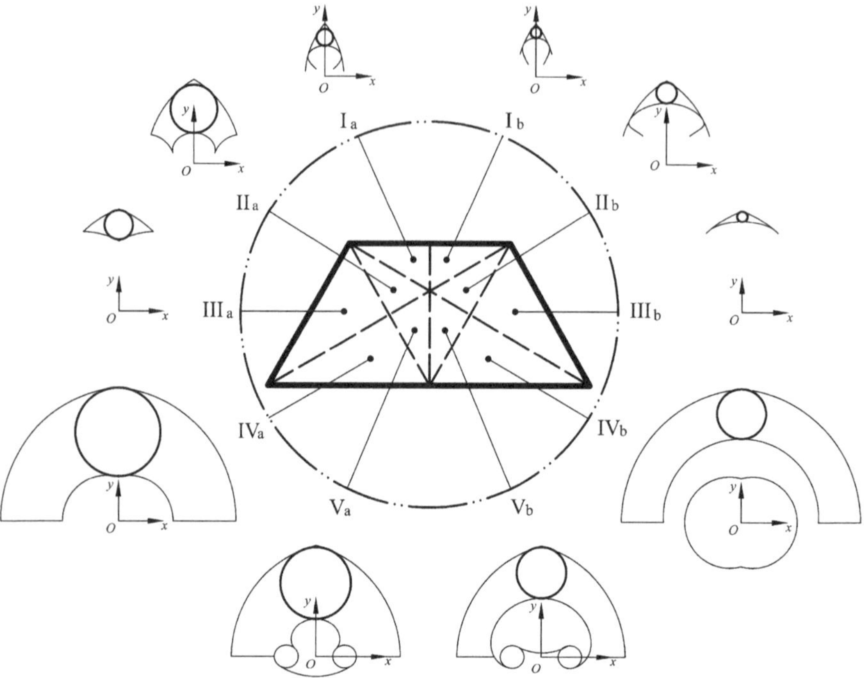

Fig. 8.15 Distribution of the MIC radius in the design space

A usable workspace may be seen as an irregular region. Within the usable workspace, many points cannot be used by the mechanism in practical applications because these points are near the singular loci. The MIC is a regular region, and its area is $\pi\, r_{\mathrm{MIC}}^2$ characterized by radius r_{MIC}. Such a region circled by the MIC is referred to as the MIW.

Distribution of the MIC Radius in the Design Space and Atlas of the MIW Area

Figure 8.15 shows the distribution of the MIC radius in the design space. Figure 8.16 illustrates the atlas of MIC radius r_{MIC}.

As shown in Fig. 8.16, line $r_1 + r_3 = r_2$ divides the design space into two parts. On the right side, the MIC radius is proportional to r_2 when r_1 is specified. On the left side, a specified r_2 yields an MIC radius that is proportional to r_1. The MIC radius reaches a maximal value of 1.5 when $r_1 = r_2 = 1.5$.

Although the MIW cannot reflect the exact value of the usable workspace area, its radius atlas (Fig. 8.16) indicates which mechanisms have large and small usable workspaces; these mechanisms are identified with the help of the workspace shape characteristics of the MIW (Fig. 8.14). In addition, the MIW always stands for the thickest part of the usable workspace. In this region, therefore, more points are

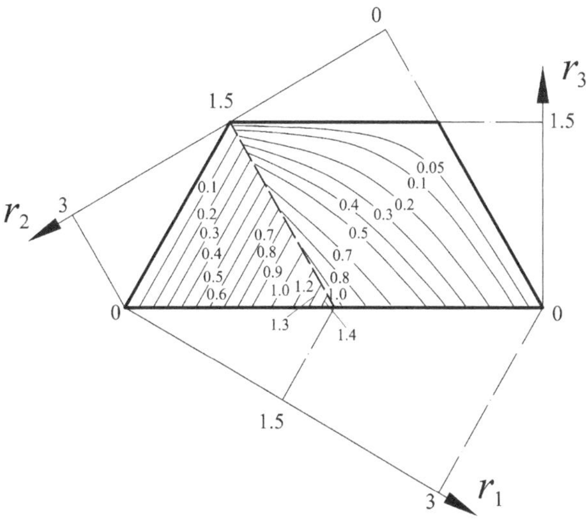

Fig. 8.16 Atlas of the MIC radius

located far from the singular curve. The practical workspace of a device uses the majority of this region. For this reason, the MIW can characterize the workspace of a practical mechanism. In performance analysis, we are usually concerned with the comparative result, and not the exact value, which is why we provide the atlas of the MIC radius instead of the atlas of the usable workspace area. Given that the definition of most global performance indices depends on the workspace, the MIW is used to evaluate performance.

The comparison of Figs. 8.14, 8.15, and 8.16 shows that the MIC radius atlas is not exactly identical to the usable workspace area atlas, which is not plotted in this chapter. For example, two mechanisms may exist in subregions II_a and IV_b in Fig. 8.16, and they can have the same MIC radii. However, the usable workspace areas in Figs. 8.14 and 8.15 are different. Therefore, to generate a result for the usable workspace performance of a 5R parallel mechanism, Figs. 8.14, 8.15, and 8.16 should be used simultaneously.

Atlas of the GCI

For the mechanism studied here, workspace W in Eq. (7.3) can be the MIW. The relationship between the GCI and three normalized parameters r_i $(i = 1, 2, 3)$ can be studied in the design space. The corresponding atlas is shown in Fig. 8.17. To plot the atlas of Fig. 8.17, we first calculate every GCI value of each non-dimensional mechanism with r_1, r_2, and r_3. Each of these mechanisms is included in the design space. Using Eq. (8.11), we can then obtain the relationship between the GCI and orthogonal coordinates s and t. Using this relationship to plot the atlas in the planar

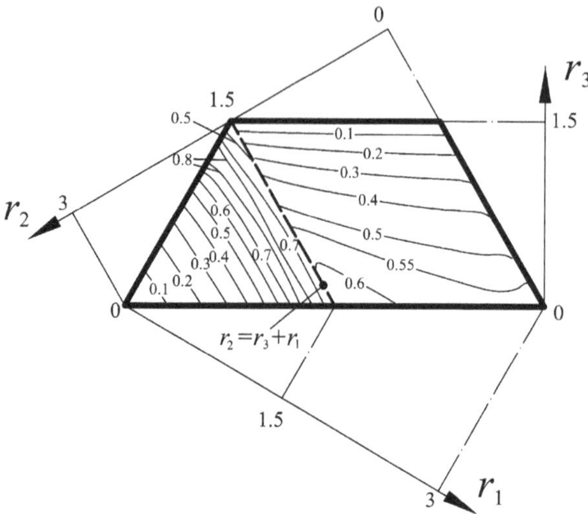

Fig. 8.17 Atlas of the GCI

system respect with s and t is a practical approach. The other atlases are also plotted using the same method. Figure 8.17 shows not only the relationship between the GCI and two orthogonal coordinates but also that between the GCI and three non-dimensional parameters. Our primary focus is the latter, for which s and t do not appear in the figure.

Figure 8.17 shows that:

- Line $r_1 + r_3 = r_2$ divides the design space into two parts.
- On the right side, specifying r_1 yields a GCI value that is inversely proportional to r_3.
- The mechanism with the best GCI is on the left side; in the region where $r_2 \in [1.56,\ 1.9]$, the GCI (>0.6) is better.

Atlas of the GSI

In Sect. 7.1.3, a GSI η_D is defined by Eq. (7.27). In the planar 5R parallel mechanism, setting W in Eq. (7.27) as a circular workspace with the radius $r_{\mathrm{MIC}}/2$ yields the GSI value for each mechanism in the design space. Figure 8.18 shows the plotted atlas, which indicates the following:

- Line $r_1 + r_3 = r_2$ divides the design space into two parts.
- On the right side of the atlas for η_D, the η_D value is proportional to r_1 when r_3 is specified.

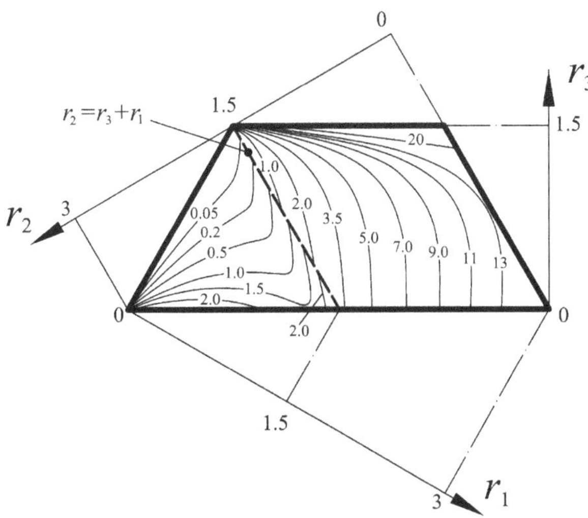

Fig. 8.18 Atlas of the GSI η_D

- On the left side of the atlas for η_D, a specified r_2 generates an η_D value that is proportional to r_1.
- All in all, when r_3 is specified, a longer r_1 leads to a larger η_D, i.e., worse stiffness performance.
- Figure 8.18 shows that in the region where $r_1 + r_3 < r_2$, the stiffness of a mechanism is better. A shorter r_1 produces better stiffness in a mechanism. In particular, $r_1 = 0$ produces the best stiffness in a mechanism and results in a triangular mechanism, which exhibits the most excellent stability.
- The figure also shows that the stiffness of a mechanism where $r_3 = 1.5$ ($r_3 = r_1 + r_2$) can be at its worst level along one direction and at its best along another direction. The mechanism where $r_3 = r_1 + r_2$ is such a mechanism, in which the five bars are collinear. It can be assembled but cannot be moved. When an external force acts on end-effector point P along the direction from A_1 (A_2) to A_2 (A_1), the deformation is at its smallest level. Along the direction normal to A_1A_2, the deformation is at its largest.

8.2.1.2 Performance Chart of 3-PRS Parallel Mechanisms

Parameter Design Space of 3-PRS Parallel Mechanisms

The PDS of the 3-PRS parallel mechanism is shown in Fig. 8.4. For the 3-PRS parallel mechanism studied here, there are three geometric parameters, r, L, and R, which are all measured in length units. Its PDS can be established, as shown in

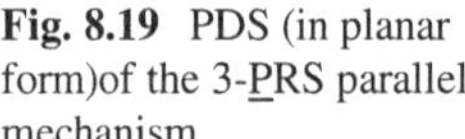
Fig. 8.19 PDS (in planar form)of the 3-PRS parallel mechanism

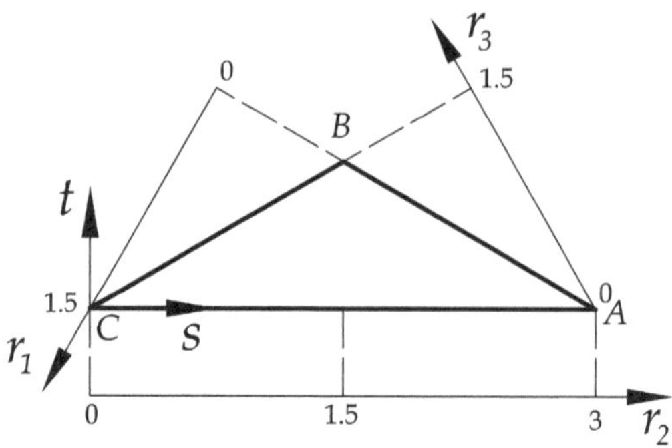

Fig. 8.4; triangle ABC is our target. The PDS can take on another form (Fig. 8.19), called the planar-closed configuration of the PDS. This configuration enables easier plotting of an atlas. For convenience, two orthogonal coordinates s and t are used to express r_1, r_2, and r_3. Thus, using

$$\begin{cases} s = r_2 \\ t = \sqrt{3} - 2r_1 \big/ \sqrt{3} - r_2 \big/ \sqrt{3}, \end{cases} \tag{8.15}$$

coordinates r_1, r_2, and r_3 can be transformed into s and t, turning Eq. (8.15) into a useful formula for constructing an atlas.

Atlases of Orientation Capability and Error Indices

The *orientation capability index* (OCI) is defined as the maximum tilt angle that can be achieved in any direction (for any ϕ), i.e., the minimum of orientation capability. For example, the OCI of the 3-PRS in Sect. 6.4 is 56° (Fig. 6.12). However, to avoid crossing singularity due to input errors, our OCI does not represent the angle from the actual singular orientation but that from the smallest orientation. From the latter, input errors can lead the mechanism to a singular configuration.

The atlas illustrating the relationship between normalized parameters r_i $(i = 1, 2, 3)$ and the OCI is shown in Fig. 8.20a. In this relationship, the index is inversely proportional to parameter r_3.

For comparison, the error index is defined as the largest value (denoted as E_{e_max}) of all maximum errors E_e of point E at the z-axis when $\varepsilon = 10$ μm, $\varphi \in [-90°, \ 90°]$, and $\theta \in [-\text{OCI}, \ \text{OCI}]$. Setting $l = 1$, the atlas of the error index defined in Sect. 7.2 is plotted (Fig. 8.20b). It can be indicated from the figure that a small r_2 usually leads to poor accuracy.

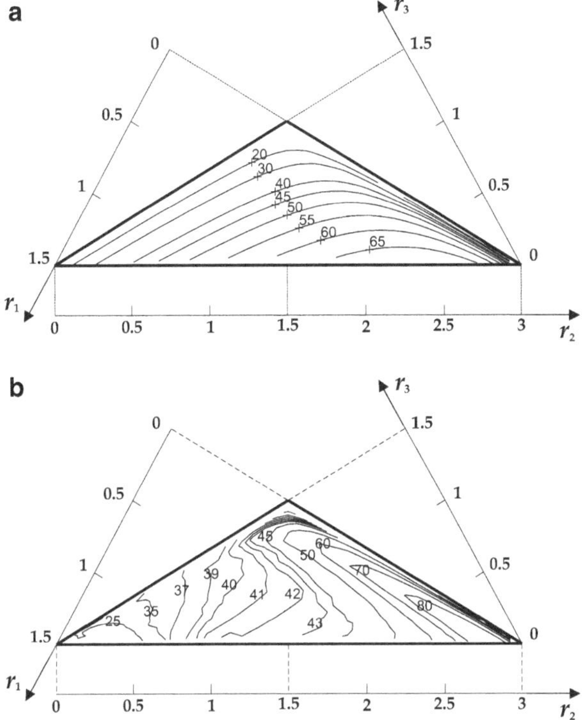

Fig. 8.20 Atlases of the 3-PRS parallel mechanism: (**a**) OCI; (**b**) error index

8.2.2 *LTI-Related Performance Chart*

8.2.2.1 Planar 5R Parallel Mechanism

Distribution of the GTW Shape in the Design Space

For the planar 5R parallel mechanism, the LTI can be written as $\chi = \min\{|\sin\gamma_1|, |\sin\gamma_2|, \ |\sin\mu|\}$, where

$$\mu = \cos^{-1}\left(\frac{2r_2^2 - |\boldsymbol{a}_{\Re}\boldsymbol{c}_{\Re}|^2}{2r_2^2}\right) \text{ and } \gamma_i = \cos^{-1}\left(\frac{r_1^2 + r_2^2 - |\boldsymbol{o}_{i\,\Re}\boldsymbol{p}|^2}{2r_1r_2}\right), \quad i = 1, 2. \tag{8.16}$$

$\boldsymbol{p} = (x \quad y)^{\mathrm{T}}$ and $\boldsymbol{o}_{i\,\Re}$ $(i = 1,2)$ are the position vectors of points O_i. Setting $\gamma \geq \sin(\pi/4)$ enables the numerical derivation of the GTW of a 2-DOF 5R parallel

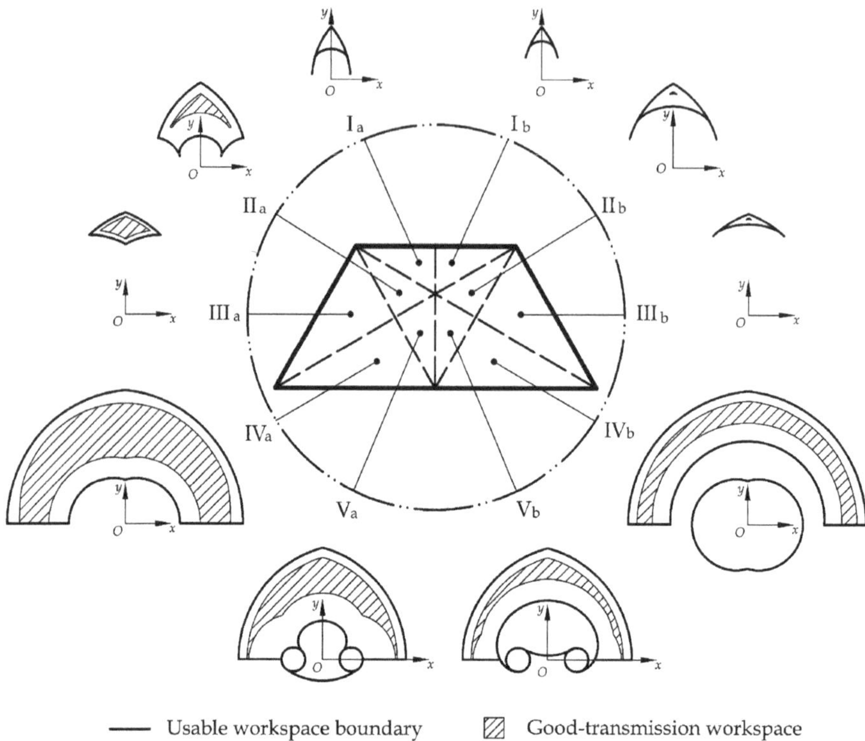

Fig. 8.21 GTW shape in the design space of planar 5R parallel mechanisms

mechanism. The GTW for the 5R parallel mechanisms is classified in the design space, as shown in Fig. 8.21. From the distribution, one sees that:

- In Fig. 8.21, each of the hatched regions is the GTW of the mechanism, characterized by both the "+ −" working mode and up-configuration.
- Some mechanisms, especially those in subregions I_a, I_b, II_b, and III_b, have a GTW of zero.
- Each GTW is symmetrical about the y-axis.
- The distribution in the design space is not symmetrical about $r_1 = r_2$.
- The thickest part in the workspace is always located at the y-axis.

Performance Chart of the GTW

The GTW of any one non-dimensional mechanism in the PDS can be calculated. The performance chart illustrating the relationship between normalized parameters r_i $(i = 1, 2, 3)$ and the GTW, in which $\gamma \geq \sin(\pi/4)$, is illustrated in Fig. 8.22. The figure shows that (a) the index is inversely proportional to parameter r_3 when r_2 is specified and (b) the GTW of some mechanisms in the design space is zero.

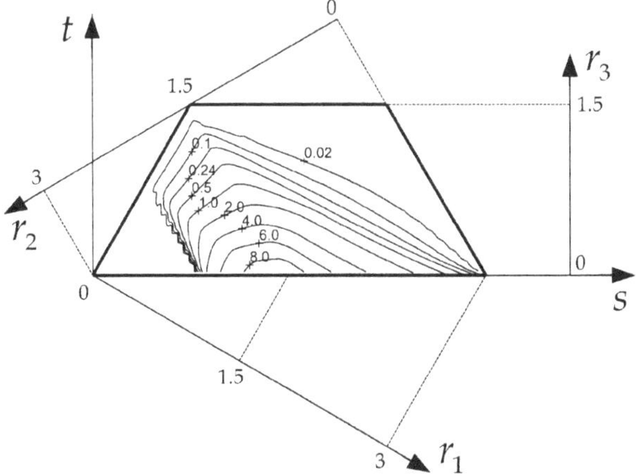

Fig. 8.22 Performance chart of the GTW of planar 5R parallel mechanisms

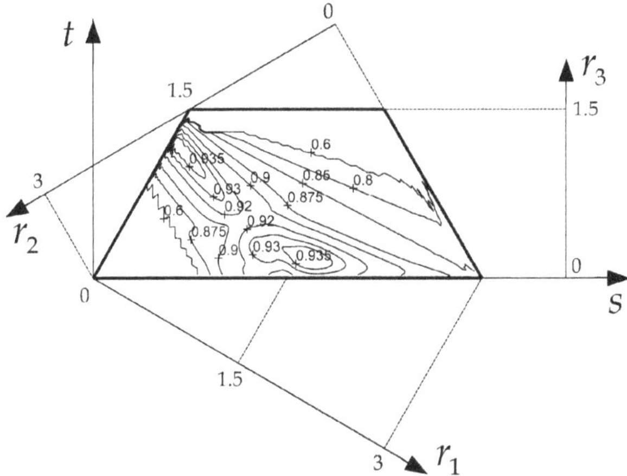

Fig. 8.23 Performance chart of the GTI

Performance Chart of the GTI

Using Eq. (7.60) and the GTW result, we can represent the relationship between the GTI and non-dimensional mechanisms in the PDS. The corresponding performance chart is illustrated in Fig. 8.23, which shows that in the design space, the mechanisms near the area $r_2 = 1.69$ usually exhibit better GTI performance.

Performance Chart of the Fatness Index of the GTW

Figure 8.21 demonstrates that the GTW shape may differ with different link lengths. The GTW may be of narrow shape for some mechanisms but broad for others. The

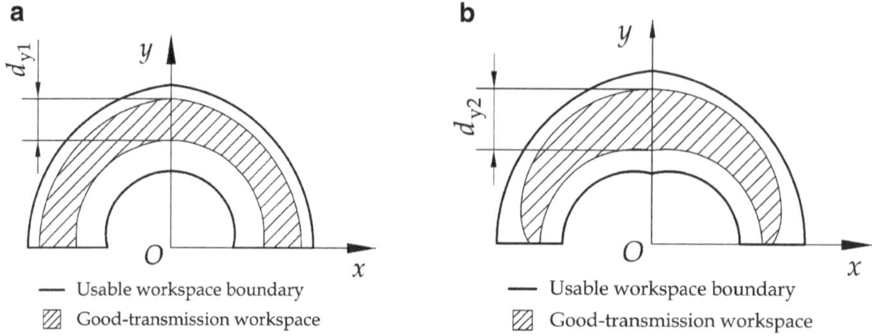

Fig. 8.24 GTW and FI of two mechanisms where (**a**) $r_1 = 1.57$, $r_2 = 1.161$, and $r_3 = 0.269$; and (**b**) $r_1 = 0.795$, $r_2 = 1.93$, and $r_3 = 0.275$

comparison between Figs. 8.21 and 8.22 shows that a GTW of large volume does not necessarily indicate broadness, whereas a GTW of small volume may be broad.

In some applications, a narrow or circular workspace is required. Certain industrial applications, however, require a regular workspace, such as the rectangular or foursquare type. In this case, a broader workspace is desirable. For this purpose, we introduce an index to evaluate the fatness of the GTW. Figure 8.21 shows that the fattest part in the GTW is always located at the y-axis. In this chapter, the GTW length along the y-axis, denoted as d_y, is defined as the *thickness index* (FI) of the GTW. The larger the FI value, the broader the GTW.

Two mechanisms with similar GTW volumes may have different FIs. For example, Fig. 8.24a, b show the GTW of a mechanism with $r_1 = 1.57$, $r_2 = 1.161$, and $r_3 = 0.269$ and that of a mechanism with $r_1 = 0.795$, $r_2 = 1.93$, and $r_3 = 0.275$, respectively. Both have the same volume (4.0); however, their FIs ($d_{y1} = 0.7$ and $d_{y2} = 0.95$) are different.

Figure 8.25 illustrates the FI performance chart of the 2-DOF 5R parallel mechanisms. The index is inversely proportional to parameter r_3 when r_2 is specified. The performance chart can be used in the optimization of a mechanism when a regular workspace is required in practical design.

8.2.2.2 Spatial 3-PRS Parallel Mechanism

Good-Transmission Orientation Capability and GTI

With the limit of the LTI (i.e., 0.7), we can identify an orientation capability with good motion/force transmission for the studied mechanism. The corresponding orientation capability is referred to as the *good-transmission orientation capability* (GTOC), defined as the subset of the MOC when the LTI for every transmission angle is greater than 0.7. That is, at every pose when $-\text{GTOC} \le \theta \le \text{GTOC}$, the LTI is subject to LTI ≥ 0.7. Tilting range $\pm$GTOC is referred to as the *good-transmission orientational workspace* of the mechanism.

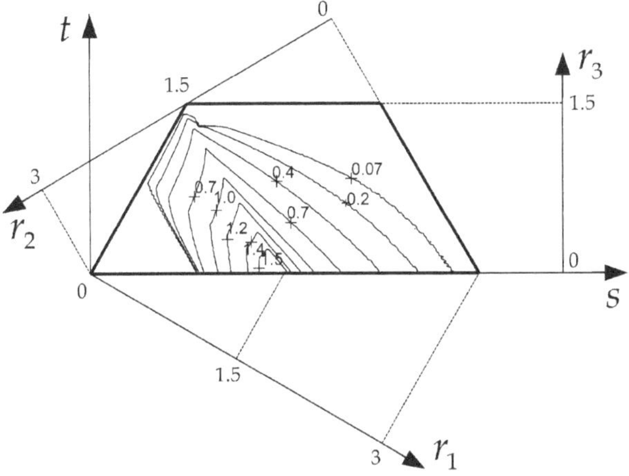

Fig. 8.25 Performance chart of the *thickness index* for the GTW

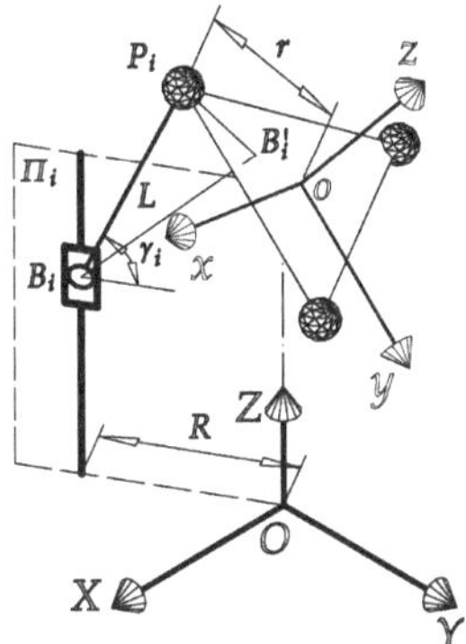

Fig. 8.26 Inverse transmission angle of the 3-PRS parallel mechanism

The mobile platform of a 3-PRS parallel mechanism has three independent output DOFs, i.e., two rotations and one translation. To restrict the other three DOFs, each leg produces a constraint wrench for the mobile platform. As shown in Fig. 8.26, the constraint wrench of the *i*th leg is a pure force, whose axis passes through center point P_i and is perpendicular to restricting plane Π_i. Hence, with respect to reference frame $\Re$: *O-XYZ*, the unit constraint wrench of the *i*th leg can be represented by

$$\$_{Ti}^{r} = (\boldsymbol{u}_i;\quad \boldsymbol{u}_i \times \boldsymbol{p}_i), \tag{8.17}$$

where $\boldsymbol{u}_i = \left[\cos(i \times 120° - 90°), \sin(i \times 120° - 90°), 0\right]^{\mathrm{T}}$ and $\boldsymbol{p}_i$ stands for the vector from origin O to point P_i.

In addition, to activate the mobile platform, each leg provides a transmission wrench, which is a pure force along the strut. With respect to reference frame $\Re$: *O-XYZ*, the unit transmission wrench of the *i*th leg can be expressed by

$$\$_{Ti} = \left(\overrightarrow{\mathrm{B}_i\mathrm{P}_i};\quad \overrightarrow{\mathrm{B}_i\mathrm{P}_i} \times \boldsymbol{p}_i\right), \tag{8.18}$$

where $\overrightarrow{\mathrm{B}_i\mathrm{P}_i} = \left[-\cos\gamma_i\cos(i \times 120°),\ -\cos\gamma_i\sin(i \times 120°),\ \sin\gamma_i\right]^{\mathrm{T}}$, which stands for the unit vector along the line passing through points B_i and P_i.

The slider in each leg can move up and down along the corresponding vertical slideway; hence, the unit input twist screw of the *i*th leg can be represented by

$$\$_{Ii} = (0,\ 0,\ 0;\ 0,\ 0,\ 1). \tag{8.19}$$

Therefore, according to Eq. (7.35), the reciprocal product between $\$_{Ti}$ and $\$_{Ii}$ can be obtained as follows:

$$\$_{Ti} \circ \$_{Ii} = \sin\gamma_i. \tag{8.20}$$

On the basis of Eq. (7.49), the input transmission index of the *i*th leg can be obtained as

$$\lambda_i = \frac{|\$_{Ti} \circ \$_{Ii}|}{|\$_{Ti} \circ \$_{Ii}|_{\max}} = \frac{|\sin\gamma_i|}{|\sin\gamma_i|_{\max}} = |\sin\gamma_i| \tag{8.21}$$

When only the slider in the *i*th leg is driven and the other two sliders are fixed, only the transmission wrench represented by $\$_{Ti}$ can contribute to the mobile platform. The other two transmission wrenches become two constraint wrenches for the mobile platform. In this case, the 3-DOF parallel mechanism becomes a single-DOF mechanism, and the unit instantaneous motion of the mobile platform can be represented by a unit twist screw $\$_{Oi}$, i.e.,

$$\$_{Oi} = \left(s_{i,1}, s_{i,2}, s_{i,3};\quad s_{i,4}, s_{i,5}, s_{i,6}\right). \tag{8.22}$$

Considering that $\$_{Oi}$ is a unit screw, we can obtain

$$s_{i,1}^2 + s_{i,2}^2 + s_{i,3}^2 = 1. \tag{8.23}$$

Equation (8.22) has six undetermined elements; thus, another five conditions are needed to determine unit twist screw $\$_{Oi}$. According to Eq. (7.52), the five equations can be obtained as follows:

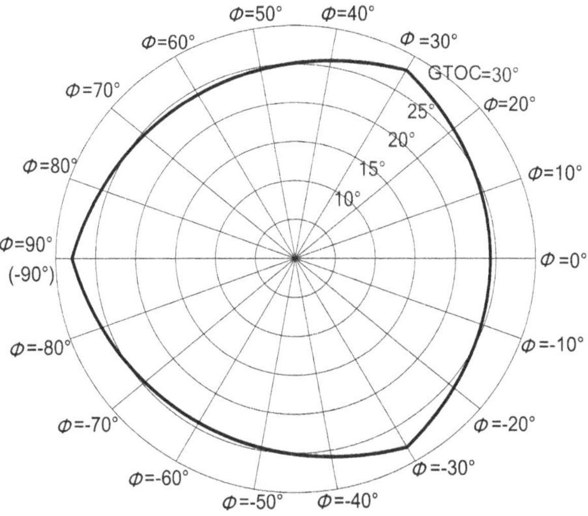

Fig. 8.27 Good-transmission orientation capability of a 3-PRS parallel mechanism

$$\$_{Tj} \circ \$_{Oi} = 0 \quad \left(j \in \{1,2,3\}, \text{and } j \neq i \right), \tag{8.24}$$

and

$$\$^r_{Tk} \circ \$_{Oi} = 0 \quad (k = 1,2,3). \tag{8.25}$$

Unit twist screw $\$_{Oi}$ can be determined on the basis of Eqs. (8.23), (8.24), and (8.25).

According to Eq. (7.57), the output transmission index of the ith leg can be obtained by

$$\eta_i = \frac{\left| \$_{Ti} \circ \$_{Oi} \right|}{\left| \$_{Ti} \circ \$_{Oi} \right|_{\max}}. \tag{8.26}$$

Therefore, the LTI of the 3-PRS tool head can be determined by $\chi = \min\{\lambda_i, \eta_i\}, \quad (i = 1, 2, 3)$.

The GTOC of the parallel mechanism can then be obtained by setting $\chi \geq 0.7$. Let us take the 3-PRS parallel mechanism with parameters $R = 2.5$, $r = 1.0$, and $L = 4.5$ as an example. By setting $\chi \geq 0.7$, we can numerically obtain the GTOC of the parallel mechanism. As shown in Fig. 8.27, the GTOC is much lower than the MOC, and the performance of the former is symmetrical. Similar to the MOC, the GTOC reaches its maximum at $\phi = 30°, -30°$, and $\pm 90°$ and its minimum at $\phi = 0, -60°$, and $60°$.

LTI χ can be used to evaluate only the effectiveness of the motion/force transmission of a 3-PRS parallel mechanism at a single pose. Usually, a 3-PRS parallel mechanism performs a task in a set of poses, not in a single pose. In practical design, whether a parallel mechanism is good should be decided by taking into account the behavior within the set of poses. To measure the global behavior of the motion/force transmission with the entire set of poses, we define the GTI following the definition of the GCI suggested by Liu et al. (2002):

$$\Gamma = \frac{\int_{\theta}\int_{\phi}\chi\,\mathrm{d}\phi\,\mathrm{d}\theta}{\int_{\theta}\int_{\phi}\mathrm{d}\phi\,\mathrm{d}\theta} \tag{8.27}$$

where n is the number of transmission angles, $-\text{GTOC} \le \theta \le \text{GTOC}$, and $-90^\circ \le \phi \le 90^\circ$ for the studied parallel mechanism; and $\Gamma_{\min} < \Gamma < 1$ ($\Gamma_{\min}$ equals 0.7). The GTI is also independent of any coordinate frame. A larger Γ usually indicates better motion/force transmissibility for all orientations.

Using the condition $\chi \ge 0.7$, we can identify an orientational workspace (i.e., the GTOC) that exhibits good force transmission and is far from singularity. The GTOCs of the 3-PRS parallel mechanisms with different link lengths may be the same. When such a case occurs, we cannot evaluate which mechanism is better with respect to the GTOC index itself. However, with the same GTOCs, their GTIs may be different. Indices GTOC and GTI in concern facilitate the optimal design of the parallel mechanism, as discussed in the succeeding section.

Atlases of the GTOC and GTI

According to the definition of the GTOC in Sect. 4.2, the GTOC of any non-dimensional mechanism in the PDS (Fig. 8.19) can be calculated. The atlas illustrating the relationship between normalized parameters r_i ($i = 1, 2, 3$) and the GTOC, in which $\chi \ge 0.7$, is illustrated in Fig. 8.28a. The figure shows that the index is inversely proportional to parameter r_3.

Figure 8.28b shows the GTI atlas, which indicates that a smaller parameter r_3 usually leads to a higher GTI, that is, better motion/force transmissibility for all orientations.

8.3 A General Procedure for Dimensional Synthesis

As pointed out in Sect. 8.1, every SM in mechanism space Π has its unique BSM in the PDS; one BSM in the PDS corresponds to infinite SMs (with different factor D) in space Π. The analysis results mentioned above show that the SMs and BSM are similar not only in size but also in performance. Regardless of how many parameters exist, the number of parameters can be reduced from n to $n-1$ using the PFNM.

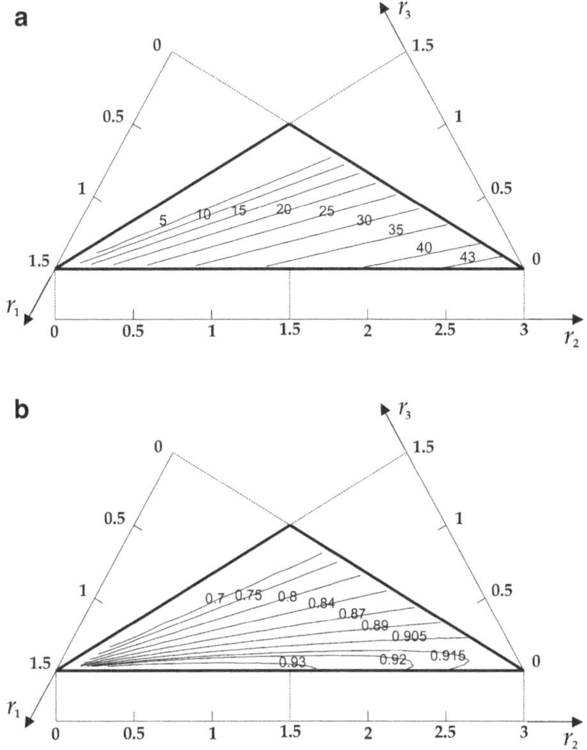

Fig. 8.28 Performance atlases of the 3-PRS parallel mechanism: (**a**) for the GTOC; (**b**) for the GTI

Moreover, the $n-1$ normalized parameters can be limited. These features provide useful information on the parameters regardless of the kind of design methodology the designer uses. Therefore, the PFNM can be applied to the kinematic design of a mechanism. For classical objective function-based optimal design, the method solves the problem of parameter limitation. Such a resolution can substantially reduce the costs involved in this optimal method. However, a detailed discussion of this matter is not provided in this chapter because of space limitations.

The PDS (or the PDS section) of a mechanism with fewer than five parameters can be expressed in a 2-dimensional space; hence, the relationship between a performance index and parameters (or all possible BSMs) can be illustrated graphically. Therefore, this mechanism can always be studied completely in a finite space, motivating us to propose a new design methodology for the aforementioned mechanism. We refer to this methodology as the *Performance Chart-Based Design Methodology* (PCbDM) and describe it as follows:

S1. All linear characteristic parameters $L_1, \ L_2, \ldots, \ L_n$ are identified from the Jacobian matrix.

S2. The involved design parameters are normalized using the PFNM proposed in this chapter, and the $(n-1)$-dimensional PDS is established.

S3. The relationships between the involved performance indices, especially that between the GCW and GCI over the GCW or that between the GTW and GTI over the GTW, are illustrated. The normalized parameters are also graphically depicted in the PDS.

S4. Performance charts are used to identify the optimum region that satisfies kinematic performance constraints.

S5. A candidate, that is, the BSM $(l_1, l_2, \ldots, l_n)$, is selected from the optimum region.

S6. The GCW or GTW of the BSM is determined.

S7. The normalization factor is calculated by comparing the GCW or GTW and the desired task workspace.

S8. The design parameters are determined using the normalization factor. The corresponding SM can then be constructed.

S9. The input parameters of the SM are calculated with respect to the task workspace.

S10. The involved performance indices and input parameters (when commercially available actuators are used) of the SM are validated. If all the performance indices are available and the input parameters are subject to the actuators, all the parameters can be determined. If not, we proceed to step **S5** and repeat steps **S5–S10**.

Notably, this design methodology is based on the performance chart. The optimum region in step **S4** is the intersection of several charts. Every designer has different design conditions, which cannot be predicted beforehand. Thus, the optimum region is nonunique. For the same reasons, the designer can adjust the BSM in step **S5** because the optimum region provides a nonunique solution to a design problem. Therefore, human interactions are inevitably used in the design process. All the design conditions in an integrated equation cannot be considered; compared with the mathematical optimal result or that generated by any automatic approach, only the design result in an application sense is truly optimal. Thus, human interactions are necessary in practical design. The PCbDM possesses the advantage of human interaction; it is described schematically in Fig. 8.29.

Note that in the PCbDM, the input parameter is not the optimized parameter. Input parameters determine the workspace. In most developed optimal design methods, therefore, input parameters are usually involved in the design process, thereby increasing the parameter number. In the optimal design, the workspace is not the only criterion. A GCW or GTW, which is part of the theoretical workspace, should be identified. Only such a workspace is acceptable in practice. Normally, the input parameters are not contained in the final Jacobian matrix or index expression. Thus, viewing the input parameters as the design parameters is unnecessary. Finally, the input parameters should be provided, a task that can be accomplished with respect to the prescribed task workspace after the characteristic parameters are optimized and determined. The theoretical workspace of a mechanism with variable-length rods (e.g., the mechanisms in Figs. 8.1b and 8.2b) is an infinite

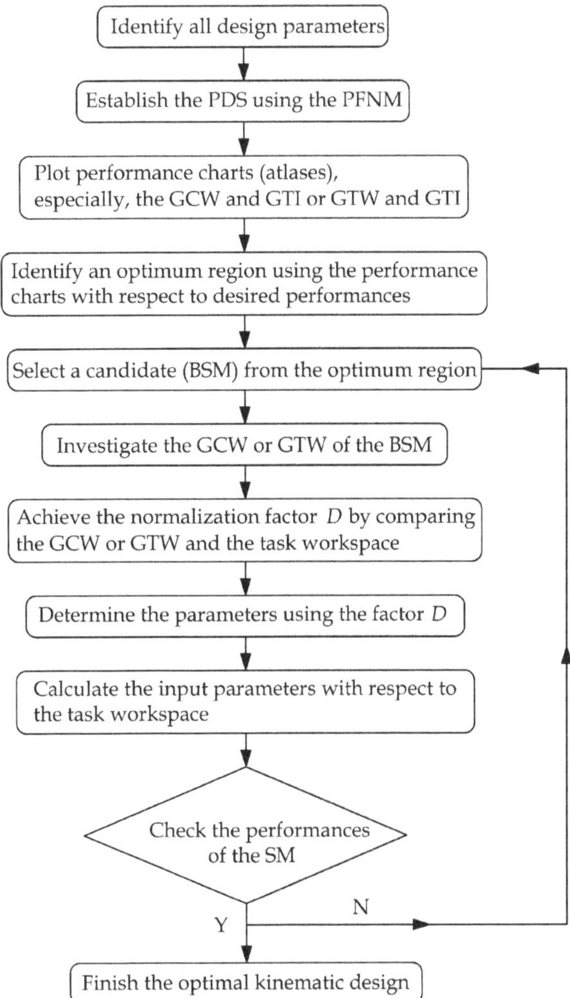

Fig. 8.29 Flowchart describing the PCbDM

region. The GCW or GTW of such a mechanism is possibly infinite. Experimentally, maximum length $\rho_{\max}$ should be less than three times minimum length $\rho_{\min}$ of a variable-length rod, considering the stiffness. Hence, for such a mechanism, minimum length $\rho_{\min}$ should first be determined with the nearest pose where the LCI is equal to the specified LCI. The GCW or GTW can be achieved by setting $\rho_{\max} = 3\rho_{\min}$.

In the PCbDM, steps **S6–S10** are applicable in the objective function-based optimal design after a BSM is achieved using the algorithm introduced by Stock and Miller (2004), Ryu and Cha (2003), and Chablat and Wenger (2003), among others.

For some parallel mechanisms, the angular parameter may be related to performance. For such a case, the angular parameter is also a design parameter. Given that the angular parameter cannot be normalized, the proposed design procedure does not include the design of such a parameter. However, this does not mean that the design method does not enable the optimal design of a parallel mechanism with angular parameters. From the design procedure, we can see that performance charts are highly important in implementing the optimal design. For the parallel mechanisms with angular parameters, we can use the method introduced by Liu et al. (2002) to illustrate the performance charts. Therefore, the design methodology is also applicable to the design of these mechanisms. Additionally, because it is possible to illustrate the performance chart of a serial robot after the design parameters are normalized, the PCbDM can also be applied to serial robots. All in all, the PCbDM can be used to design an optimal mechanism as long as the performance charts of a mechanism can be presented, and the performance levels of the original mechanism and its normalized mechanism are similar to each other.

A performance chart can globally and visually show the relationship between a performance index and associated design parameters in a limited space. Hence, performance charts can illustrate how antagonistic the involved criteria are. Therefore, the performance chart method can be regarded as an ideal method. Compared with the result achieved from the objective unction-based method, the optimum result is fuzzy. However, the performance chart method is more flexible because the optimal design method does not provide a unique solution to a design problem. Thus, the designer can adjust the optimum result according to his design conditions.

8.4 Examples

8.4.1 Jacobian-Matrix-Based Design

8.4.1.1 Planar 5R Parallel Mechanism

The relationships between the performance indices and link lengths of the planar 5R parallel mechanism have been examined in this chapter. The results are illustrated by their atlases, from which one can visually determine which mechanism can exhibit better performance and which cannot. Determining mechanism performance is highly important in identifying a global optimum mechanism for a specified application. In this section, the optimum region with respect to possible performance is shown first.

Before presenting the determination of mechanism dimensions, we explain the reasonability of the definition and application of two issues: the usable workspace and the importance of the MIW. In Sect. 6.1, a singularity-free region above the x-axis is defined as the *usable workspace* of a planar 5R parallel mechanism. According to the definition, a usable workspace is not the workspace used in practice

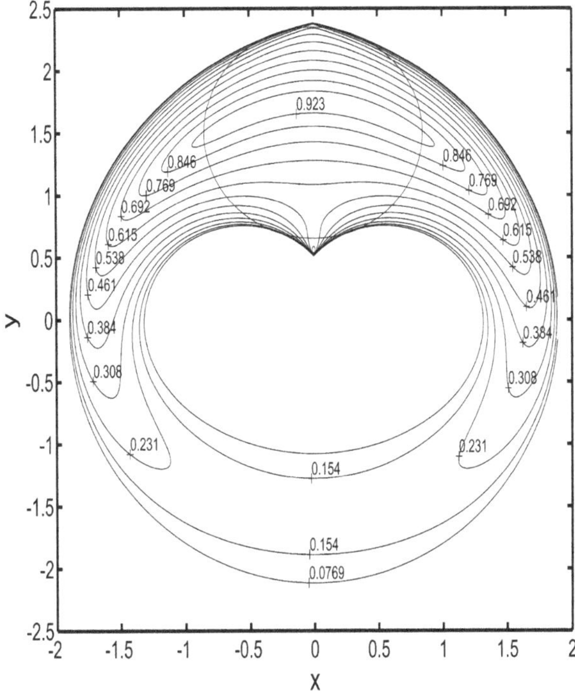

Fig. 8.30 Sample distribution of the LCI in the entire workspace

by a device. In fact, only part of the usable workspace can be available to a device. This workspace can be referred to as the *practical workspace* or GCW, which can be restricted by the LCI. The specified index value can be determined by the designer. Figure 8.30 shows the distribution of the LCI on the entire singularity-free workspace of the mechanism with $r_1 = 0.85$, $r_2 = 1.6$, and $r_3 = 0.55$. The figure illustrates that the LCI within the region where $x < 0.0$ is worse. Hence, if the region restricted by $1/\kappa_J = 0.5$ is considered the GCW, the space when $x < 0.0$ will be excluded. Most mechanisms (except those with $r_3 = 0$) in subregions IV_a and V_a have similar characteristics. Although there are two or more good-conditioning regions, these are separated by points with worse LCIs for the mechanisms in subregions IV_b and V_b. Thus, the usable workspace defined as the region where $x > 0.0$ is closer to the practical workspace than to the entire workspace.

As for the importance of the MIW, the atlas of the MIC radius instead of that of the usable workspace is presented in Sect. 8.2. For the 5R parallel mechanism, the global indices are also defined with respect to the MIW, not the usable workspace. To elucidate, let us examine two 5R parallel mechanisms. The first has the parameters $r_1 = 0.4$, $r_2 = 2.33$, and $r_3 = 0.27$, in which the usable workspace has an area of 3.3453 (Fig. 8.31a). Its MIC radius is $r_{\text{MIC}} = 0.4$. The

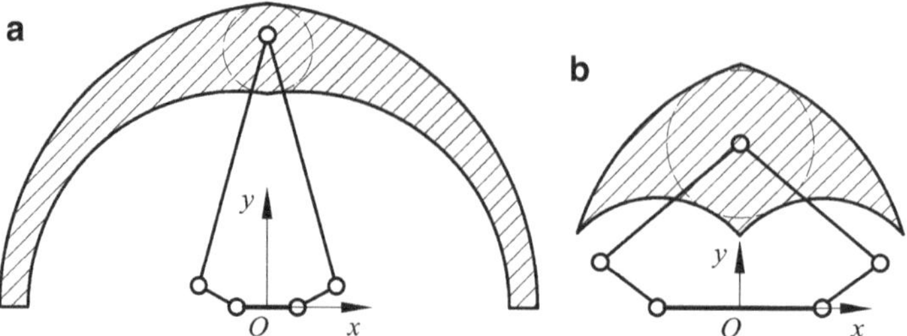

Fig. 8.31 Workspaces of two sample mechanisms

second mechanism has the parameters $r_1 = 0.65$, $r_2 = 1.62$, and $r_3 = 0.73$. Its usable workspace is shown in Fig. 8.31b, its area is 2.2390, and its MIC radius is $r_{\mathrm{MIC}} = 0.65$. The first mechanism has a larger usable workspace area than does the second mechanism. However, the MIC radius of the first mechanism is shorter than that of the second. For the two mechanisms, the distribution of the LCI on the usable workspace is shown in Fig. 8.32a, b, respectively. The figures clearly show that the GCW (restricted by $1/\kappa_J = 0.5$) of the second mechanism is larger than that of the first. In fact, there is no such GCW for the first mechanism. Generally, a flat workspace means that most points are far from the singular loci. A thin one indicates that most points are near singularity. Our concern is how useful the workspace is, but not how large the area is. These two examples show that a large usable workspace does not translate to a large GCW. Nevertheless, if the MIW is large, the GCW will be large as well. Therefore, the MIC or the MIW can characterize the workspace performance, which is why we present the atlas of the MIW and define the global indices with respect to the workspace. Although the MIW is more useful, the usable workspace cannot be replaced and disregarded. This is our research object as well because the final GCW of an actual mechanism is part of the usable workspace.

Additionally, Figs. 8.30 and 8.32 show that the distribution of the LCI is continuous. The MIW includes all points with the LCI values from the minimum (zero) to the maximum. The point with the maximum (best) LCI is always within the MIW. Because of this continuous distribution, we can conclude that an index is good within the region near the MIW when it exhibits good performance within it. This is why the MIW can be used as a reference in globally evaluating performance.

Optimum Region Considering the Workspace and GCI

Most designs usually consider the workspace and GCI. On the basis of the atlas of the MIC radius (Fig. 8.16) and the distribution chart of the MIC in the design space (Fig. 8.15), we can see that the workspace of a mechanism in subregions $\mathrm{IV_a}$, $\mathrm{IV_b}$, $\mathrm{V_a}$, and $\mathrm{V_b}$ can be large. The atlas of the GCI (Fig. 8.17) shows that the mechanism can have a better GCI when $r_2 \in [1.56,\ 1.9]$. When the MIC radius,

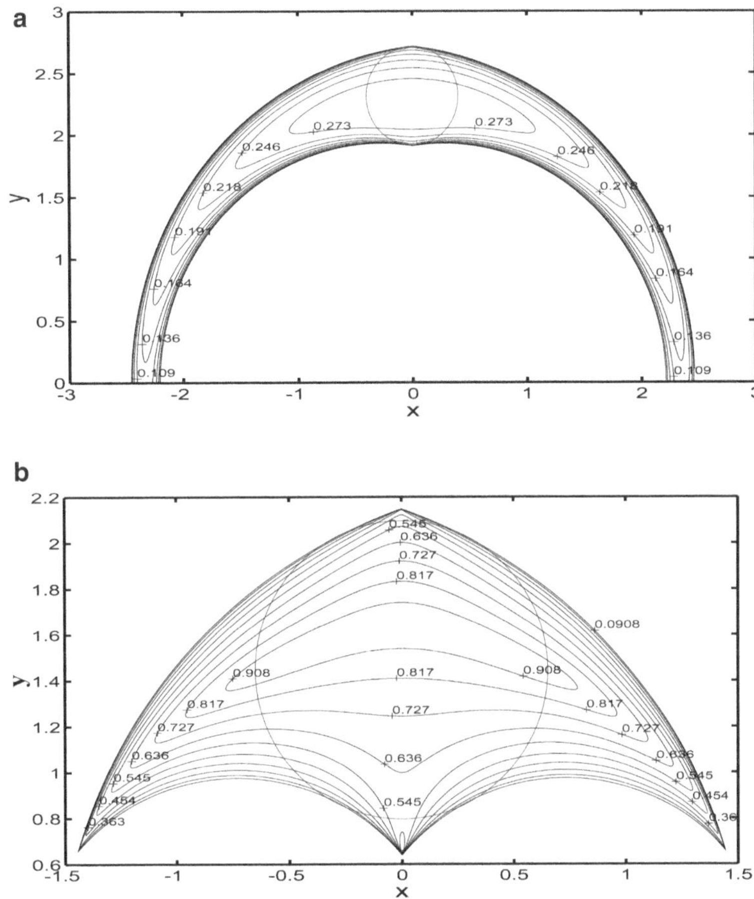

Fig. 8.32 LCI distribution for two comparative mechanisms with usable workspace areas of (**a**) 3.3453 and (**b**) 2.2390

denoted as r_{MIC}, is greater than 0.8 ($r_{\mathrm{MIC}} > 0.8$) and the GCI is greater than 0.6, the optimum region in the design space can be obtained (see Fig. 8.33a). The region is denoted as $\Omega_{\mathrm{W-GCI}} = [(r_1,\ r_2,\ r_3)\,|r_{\mathrm{MIC}} > 0.8 \text{ and } \eta_J > 0.6]$, with performance restriction. One can also obtain an optimum region with a better workspace and GCI. An example is region $\Omega'_{\mathrm{W-GCI}}$ where $r_{\mathrm{MIC}} > 1.0$ and $\eta_J > 0.7$ (Fig. 8.33b). To derive a better result, we can decrease the optimum region with more stringent restrictions. Such a region comprises all possible optimum results.

After the optimum region is identified, there are two ways to achieve the optimum design result with non-dimensional parameters. One is to search for the most optimal result within region $\Omega_{\mathrm{W-GCI}}$ or $\Omega'_{\mathrm{W-GCI}}$ using a classical searching algorithm based on an established objective function. This method yields a unique

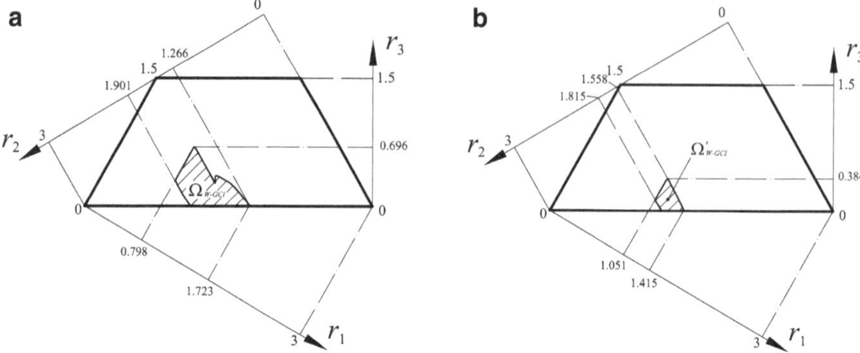

Fig. 8.33 Optimum region with a specified MIC and good GCI

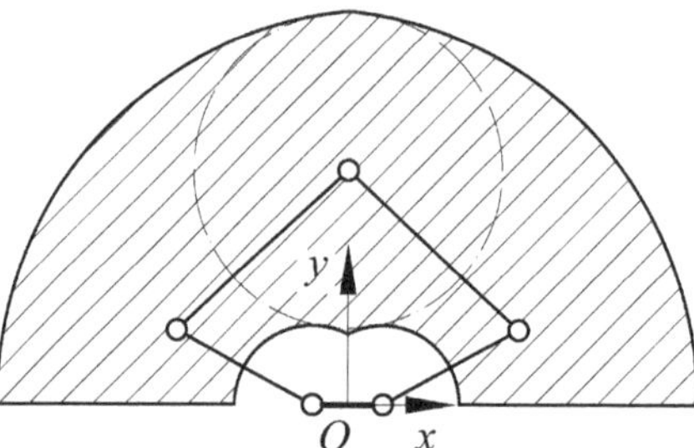

Fig. 8.34 Example in optimum region Ω'_{W-GCI}

solution, but is not the focus of this chapter. The other approach is to select a mechanism within the obtained optimum region. For example, the mechanism with $r_1 = 1.1$, $r_2 = 1.65$, and $r_3 = 0.25$ can be selected as the candidate if only the workspace and GCI are involved in the design. Its MIC radius and GCI are $r_{MIC} = 1.1$ and 0.7929, respectively. The mechanism is in subregion IV_a because $r_2 > r_1 + r_3$ and $r_1 > r_3$. No singularity is observed, that is, no collinearity among points B_1, P, and B_2 is found in the workspace. The mechanism and its workspace are shown in Fig. 8.34. The advantage of the second method is that the designer can adjust the design result appropriately by selecting another candidate in the optimum region.

Optimum Region Considering the Workspace, GCI, and GSI

Because a mechanism with smaller $\eta_{D\max}$ and $\eta_{D\min}$ values usually exhibits better stiffness, the mechanisms in subregions III_a and IV_a, where $r_2 > r_1 + r_3$, have better performance. Some mechanisms in subregion III_a are excellent in terms of stiffness and the GCI but perform poorly when it comes to workspace.

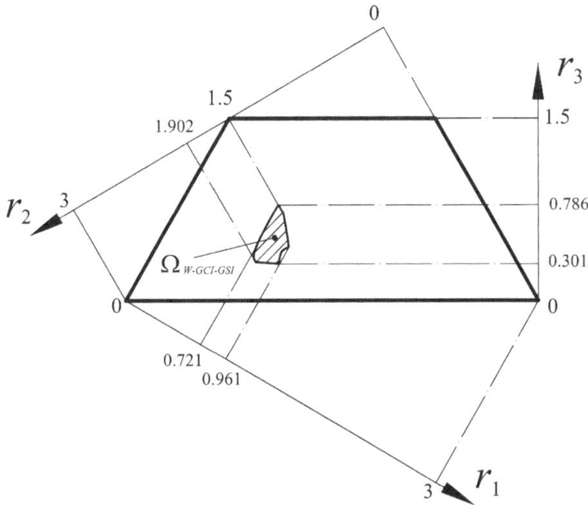

Fig. 8.35 Optimum region with a specified MIC, GSI, and good GCI

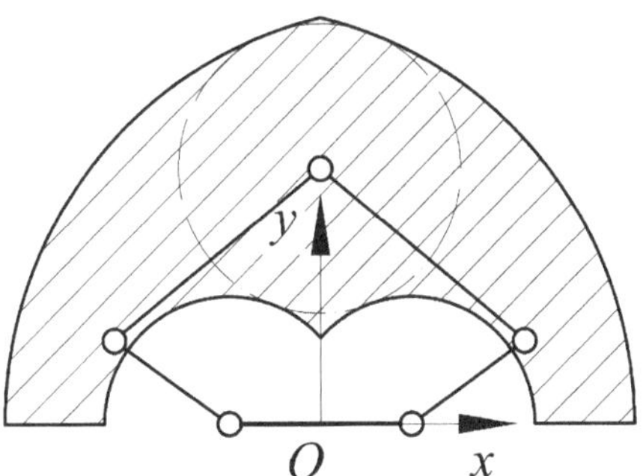

Fig. 8.36 Example in optimum region $\Omega_{\text{W-GCI-GSI}}$

These mechanisms can be employed in field applications that require a small workspace. If the workspace, conditioning index, and stiffness are all considered in the design, the mechanism can be selected from an optimum region $\Omega_{\text{W-GCI-GSI}} = [(r_1,\ r_2,\ r_3)\,|r_{MIC}\,(r_1,\ r_2,\ r_3) > 0.7,\ \ \eta_J > 0.6,\ \ \eta_{D\min} < 0.7$ and $\eta_{D\max} < 1.0]$, as shown in Fig. 8.35. The non-dimensional parameters of one example within the region are $r_1 = 0.85$, $r_2 = 1.6$, and $r_3 = 0.55$. Its indices are $r_{\text{MIC}} = 0.85$, $\eta_J = 0.7509$, $\eta_{D\max} = 0.8133$, and $\eta_{D\min} = 0.5475$. Because the mechanism is in subregion IV_{a}, no singularity occurs (i.e., the workspace does not exhibit collinearity among points B_1, P, and B_2). Figure 8.36 shows the mechanism and its workspace.

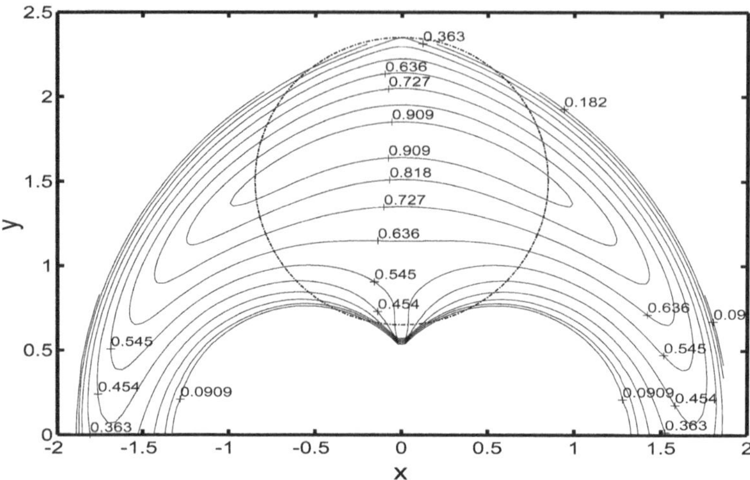

Fig. 8.37 LCI distribution on the usable workspace of the optimum mechanism where $r_1 = 0.85$, $r_2 = 1.6$, and $r_3 = 0.55$

Dimension Determination

The final objective of optimum design is to determine the link lengths of a mechanism. In the previous section, some optimum regions in the design space have been presented as examples. These regions consist of mechanisms with non-dimensional parameters. The selected optimum mechanisms with non-dimensional parameters show comparative results. They are not the final optimum design results. Their workspaces may be so small that they cannot be used in practical applications. Moreover, the defined MIW still consists of singular points. In the design process, the points where the desired indices perform poorly should be excluded from the workspace. In this section, a desired workspace is provided beforehand. The link lengths of a mechanism that can work within this workspace with better (best) performance are determined.

As an example on how to determine real dimensional parameters with respect to a non-dimensional optimum mechanism, we consider the mechanism with $r_1 = 0.85$, $r_2 = 1.6$, and $r_3 = 0.55$. This mechanism belongs to optimum region $\Omega_{\mathrm{W-GCI-GSI}}$, where the workspace, conditioning index, and stiffness are considered in the design objective. The MIC radius of the mechanism is $r_{\mathrm{MIC}} = 0.85$. As mentioned, the MIW consists of singular points and their neighborhoods, where control over the mechanism is lost. For the example, only stationary singularity is observed in the usable workspace. As shown in Fig. 8.37, the LCI is zero at the singular points and very small near these points. However, the stiffness of the mechanism at these points is high (Fig. 8.38). Therefore, in determining the

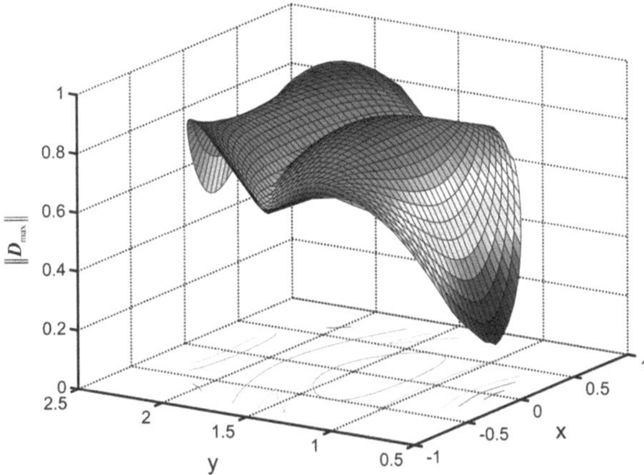

Fig. 8.38 Distribution of $\|D_{\max}\|$ on the MIW of the optimum mechanism where $r_1 = 0.85$, $r_2 = 1.6$, and $r_3 = 0.55$

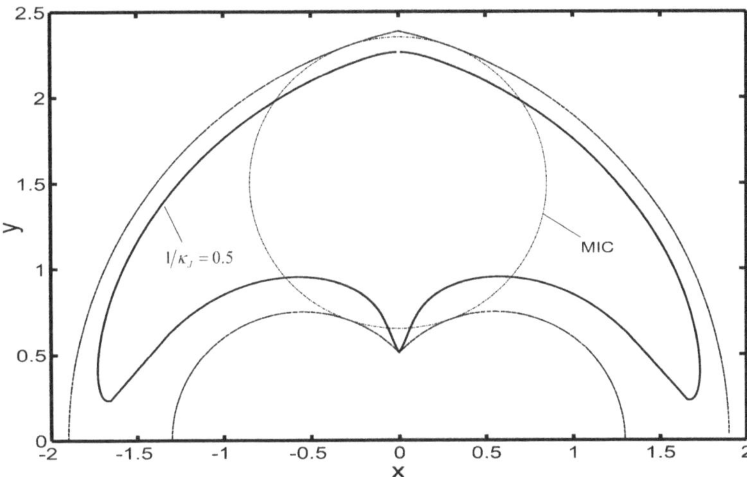

Fig. 8.39 GCW restricted by a specified LCI ($1/K_J = 0.5$) for an optimum mechanism where $r_1 = 0.85$, $r_2 = 1.6$, and $r_3 = 0.55$

dimensional parameters with respect to a desired workspace, the LCI can be the only performance index considered.

The process involved in determining the dimensions of the non-dimensional optimum mechanism with respect to a desired workspace can be summarized as follows:

Step 1: The GCW with a specified LCI is obtained. The example in Fig. 8.39 shows a given LCI of $1/\kappa_J = 0.5$, and the GCW of the mechanism is bounded

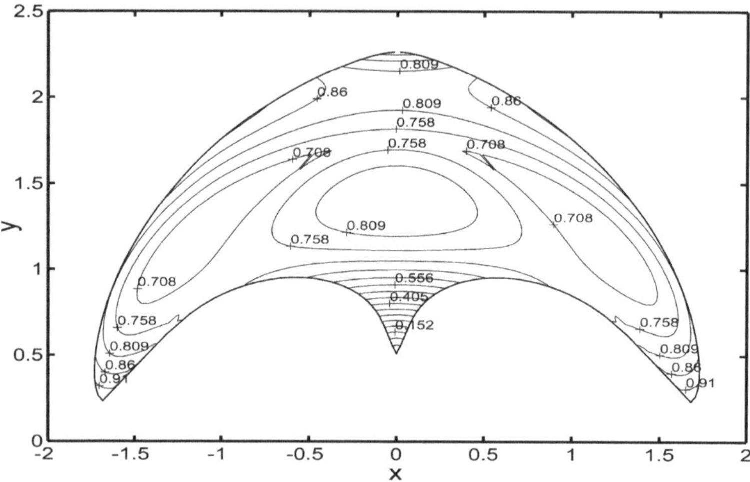

Fig. 8.40 Distribution of $\|D_{\max}\|$ on the GCW for an optimum mechanism where $r_1 = 0.85$, $r_2 = 1.6$, and $r_3 = 0.55$

by curve $1/\kappa_J = 0.5$. The area of the workspace is 3.4817. The GCI within this workspace is $\eta_J = 0.6686$. This step also allows for the verification of stiffness performance. Figure 8.40 shows the distribution of $\|\mathbf{D}_{\max}\|$ in the workspace. With respect to this workspace, $\eta_D = 0.7656$.

Step 2: Dimensional factor D, which is used to change the dimensional geometric parameters of a mechanism to non-dimensional types, is determined. The ratio of workspace W_{Dim} of a dimensional mechanism to $W_{\text{Non-Dim}}$ of a non-dimensional mechanism is D^2, that is,

$$W_{\text{Dim}} = D^2\, W_{\text{Non-Dim}}. \tag{8.28}$$

The GCW of the non-dimensional optimum mechanism is obtained in *Step 1*. If the workspace of a dimensional mechanism is given with respect to the design specification, factor D can easily be derived from Eq. (8.28). For example, if the objective workspace has a shape identical to that of the GCW (Fig. 8.39) and the workspace area is 150 mm^2, factor D can be obtained as $D = \sqrt{150/3.4817} \approx$ 6.6mm.

Step 3: The dimensional parameters of the optimum mechanism are derived using dimensional factor D. The relationship between a dimensional parameter and a non-dimensional one is $R_i = D\, r_i$. When D is determined, R_i can be achieved. For the example above, $R_1 = 5.61$ mm, $R_2 = 10.56$ mm, and $R_3 = 3.63$ mm. This step also allows for the verification of dimensional mechanism performance. Figure 8.41 shows the distribution of the LCI and stiffness index in the workspace, illustrating that the distribution is similar to that when the mechanism is non-dimensional. No change in the conditioning index is observed, and the GCI is still equal to 0.6686.

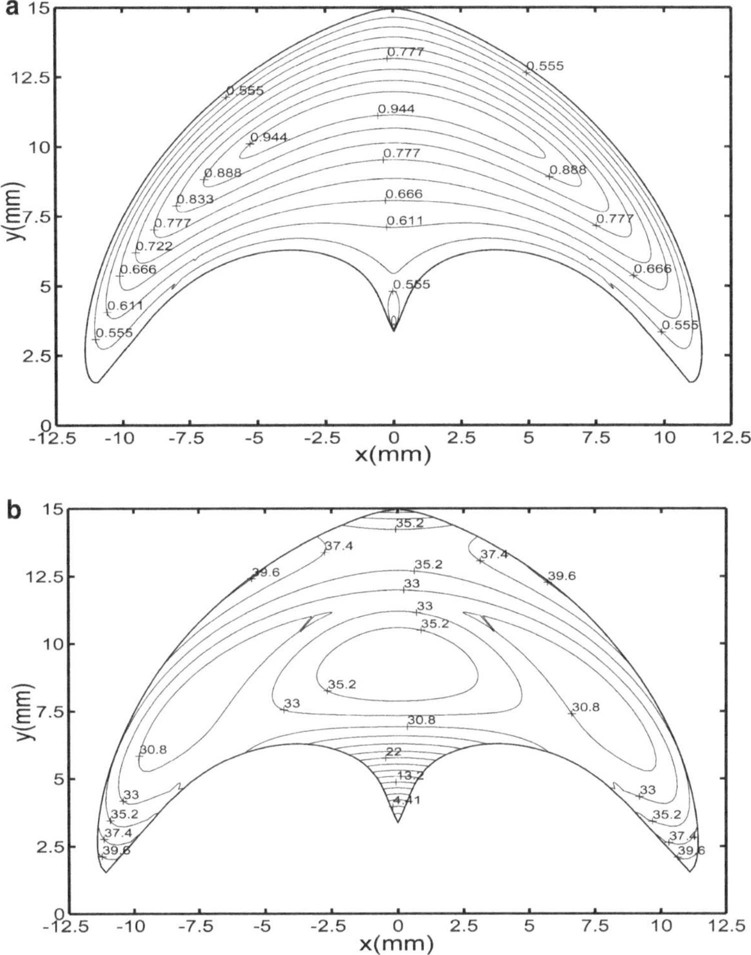

Fig. 8.41 Distribution of performance indices on the desired workspace for the dimensional optimum mechanism: (**a**) $1/K_J$; (**b**) $\|D_{\max}\|$

The stiffness index value is different, i.e., $\eta_D = 33.3475$. Hence, factor D changes the value of the stiffness index, but not the conditioning index and distribution of all the indices.

Step 4: The input limit for θ_1 and θ_2 is calculated. These can be computed from the inverse kinematic equation. For the example, the limits are $\theta_1 \in [36.40°, \ \ 274.97°]$ and $\theta_2 \in [-94.97°, \ \ 143.60°]$.

In *Step 2*, the desired workspace can be any shape and should be determined by the design specification. Regardless of workspace shape, we can always identify a similar downsized workspace. The downsized workspace, which is symmetrical about the y-axis, should be the maximal workspace included in

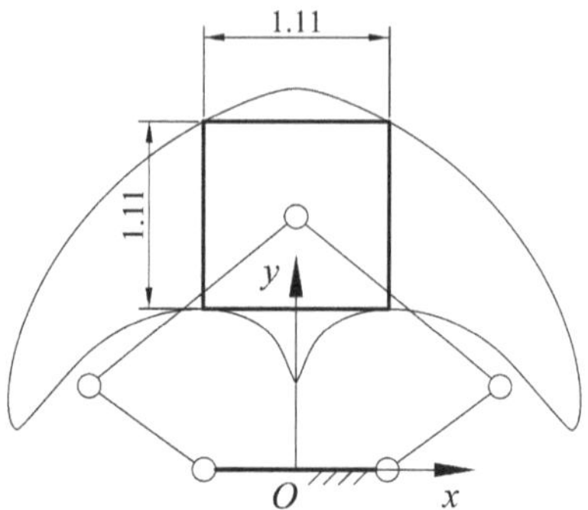

Fig. 8.42 Maximum square included in the GCW

the obtained GCW. For example, if the desired workspace is a square with an area of 10 mm × 10 mm = 100 mm^2, we can find a maximal square with an area of 1.11 × 1.11 = 1.2321 (Fig. 8.42) in the GCW. Factor D should be 9.1 mm. The dimensional parameters of the design are $R_1 = 7.74$ mm, $R_1 = 14.56$ mm, and $R_1 = 5.0$ mm. The input limits are $\theta_1 \in [84.65°, 216.0°]$ and $\theta_2 \in [-36.0°, 95.35°]$.

This section shows that all mechanisms with parameters $R_i = D\,r_i$ exhibit performance similar to that with non-dimensional parameters r_i. For example, the workspace of the mechanism with R_i is D^2-times that of the mechanism with r_i. Within the similar workspaces, the GCI values are the same as those of the other workspaces. The other performance indices are the same in terms of distribution in the workspaces and differ only in quantity. The mechanisms that exhibit such performance similarity are referred to as SMs. Hence, the ratios of geometric parameters R_i of all SMs are constant. Any one of the non-dimensional mechanisms in the established design space stands for all of its possible SMs. After normalization, a mechanism with any kind of link length can find its position in the design space. For this reason, the design space is a useful tool for presenting a global comparative result in terms of the performance of all mechanisms. This feature also guarantees a global optimum result in the design issue.

8.4.1.2 Spatial 3-PRS Parallel Mechanism

Suppose that an application requires a parallel mechanism to have a desired orientation capability of 40° with any axis in the *o*-*xy* plane and higher accuracy. The optimization process based on the atlases in Fig. 8.20 can be summarized as follows.

Step 1: An optimum region in the PDS is identified. According to the definition of the OCI, some orientations near singularity cause loss of control over a mechanism, and the accuracy near these orientations is poorer. This definition indicates that if we use the desired orientation capability to identify a mechanism in the atlas in Fig. 8.20a, the parallel mechanism will be unable to reach this capability when used

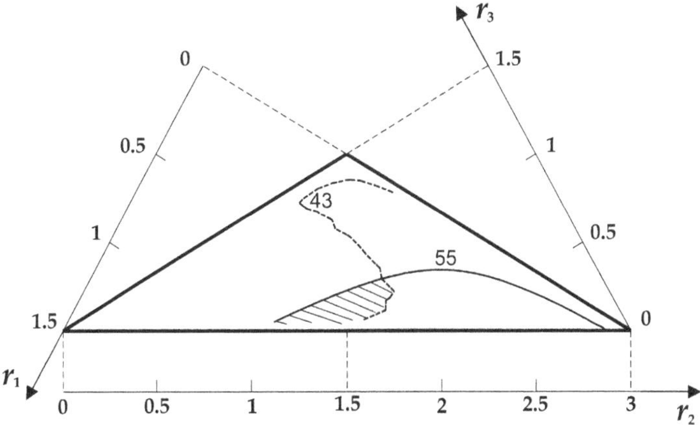

Fig. 8.43 Identified optimum region when OCI >55° and $E_{e\_max} < 43$ μm

in practical situations. Consequently, we use the constraint OCI > 55° (we allow for a 15° safety buffer) to identify the optimum region. The error constraint is defined as $E_{e\_max} < 43$ μm. With the OCI and error constraints, an optimum region, denoted as Ω_{OCI-E}, can be identified (shown as the hatched region in Fig. 8.43) using the atlases in Fig. 8.20. The optimum region contains all possible solutions with parameters r_i ($i = 1, 2, 3$) that are subject to the design requirement.

Step 2: A solution candidate is selected from the optimum region. This region contains all possible solutions for the design. Given that no best solution exists, only a comparatively better one, we can choose any non-dimensional mechanism from this region. An example is the mechanism with $r_1 = 0.65$, $r_2 = 1.5$, and $r_3 = 0.85$. The OCI and error index E_{e_max} of this mechanism are 58.55° and 42.5 μm, respectively.

Step 3: Dimensional parameters r, L, and R are determined. We first need to determine normalization factor D, which can be obtained by comparing the desired workspace of a design problem to the workspace of the non-dimensional mechanism selected from the optimum region. For the 3-PRS parallel mechanism considered in this chapter, any performance is independent of the positional workspace along the z-axis; the orientational workspace is dependent on the ratio of related linear parameters, but not on any one parameter. Therefore, we cannot determine factor D with respect to the orientational workspace. In this case, we should first determine parameter r according to practical applications and ensure that the parameter is as small as possible to reduce the size of the mechanism. We assume that $r = 200.00$ mm. Thus, we obtain $D = r/r_1 = 200/0.65 \approx 307.70$ mm. Accordingly, we derive $L = Dr_2 \approx 461.55$ mm and $R = Dr_3 \approx 261.55$ mm.

Step 4: The error of the dimensional mechanism is examined, and if necessary, the design solution is revised. In the previous step, the dimensional parameters ($r = 200.00$ mm, $L = 461.55$ mm, and $R = 261.55$ mm) are determined. In this step, we analyze the error/accuracy of the solution for the desired orientation capability.

Before doing so, however, distance l between point E and o should be determined. We can assume that point E is the end point of the tool mounted on the mobile platform. In our design problem, the desired orientation capability is 40°. For $\varphi \in [-90°,\ 90°]$ and $\theta \in [-40°,\ 40°]$, numerical calculations show that error index E_{e_max} of point E for input error $\varepsilon = 10\ \mu m$ is 9.0 μm when $l = 100$ mm. If the error is acceptable, we proceed to Step 5. If the error is greater than expected, we return to Step 2, choose another group of non-dimensional parameters from the optimum region with a smaller error, and repeat Steps 3 and 4.

Step 5: The input range is calculated. The input range required to reach an orientation capability of 40° can be calculated using Eq. (8.14). For the solution obtained in Step 4, we have $\rho_i = [-570.9079\ \text{mm}, -313.7929\ \text{mm}]\ (i = 1, 2, 3)$ when $z = 0$. The final input range for each actuator should include the positional workspace along the z-axis.

Note that if the input range does not correspond to an off-the-shelf actuator, the designer can return to Step 2 and choose another non-dimensional mechanism to adjust the design, or return to Step 1 and identify another optimum region.

8.4.2 Motion/Force Transmissibility-Based Design

8.4.2.1 Planar 5R Parallel Mechanisms

The objective of optimization is to determine the geometrical parameters of a mechanism for a desired workspace. In this process, other performance indices may be considered. For the parallel mechanism studied here, the optimization, which takes into account the desired workspace and force transmissibility on the basis of the performance charts in Sect. 8.2.2.1, can be summarized as follows:

Step 1: An optimum region in the PDS is identified. The identification of the optimum region is dependent on the design requirement. Different practical designs may have varied optimum regions. Nevertheless, the mechanism candidate that has the largest possible workspace is typically the desirable candidate. In this case, we can identify an optimum region with a large GTW and better GTI. For example, setting GTW $\geq$ 5.0 and GTI $\geq$ 0.925, an optimum region denoted as $\Omega_{\text{GTW-GTI}}$ can be identified [shown as the hatched region in Fig. 8.44a) using the performance charts in Figs. 8.22 and 8.23. Optimum region $\Omega_{\text{GTW-GTI}}$ contains all possible optimum solutions with non-dimensional parameters r_i ($i = 1$, 2, 3). However, the optimum region is nonunique and can be changed under different constraints. For example, reducing the optimum region necessitates more stringent constraints, such as GTW $\geq$ 6.0 and GTI $\geq$ 0.93. The optimum region denoted as $\Omega'_{\text{GTW-GTI}}$ is shown in Fig. 8.44b.

For a practical design that requires a regular workspace, the fatness of the GTW should be considered. The corresponding optimum region can be identified using the performance charts in Figs. 8.22, 8.23, and 8.25. For example, when

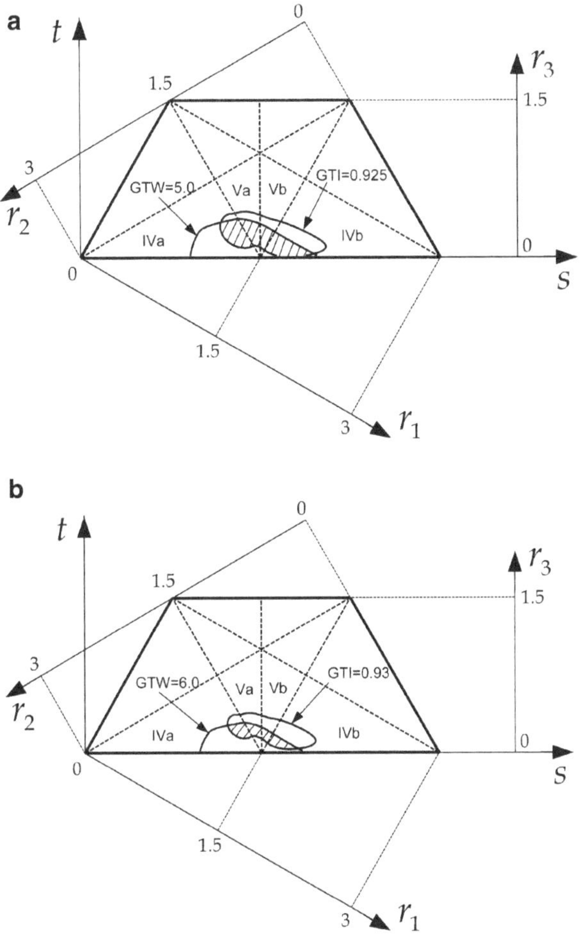

Fig. 8.44 Optimum regions for the 5R parallel mechanism when (**a**) GTW $\geq$ 5.0 and GTI $\geq$ 0.925, and (**b**) GTW $\geq$ 6.0 and GTI $\geq$ 0.93

GTW $\geq$ 6.0, GTI $\geq$ 0.93, and FI $\geq$ 1.2, an optimum region denoted as $\Omega_{\text{GTW-GTI-FI}}$ can be obtained (Fig. 8.45).

Step 2: A solution candidate, that is, the BSM, is selected from the optimum region. The optimum region obtained contains all possible solutions with better performance. Only a comparatively better solution to a design problem exists; hence, we can choose any non-dimensional mechanism from the region. For example, the BSM with parameters $r_1 = 1.25$, $r_2 = 1.49$, and $r_3 = 0.26$ is selected from optimum region $\Omega_{\text{GTW-GTI-FI}}$. The GTW, GTI, and FI of the mechanism are 6.4125, 0.9337, and 1.3911, respectively. Figure 8.45 shows the mechanism, its GTW, and the distribution of the LTI within the GTW.

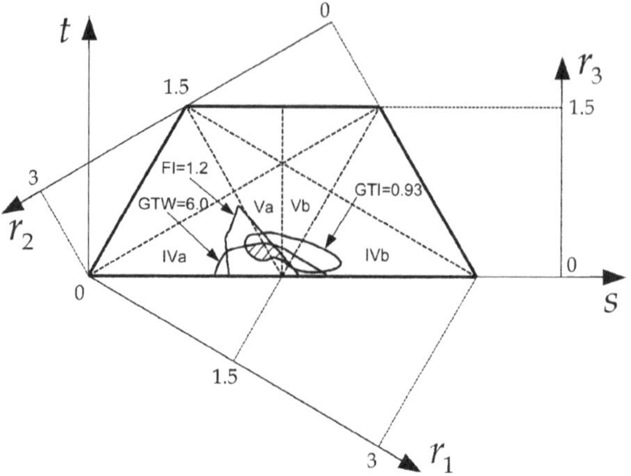

Fig. 8.45 Optimum region for the 5R parallel mechanism when GTW $\geq$ 6.0, GTI $\geq$ 0.93, and FI $\geq$ 1.2

Step 3: Dimensional parameters R_i are determined, but first, we ascertain normalization factor D. This factor can be obtained by comparing the desired workspace of a design problem with the workspace of the non-dimensional mechanism selected from the optimum region. Therefore, the magnitude of D depends on the desired workspace. The factor can be calculated once the desired workspace is given. For example, when the desired workspace is similar to the GTW shown in Fig. 8.46 and its volume $\text{GTW}_\text{D} = 200\ \text{mm}^2$ because $\text{GTW}_\text{non-D} = 6.4125$, then $D = \sqrt{\text{GTW}_\text{D}/\text{GTW}_\text{non-D}} \approx 5.5847$ mm. With $R_i = D\,r_i$, we have $R_1 = 5.5847\ \text{mm} \times 1.25 \approx 6.98$ mm, $R_2 = 5.5847\ \text{mm} \times 1.49 \approx 8.32$ mm, and $R_3 = 5.5847\ \text{mm} \times 0.26 \approx 1.45$ mm.

For the example above (Fig. 8.46), an included rectangular workspace with a volume of $2.2072(\text{length}) \times 0.9815(\text{width}) \approx 2.16$ can be identified in the GTW. Hence, a similar workspace must exist in the entire GTW of the dimensional mechanism with $R_1 = 6.98$ mm, $R_2 = 8.32$ mm, and $R_3 = 1.45$ mm. The volume of this workspace should be $D^2 \times 2.17 = (D \times 2.2072) \times (D \times 0.9815) \approx 67.56\ \text{mm}^2$, indicating that obtaining a rectangular workspace in practical design necessitates the identification of an included rectangle in the GTW of the selected BSM. The identification is carried out by considering the desired workspace.

Step 4: Whether the dimensions obtained in Step 3 are sufficient for practical design is verified, and if necessary, the design solution is revised. In the previous step, the dimensional parameters ($R_1 = 6.98$ mm, $R_2 = 8.32$ mm, and $R_3 = 1.45$ mm) are determined. In this step, the dimensions should be examined to ensure that all components are well incorporated and that any other item needed in the practical design problem is feasible. Once everything is in place, we proceed to Step 5; otherwise, we return to Step 2 and choose another group of non-dimensional parameters from the optimum region, and repeat Steps 3 and 4.

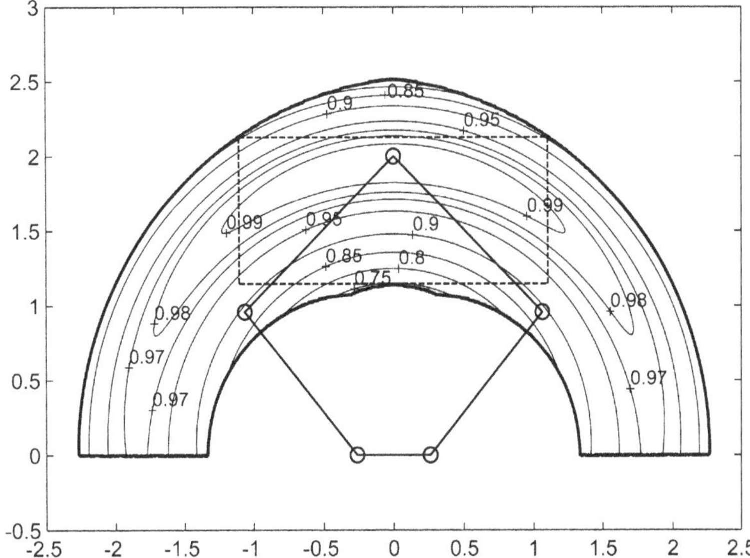

Fig. 8.46 Mechanism, GTW, LTI distribution, and included rectangular workspace of an optimum BSM

Step 5: The input range is calculated. The input range required to realize the desired workspace of the optimized mechanism can be calculated by inverse kinematics. For example, for the solution obtained in Step 3, the input ranges are $\theta_1 \in [-13.68°,\ 97.64°]$ and $\theta_2 \in [82.36°,\ 193.68°]$ for the desired workspace of $12.3265(\text{length}) \times 5.4813(\text{width}) \approx 67.56\ \text{mm}^2$.

Step 6: The solution to the optimization problem is implemented.

The design method can provide all possible solutions because the above-mentioned process not only provides more than one solution to an optimum region but also enables the designer to adjust his design solution.

8.4.2.2 Spatial 3-<u>P</u>RS Parallel Mechanism

Suppose that an application requires tilting of the mobile platform to ±40° about any axis in the *o-xy* plane and good motion/force transmissibility. The optimization process based on the atlases in Fig. 8.28 can be summarized as follows:

Step 1: An optimum region in the PDS is identified. When GTOC $\geq$ 40° and $\Gamma \geq 0.92$, an optimum region, denoted as $\Omega_{\text{GTOC-GTI}}$, can be identified (shown as the hatched region in Fig. 8.47) using the atlases in Fig. 8.28. The optimum region contains all possible solutions with non-dimensional parameters r_i $(i = 1, 2, 3)$, which are subject to the design requirement. Notably, the optimum region is nonunique and can be changed under different constraints.

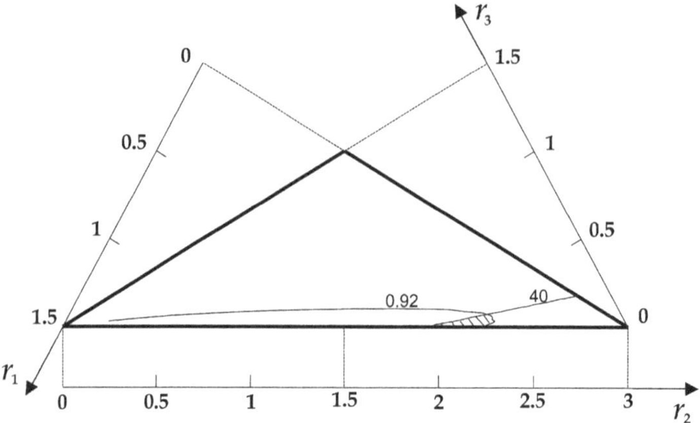

Fig. 8.47 Optimum region for the 3-PRS parallel mechanism when GTOC $\geq 40°$ and $\Gamma \geq 0.92$

Step 2: A solution candidate is chosen from the optimum region. The optimum region obtained, $\Omega_{\text{GTOC-GTI}}$, contains all possible solutions for the design. Because no best solution for a design problem exists (only a comparatively better one), we can choose any non-dimensional mechanism from this region. For example, the mechanism with parameters $r_1 = 0.467$, $r_2 = 2.033$, and $r_3 = 0.5$ is selected in this study. The GTOC and GTI Γ of the mechanism are 40.1° and 0.9238, respectively.

Step 3: Dimensional parameters *r*, *L*, and *R* are determined, but we first derive normalization factor *D*, which can be obtained by comparing the desired workspace of a design problem with the workspace of the non-dimensional mechanism selected from the optimum region. For the 3-PRS parallel mechanism considered here, any performance is independent of the positional workspace along the *z*-axis, and the orientational workspace is dependent on the ratio of related linear parameters, but not on any one parameter. Therefore, we cannot determine factor *D* with respect to the orientation. In this case, we ascertain parameter *r* according to practical applications and ensure that the parameter is as small as possible to reduce the space occupied by the mechanism. We assume that $r = 200$ mm. Hence, we have $D = r/r_1 = 200/0.467 \approx 428.26$ mm. Accordingly, we obtain $L = Dr_2 \approx 870.66$ mm and $R = Dr_3 \approx 214.13$ mm.

Step 4: Whether the dimensions obtained in Step 3 are sufficient for practical design is determined, and if necessary, the design solution is revised. In the previous step, the dimensional parameters ($r = 200$ mm, $L = 870.66$ mm, and $R = 214.13$ mm) are determined. In this step, we examine the dimensions to guarantee that all components, including the spindle and any other item needed, are well incorporated into the practical design problem. Once everything is in place, we proceed to Step 5; otherwise, we return to Step 2, select another group of non-dimensional parameters from the optimum region, and repeat Steps 3 and 4.

Step 5: The input range required to reach an orientational workspace of $\pm 40°$ for the optimized mechanism can be calculated by inverse kinematics. For the solution obtained in Step 4, $\rho_i = [-995.1252\ \text{mm}, -738.0102\ \text{mm}]$ ($i = 1, 2, 3$) when $z = 0$. The final input range for each actuator should incorporate the positional workspace along the z-axis.

If the input range is unsuitable for a commercial actuator, the designer can return to Step 2 and choose another non-dimensional mechanism to adjust the design solution, or return to Step 1 and identify another optimum region.

References

Angeles J (1995) Kinematic isotropy in humans and machines. In: Proceedings of IFToMM 9th world congress on theory of machines and mechanisms, Edizioni Unicopli, Milano, pp XLII–XLIX

Bruyninckx H (1997) The 321-HEXA: a fully-parallel manipulator with closed-form position and velocity kinematics. In: Proceedings of IEEE international conference on robotics and automation, Albuquerque, New Mexico, USA, IEEE press, Piscataway, N.J., pp 2657–2662

Chablat D, Wenger P (2003) Architecture optimization of a 3-DOF parallel mechanism for machining applications: the Orthoglide. IEEE Trans Robot Autom 19(3):403–410

Gao F, Zhang X, Zhao Y, Wang H (1996) A physical model of the solution space and the atlases of the reachable workspaces for 2-DOF parallel plane wrists. Mech Mach Theory 31(2):173–184

Hunt KH (1983) Structure kinematics of in parallel actuated robot arms. J Mech Trans Autom Des 105:705–712

Liu X-J, Wang J, Gao F, Wang L-P (2002) The mechanism design of a simplified 6-DOF 6-RUS parallel manipulator. Robotica 20(1):81–91

Liu X-J, Wang J, Pritschow G (2005) A new family of spatial 3-DOF fully-parallel manipulators with high rotational capability. Mech Mach Theory 40(4):475–494

Ma O, Angeles J (1991) Optimum architecture design of platform manipulators. In: Proceedings of the fifth international conference on advanced robotics, Pisa, IEEE press, Piscataway, N.J., pp 1130–1135

Ottaviano E, Ceccarelli M (2002) Optimal design of CaPaMan (Cassino Parallel Manipulator) with a specified orientation workspace. Robotica 20:159–166

Pierrot F, Dauchez P, Fournier A (1991) Towards a fully-parallel 6 d.o.f. robot for high speed applications. In: Proceedings of IEEE international conference on robotics & automation, Sacramento, CA, IEEE press, Piscataway, N.J., pp 1288–1293

Ryu J, Cha J (2003) Volumetric error analysis and architecture optimization for accuracy of HexaSlide type parallel manipulators. Mech Mach Theory 38:227–240

Sorli M, Ferraresi C, Kolarski M et al (1997) Mechanics of TURIN parallel robot. Mech Mach Theory 32(1):51–57

Stock M, Miller K (2004) Optimal kinematic design of spatial parallel manipulators: application to linear Tsai's robot. J Mech Des 125:292–301

Tandirci M, Angeles J, Ranjbaran F (1992) The characteristic point and characteristic length of robotic manipulators. In: Proceedings of ASME 22nd biennial conference on robotics, spatial mechanisms and mechanical systems, Scottsdale, AZ, ASME press, New York pp 203–208

Yang J-H (1987) The space model and dimensional types of the four-bar mechanisms. Mech Mach Theory 22(1):71–76

Chapter 9
Kinematic Optimal Design of a Spatial 3-DOF Parallel Manipulator

Abstract This chapter provides a case study for the kinematic optimal design of a spatial 3-DOF (two translations and one rotation) parallel manipulator. The final design of the manipulator is used for a module of a five-axis hybrid kinematic machine developed at Tsinghua University.

Keywords Kinematic design • Optimal design • Parallel manipulator • Hybrid kinematic machine

A five-axis serial-parallel kinematic machine, SPKM 165 (Fig. 9.1), has recently been developed at Tsinghua University. The machine consists of a parallel module and serial table.

As shown in Figs. 9.1b and 9.2, the mobile platform of the parallel module is connected to the base through three legs, namely, legs I, II, and III. Legs I and II are in the same plane (i.e., the *o-yz* plane shown in Fig. 9.2) and have identical chains. Each chain consists of a length-fixed link connected to the active slider through a passive revolute joint and to the mobile platform through a universal joint. Leg III is in the *o-xz* plane, perpendicular to the plane where legs I and II are located. This leg consists of another length-fixed link connected to the active slider through a passive revolute joint and attached to the mobile platform through a cylinder joint. The three sliders are then linked to the base through three prismatic joints and are actuated along the *z*-axis.

Rotating table IV in the serial table is placed on a slider and rotates around the *z*-axis. Slider V is connected to the base through a prismatic joint and is actuated along the *x*-axis.

Hence, the parallel module has three spatial DOFs: two translations in the *o-yz* plane and one rotation around the axis defined by the two universal joints attached to the mobile platform. The serial table has two DOFs: one translation along the *x*-axis and one rotation about the *z*-axis. All the five DOFs are shown in Fig. 9.2.

X.-J. Liu and J. Wang, *Parallel Kinematics: Type, Kinematics, and Optimal Design*, Springer Tracts in Mechanical Engineering, DOI 10.1007/978-3-642-36929-2_9,

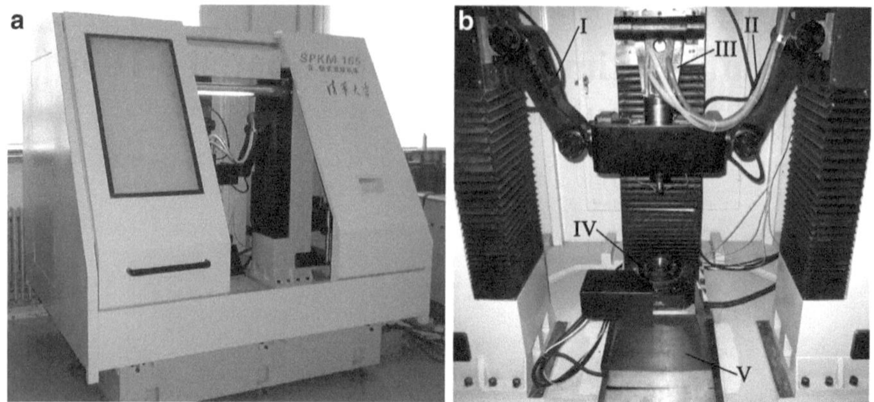

Fig. 9.1 SPKM165, a five-axis serial-parallel kinematic machine

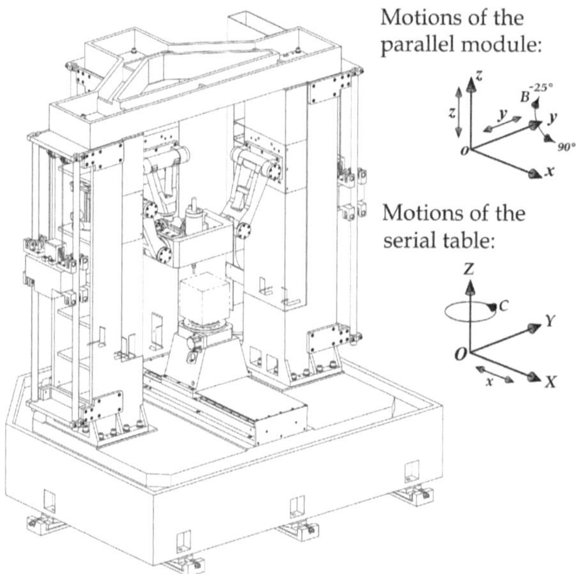

Fig. 9.2 Motions of the SPKM 165 machine

Generally, thismilling machine functions as a vertical hybrid machine. Its positional workspace is 165 mm × 165 mm × 165 mm and rotation angle β about the y-axis in the parallel module can be $\beta \in [-25°, 90°]$. Rotating table IV in the serial table can rotate continuously. Thus, this machine is capable of five-face machining and can satisfy special requirements.

In this chapter, we consider the optimal kinematic design of the parallel module.

9.1 Inverse Kinematics

The parallel module is the mechanism shown in Fig. 2.39a. A kinematical scheme of the manipulator is developed, as shown in Fig. 9.3. The vertices of the output platform are denoted as platform joints P_i ($i = 1, 2, 3$), and the central points of the three revolute joints attached to the sliders are denoted as B_i ($i = 1, 2, 3$). A fixed global reference frame O-xyz is located at the center point of line segment ab with the y-axis normal to plane abc, and the x-axis is directed along ab. Another reference frame, called moving frame O'-$x'y'z'$, is located at the center of side P_1P_2. The y'-axis is perpendicular to the mobile platform and the x'-axis is directed along P_1P_2. Given that the first and second legs are the same in terms of kinematic chains and the third leg is different, the geometric parameters of the first and second legs can be identical to each other but different from those of the third leg. Thus, the geometric parameters are $O'P_1 = O'P_2 = r$, $B_1P_1 = B_2P_2 = R_2$, $O'P_3 = L_1$, $P_3B_3 = L_2$, normal distance L_3 from point O to the straight-line path of joint point B_3, i.e., $Oc = L_3$, and $Oa = Ob = R$.

The inverse kinematic problem of this manipulator is somewhat similar to that of the ***HALF*** manipulator. It can be solved easily. Vectors $\boldsymbol{b}_i$ ($i = 1, 2, 3$) are defined as the position vectors of points B_i in reference frame O-xyz and can be written as

$$\boldsymbol{b}_1 = (-R \quad y_1 \quad 0)^{\mathrm{T}}, \quad \boldsymbol{b}_2 = (R \quad y_2 \quad 0)^{\mathrm{T}}, \quad \boldsymbol{b}_3 = (0 \quad y_3 - L_3)^{\mathrm{T}}. \tag{9.1}$$

In reference frame O-xyz, position vectors $\boldsymbol{p}_i$ ($i = 1,2,3$) of points P_i can be expressed as

$$\boldsymbol{p}_1 = (x - r \quad y \quad 0)^{\mathrm{T}}, \quad \boldsymbol{p}_2 = (x + r \quad y \quad 0)^{\mathrm{T}},$$
$$\text{and} \quad \boldsymbol{p}_3 = (0 \quad y + L_1 \sin\phi - L_1 \cos\phi)^{\mathrm{T}}, \tag{9.2}$$

where $(x, \ y, \ \phi)$ is the pose of the manipulator and ϕ is the rotating angle of the mobile platform about the x'-axis. The x coordinate of point P_3 at the fixed-length

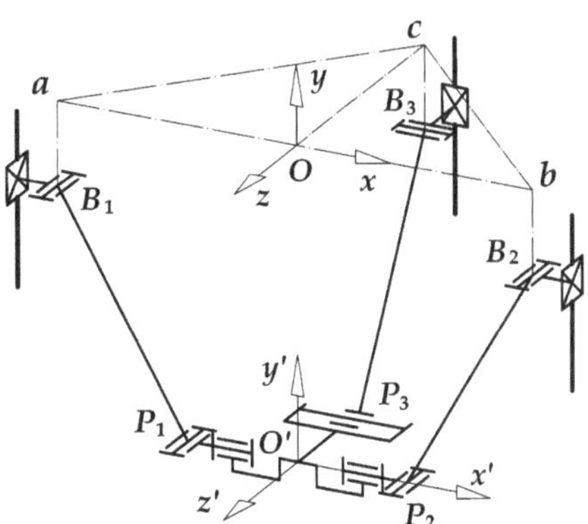

Fig. 9.3 Kinematic scheme of a spatial 3-DOF parallel manipulator

link of the third leg is always zero. The kinematic problem of the manipulator can be solved by

$$|\boldsymbol{b}_i - \boldsymbol{p}_i| = B_i P_i. \tag{9.3}$$

Hence,

$$(R - r + x)^2 + (y_1 - y)^2 = R_2^2, \tag{9.4}$$

$$(R - r - x)^2 + (y_2 - y)^2 = R_2^2, \tag{9.5}$$

$$(y_3 - y - L_1 \sin\phi)^2 + (L_3 - L_1 \cos\phi)^2 = L_2^2. \tag{9.6}$$

For a given pose $(x, \; y, \; \phi)$, input y_i ($i = 1,2,3$) can be obtained as follows:

$$y_1 = \pm\sqrt{R_2^2 - (R - r + x)^2} + y, \tag{9.7}$$

$$y_2 = \pm\sqrt{R_2^2 - (R - r - x)^2} + y, \tag{9.8}$$

$$y_3 = \pm\sqrt{L_2^2 - (L_3 - L_1 \cos\phi)^2} + y + L_1 \sin\phi. \tag{9.9}$$

Therefore, there are eight inverse kinematic solutions for the manipulator. The configuration shown in Fig. 9.3 corresponds to the solution when the "$\pm$" signs in Eqs. (9.7), (9.8), and (9.9) are all "+". In this chapter, we are concerned only with this solution. Equations (9.6) and (9.9) show that the kinematics of the third leg is independent of the x coordinate.

9.2 Optimal Design of the Parallel Manipulator

In this section, the optimal design of the manipulator is implemented using the LTI, GTI, and GTW. Setting $R - r$ in Eqs. (9.4), (9.5), (9.7), and (9.8) as R_1, i.e., $R_1 = R - r$, shows that the kinematics of the first and second legs are actually those of the PRRRP symmetrical parallel manipulator in Fig. 7.16. When the position of point O' is specified, the kinematic equation of Eq. (9.6) is that of the slider-crank mechanism shown in Fig. 7.14. Therefore, the singular configuration of the manipulator is that when one of the P̲RRRP̲ manipulators and slider-crank mechanism is in singularity. The optimal design of the manipulator can then be divided into two parts: (a) the design of the P̲RRRP̲ parallel manipulator and (b) that of the slider-crank mechanism. The transmission angle is the classical concept and tool used in the analysis and design of planar mechanisms, and the new decoupled parallel manipulator proposed in this chapter is the combination of two planar mechanisms. Thus, the LTI, GTI, and GTW can be applied to the optimal design of the new parallel manipulator. The optimal design is implemented using the method introduced in Sect. 8.3.

9.2.1 Performance Atlas

9.2.1.1 PRRRP Parallel Manipulator (First and Second Legs)

The normalization of dimensional parameters R_1 and R_2 yields two non-dimensional parameters r_1 and r_2; $r_1 + r_2 = 2$ and $0 < r_1 \leq r_2 \leq 2$. Hence, $r_1 \leq 1$, which is the PDS of the PRRRP parallel manipulator. In the design process, only parameter r_1 should be optimized. On the basis of inverse kinematics, we can express the inverse and forward transmission angles as follows:

$$\mu = \cos^{-1}\left(\frac{2r_2^2 - r_{\mathrm{AB}}^2}{2r_2^2}\right), \tag{9.10}$$

$$\gamma_i = \cos^{-1}\left(\frac{r_1^2 + r_2^2 - r_{iy}^2}{2r_1 r_2}\right), \quad i = 1, 2, \tag{9.11}$$

where $r_{\mathrm{AB}} = \sqrt{r_y^2 + 4r_1^2}$, $r_y = \sqrt{r_2^2 - (x - r_1)^2} - \sqrt{r_2^2 - (x + r_1)^2}$, and $r_{iy} = \sqrt{r_2^2 + (-1)^i 2xr_1 - r_1^2}$. Equations (9.10) and (9.11) imply that the forward and inverse transmission angles are independent of the y coordinate. For the PRRRP parallel mechanism, the LTI can be obtained thus:

$$\chi = \min\{\lambda, \eta\} = \min\{|\sin\gamma_1|, |\sin\gamma_2|, |\sin\mu|\}. \tag{9.12}$$

Hence, the LTI is unrelated to the y coordinate, indicating that in the analysis and design, we can disregard the workspace along the y-axis. If the LTI is given by $\chi \geq \sin(\pi/4)$, the minimum and maximum values of r_1 will be

$$r_{1\,\min} = \frac{2\sin(\pi/8)}{1 + \sin(\pi/8)} \approx 0.5535, \tag{9.13}$$

$$r_{1\,\max} = \frac{2\cos(\pi/4)}{1 + \cos(\pi/4)} \approx 0.8284. \tag{9.14}$$

Some manipulators when $r_1 < r_{1\,\min}$ still have a GTW along the x-axis. However, the workspace has two discontinuous regions, which are symmetrical about line $x = 0$. As an example, the GTW and some configurations of the manipulator with $r_1 = 0.54$ and $r_1 = 1.46$ are shown in Fig. 9.4. When $x = 0$, $\mu < 45°$. There

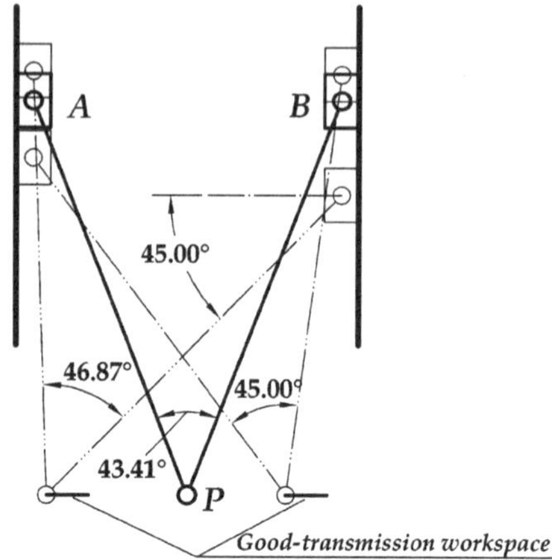

Fig. 9.4 P̲RRRP̲ parallel manipulator with two discontinuous workspace regions

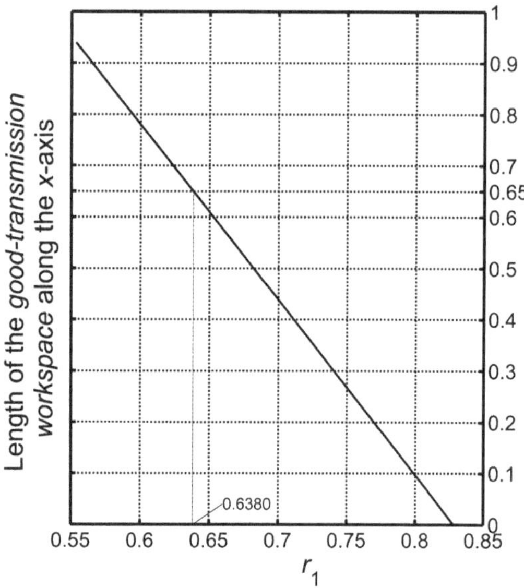

Fig. 9.5 Relationship between the GTW length along the x-axis, i.e., $W_{x-\text{GTW}}$, and normalized parameter r_1 of the PRRRP parallel manipulator (the first and second legs of the new parallel manipulator)

are two regions where the LTI is subject to constraint $\chi \geq \sin(\pi/4)$. In practice, this kind of workspace is usually undesirable. In this work, the GTW of such a manipulator is assumed to be zero.

The relationship between the GTW length along the x-axis, denoted as $W_{x-\text{GTW}}$, and normalized parameter r_1 is shown in Fig. 9.5. The figure shows that when $r_1 = r_{1\,\text{min}}$, the GTW is at its largest; the length is inversely proportional to parameter r_1.

Figure 9.6 illustrates the relationship between the GTI value and normalized parameter r_1. When $r_1 = 0.6813$, the GTI reaches its maximum value. When $r_1 < 0.6813$, the GTI value is inversely proportional to parameter r_1. The index is proportional to the parameter when $r_1 > 0.6813$.

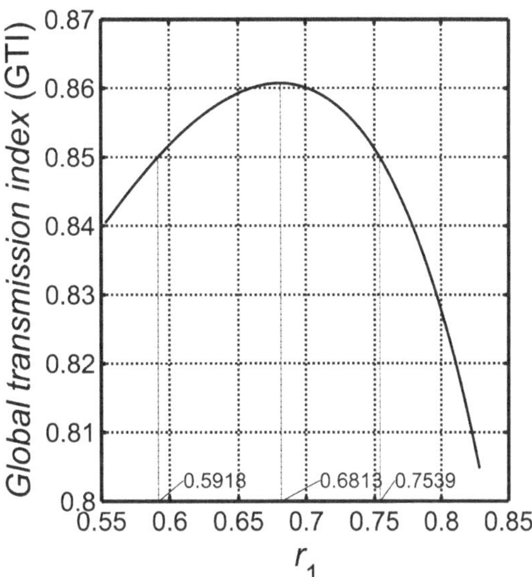

Fig. 9.6 Relationship between the GTI and normalized parameter r_1 of the PRRRP parallel manipulator (the first and second legs of the new parallel manipulator)

9.2.1.2 Slider-Crank Mechanism (The Third Leg)

As described previously, the third leg can function as the slider-crank mechanism at any moment. An example is when $y = 0$, i.e., points O' and O are coincident. Then, kinematic equation (9.6) for the third leg can be rewritten as

$$(y_3 - L_1 \sin\phi)^2 + (L_3 - L_1 \cos\phi)^2 = L_2^2. \tag{9.15}$$

This is the kinematic equation of the slider-crank mechanism in Fig. 7.14. Using the normalization introduced in Sect. 8.1, we obtain three normalization parameters l_1, l_2, and l_3 for parameters L_1, L_2, and L_3. The PDS of the third leg, i.e., the slider-crank mechanism, is constrained by

$$l_1 + l_2 + l_3 = 3, \quad l_1 \leq l_3, \quad \text{and} \quad l_1 + l_2 > l_3, \tag{9.16}$$

where $l_i = L_i/D$ and $D(L_1 + L_2 + L_3)/3$ is the normalization factor. The PDS is the same as that of the 3-PRS parallel mechanism in Fig. 8.19.

As shown in Fig. 7.14, because $l_1 \leq l_3$ the forward and inverse transmission angles of the normalized mechanism can be obtained as

$$\mu = \cos^{-1}\left(\frac{l_1^2 + l_2^2 - |\boldsymbol{cb}_3|^2}{2l_1 l_2}\right), \tag{9.17}$$

$$\gamma = \cos^{-1}\left(\frac{l_3 - |l_1 \cos\phi|}{l_2}\right), \tag{9.18}$$

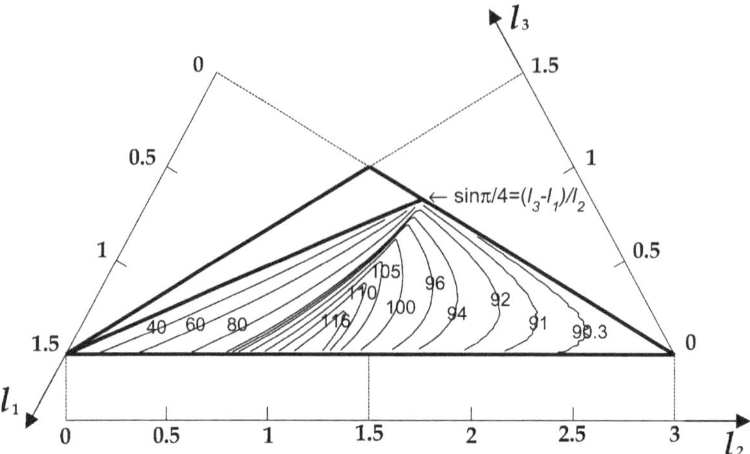

Fig. 9.7 Atlas of the GTW for the orientation of the slider-crank mechanism (the third leg)

where $\boldsymbol{c}$ is the position vector of point O' in frame O-xyz, $|\boldsymbol{c}\boldsymbol{b}_3| = \sqrt{y_3^2 + l_3^2}$, and y_3 can be obtained from the kinematic equation for a given orientation ϕ. The transmission angles are relative to orientation ϕ.

For the slider-crank mechanism, the LTI can be written as

$$\chi = \min\{\lambda, \eta\} = \min\{|\sin\gamma|, |\sin\mu|\}. \tag{9.19}$$

Suppose that when $\phi \in (\phi_{\min}, \phi_{\max})$, LTI $\chi \geq \sin(\pi/4)$. Here, $\phi_{\min}$ and $\phi_{\max}$ are the orientations when $\chi \geq \sin(\pi/4)$. Hence, the GTW for the orientation is defined as the relative angle between orientations $\phi_{\min}$ and $\phi_{\max}$. If the orientational workspace is denoted as W_{ϕ_GTW}, we obtain

$$W_{\phi\_\mathrm{GTW}} = \phi_{\max} - \phi_{\min}, \tag{9.20}$$

which is the orientation capability of the mechanism.

Figure 9.7 depicts the relationship between W_{ϕ_GTW} and normalized parameters l_1, l_2, and l_3 when $\chi > \sin(\pi/4)$. When l_2 is near 1.5 and l_3 is less than 1.0, the mechanisms have a large GTW for the orientation (orientation capability).

Figure 9.8 illustrates the relationship between the GTI and normalized parameters l_1, l_2, and l_3 when $\chi \geq \sin(\pi/4)$. Thus, the GTI is inversely proportional to parameter l_3 and proportional to parameter l_1.

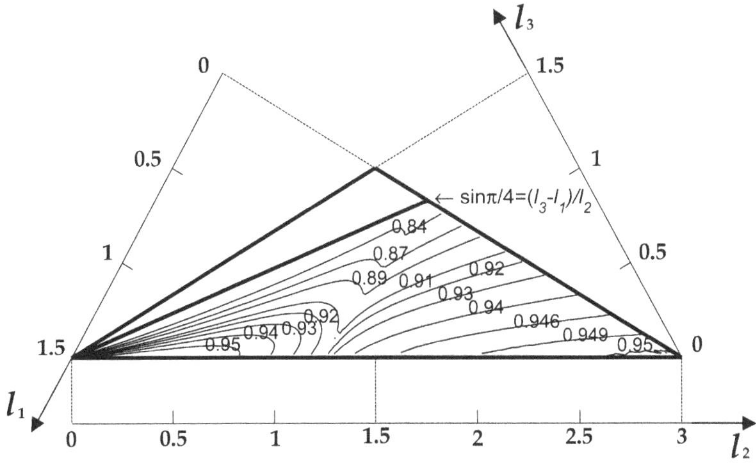

Fig. 9.8 Atlas of the GTI of the slider-crank mechanism (the third leg)

9.2.2 *Dimensional Synthesis Using the Performance Atlases*

The optimal design in which performance atlases are used has been introduced and discussed in detail in Sect. 8.3. The design process for the studied manipulators is the same. We consider the design problems encountered in the SPKM 165 machine: (a) the desired positional workspace is $x \times y = 165$ mm $\times$ 165 mm in the O-xy plane and (b) the desired rotational workspace is as much as $-90° \sim 25°$ of the mobile platform at every point in the workspace. The optimization process based on the atlases in Figs. 9.5, 9.6, 9.7, and 9.8 can be summarized as follows:

Step 1: *Identification of an optimum region in the PDS.*

Because all design conditions cannot be the same, the identification of an optimum region is the responsibility of the designer. We provide an example in this chapter.

For the first and second legs, an optimum region Ω_{1-2} can be identified using the atlases in Figs. 9.5 and 9.6 with performance constraints, such as $W_{x\text{-GTW}} > 0.65$ and $\Gamma > 0.85$. Figure 9.5 shows that when $r_1 < 0.6380$, the length of the GTW is greater than 0.65. Figure 9.6 illustrates that when $0.5918 < r_1 < 0.7539$, the GTI is no less than 0.85. Hence, the optimum region for r_1 is $\Omega_{1-2} = [0.5918 < r_1 < 0.6380 | W_{x\text{-GTW}}(r_1) > 0.65 \quad \text{and} \quad \Gamma(r_1) > 0.85]$.

The third leg is related to the orientation capability, and an optimum region Ω_3 for this leg should be determined with respect to $W_{\phi\text{\_GTW}} \geq 115°$ and a better GTI (e.g., $\Gamma > 0.91$). This region is shown as the hatched parts in the PDS shown in Fig. 9.9.

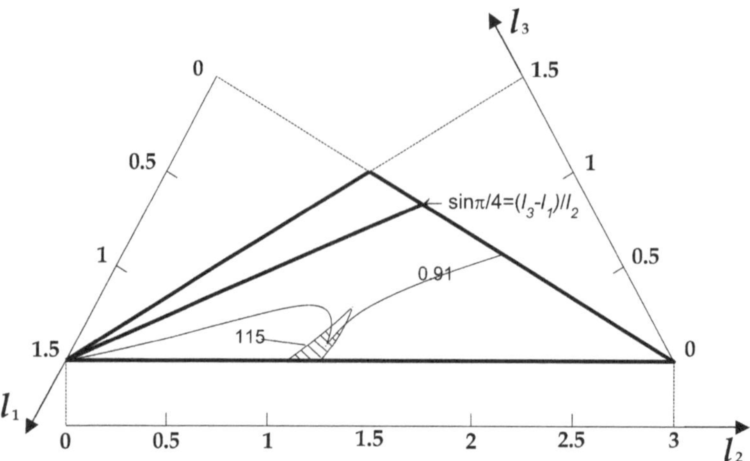

Fig. 9.9 Optimum region for the slider-crank mechanism (the third leg) when $W_{\phi\_GTW} \geq 100°$ and $\Gamma > 0.93$

The identified optimum regions contain all possible non-dimensional solutions for the design problem. Given that the non-dimensional manipulator in the PDS and all its corresponding dimensional manipulators are similar in performance, the final design result that is based on the manipulator in the optimum region is optimal. Therefore, the next step is to identify an acceptable solution candidate in the optimum region.

Step 2: *Selection of a solution candidate from the optimum region.*

Obtained optimum regions Ω_{1-2} and Ω_3 contain all possible solutions for the design. No best solution, only a comparatively better one, exists for a design problem. Thus, we can choose any one non-dimensional manipulator from the regions.

For the first and second legs, r_1 ($r_1 = 0.6$) is selected from the optimum region Ω_{1-2}. Hence, $r_2 = 2 - r_1 = 1.4$. The GTW length along the x-axis and the GTI of the non-dimensional manipulator with $r_1 = 0.6$ and $r_2 = 1.4$ are $W_{x-\mathrm{GTW}} = 0.7798$ and $\Gamma > 0.8517$, respectively.

For the third leg, we choose the non-dimensional manipulator with parameters $l_1 = 0.75$, $l_2 = 1.3$, and $l_3 = 0.95$ from optimum region Ω_3. For the selected manipulator, $W_{\phi\_\mathrm{GTW}} = 118.85°$ and $\Gamma > 0.9113$. The orientation range for the manipulator is defined by $\phi_{\min} = -87.60°$ and $\phi_{\max} = 31.25°$.

Step 3: *Determination of dimensional parameters r, R, R_2, L_1, L_2, and L_3 with respect to the desired workspace.*

According to the normalization method introduced in Sect. 8.1, the normalization factor D should first be determined. This factor can be obtained by comparing the desired workspace of a design problem and the workspace of the non-dimensional manipulator selected from the optimum region.

For the first and second legs of the manipulator, factor $D' = 165$ mm/0.7798 ≈ 211.59 mm because the desired workspace along the x-axis is 165 mm. Therefore, $R_1 = D'r_1 \approx 127.0$ mm and $R_2 = D'r_2 \approx 296.0$ mm. The GTIs for the non-dimensional and dimensional manipulators are clearly the same.

An angle is dependent only on the ratio of related linear parameters, but not on any one of the parameters. Thus, we cannot determine normalization factor D of the third leg with respect to the $\pm 50°$orientational workspace. In this case, we should first determine parameter L_1 of the mobile platform according to practical applications and ensure that the parameter is as small as possible to reduce the space occupied by the manipulator. We take $L_1 = 180$ mm and accordingly obtain $D'' = L_1/l_1 = 180$ mm/0.75 $= 240$ mm, $L_2 = D''l_2 = 312$ mm, and $L_3 = D''l_3 = 228$ mm. Considering the attachment of the spindle, we take $r = 130$ mm and derive $R = R_1 + r = 130 + 127 = 257$ mm for the first and second legs.

Step 4: *Calculation of the input range.*

The input range required to reach the desired workspace can be calculated using inverse kinematic Equations (9.7), (9.8), and (9.9) of the spatial 3-DOF parallel manipulator. For the solution obtained in Step 3, $y_1 = y_2 = [-162.18\text{ mm}, -78.65\text{ mm}]$ and $y_3 = [-330.36\text{ mm}, \ 25.16\text{ mm}]$. The final input range for each actuator should incorporate the positional workspace along the y-axis. Hence, $y_1 = y_2 = [-327.18\text{ mm}, -78.65\text{ mm}]$ and $y_3 = [-495.36\text{ mm}, \ 25.16\text{ mm}]$ when the entire positional workspace is under the x-axis.

Step 5: *Examining the design result and adjusting the design solution if necessary.*

In this step, determining which items require examination depends on the application. For example, the designer may check whether the input range is suitable for his preferred commercial actuator. Regardless of the item checked, once the solution is unsatisfactory, the designer can return to Step 2 and choose another non-dimensional manipulator, or return to Step 1 to identify another optimum region and make adjustments until a satisfactory design solution is derived. This is the advantage of this design method.

Note that we obtain orientational workspace W_{ϕ_GTW} about orientations ϕ_{min} and ϕ_{max}, as given in Eq. (9.20). When the desired orientation range is $-90° \sim 25°$, original orientation ϕ_0 of the mobile platform should be adjusted to fit the requirement because ϕ_{min} is usually not equal to ϕ_{max}. For the example, original orientation ϕ_0 can be specified as $\phi_0 = 6.0°$ because $\phi_{min} = -87.60°$ and $\phi_{max} = 31.25°$.

Index

X.-J. Liu and J. Wang, *Parallel Kinematics: Type, Kinematics, and Optimal Design*, Springer Tracts in Mechanical Engineering, DOI 10.1007/978-3-642-36929-2,